Plant Pathogenic Microorganisms: State-of-the-Art Research in Spain

Plant Pathogenic Microorganisms: State-of-the-Art Research in Spain

Editors

Elvira Fiallo-Olivé
Soledad Sacristán
Ana Palacio-Bielsa

MDPI • Basel • Beijing • Wuhan • Barcelona • Belgrade • Manchester • Tokyo • Cluj • Tianjin

Editors
Elvira Fiallo-Olivé
Instituto de Hortofruticultura
Subtropical y Mediterránea
"La Mayora"
Consejo Superior de
Investigaciones Científicas
Málaga
Spain

Soledad Sacristán
Centro de Biotecnología y
Genómica de Plantas
Universidad Politécnica de
Madrid (UPM)
(INIA/CSIC)
Madrid
Spain

Ana Palacio-Bielsa
Departamento de Sistemas
Agrícolas, Forestales y
Medio Ambiente
Centro de Investigación y
Tecnología Agroalimentaria
de Aragón (CITA)
Zaragoza
Spain

Editorial Office
MDPI
St. Alban-Anlage 66
4052 Basel, Switzerland

This is a reprint of articles from the Special Issue published online in the open access journal *Microorganisms* (ISSN 2076-2607) (available at: www.mdpi.com/journal/microorganisms/special_ issues/plant_pathogenic_microorganisms_spain).

For citation purposes, cite each article independently as indicated on the article page online and as indicated below:

LastName, A.A.; LastName, B.B.; LastName, C.C. Article Title. *Journal Name* **Year**, *Volume Number*, Page Range.

ISBN 978-3-0365-7339-7 (Hbk)
ISBN 978-3-0365-7338-0 (PDF)

Cover image courtesy of Elvira Fiallo-Olivé

Contents

About the Editors . . . . **vii**

Elvira Fiallo-Olivé, Ana Palacio-Bielsa and Soledad Sacristán
Plant Pathogenic Microorganisms: State-of-the-Art Research in Spain
Reprinted from: *Microorganisms* 2023, *11*, 816, doi:10.3390/microorganisms11030816 . . . . **1**

María Quintana, Leandro de-León, Jaime Cubero and Felipe Siverio
Assessment of Psyllid Handling and DNA Extraction Methods in the Detection of '*Candidatus*
Liberibacter Solanacearum' by qPCR
Reprinted from: *Microorganisms* 2022, *10*, 1104, doi:10.3390/microorganisms10061104 . . . . **5**

Elvira Fiallo-Olivé, Ana Cristina García-Merenciano and Jesús Navas-Castillo
Sweet Potato Symptomless Virus 1: First Detection in Europe and Generation of an Infectious
Clone
Reprinted from: *Microorganisms* 2022, *10*, 1736, doi:10.3390/microorganisms10091736 . . . . **23**

Ana M. Fernández-Sanz, M. Rosario Rodicio and Ana J. González
Biochemical Diversity, Pathogenicity and Phylogenetic Analysis of *Pseudomonas viridiflava* from
Bean and Weeds in Northern Spain
Reprinted from: *Microorganisms* 2022, *10*, 1542, doi:10.3390/microorganisms10081542 . . . . **33**

**David Hernández, Omar García-Pérez, Santiago Perera, Mario A. González-Carracedo, Ana
Rodríguez-Pérez and Felipe Siverio**
Fungal Pathogens Associated with Aerial Symptoms of Avocado (*Persea americana* Mill.) in
Tenerife (Canary Islands, Spain) Focused on Species of the Family Botryosphaeriaceae
Reprinted from: *Microorganisms* 2023, *11*, 585, doi:10.3390/microorganisms11030585 . . . . **45**

Belén Álvarez, María M. López and Elena G. Biosca
Ralstonia solanacearum Facing Spread-Determining Climatic Temperatures, Sustained
Starvation, and Naturally Induced Resuscitation of Viable but Non-Culturable Cells in
Environmental Water
Reprinted from: *Microorganisms* 2022, *10*, 2503, doi:10.3390/microorganisms10122503 . . . . **67**

Marta Sena-Vélez, Elisa Ferragud, Cristina Redondo, James H. Graham and Jaime Cubero
Chemotactic Responses of *Xanthomonas* with Different Host Ranges
Reprinted from: *Microorganisms* 2022, *11*, 43, doi:10.3390/microorganisms11010043 . . . . **85**

Iván Martín-Hernández and Israel Pagán
Gene Overlapping as a Modulator of *Begomovirus* Evolution
Reprinted from: *Microorganisms* 2022, *10*, 366, doi:10.3390/microorganisms10020366 . . . . **103**

Jordi Cabrefiga, Maria Victoria Salomon and Pere Vilardell
Improvement of Alternaria Leaf Blotch and Fruit Spot of Apple Control through the
Management of Primary Inoculum
Reprinted from: *Microorganisms* 2022, *11*, 101, doi:10.3390/microorganisms11010101 . . . . **119**

**Bàrbara Quetglas, Diego Olmo, Alicia Nieto, David Borràs, Francesc Adrover and Ana
Pedrosa et al.**
Evaluation of Control Strategies for *Xylella fastidiosa* in the Balearic Islands
Reprinted from: *Microorganisms* 2022, *10*, 2393, doi:10.3390/microorganisms10122393 . . . . **131**

Esther Badosa, Marta Planas, Lidia Feliu, Laura Montesinos, Anna Bonaterra and Emilio Montesinos
Synthetic Peptides against Plant Pathogenic Bacteria
Reprinted from: *Microorganisms* **2022**, *10*, 1784, doi:10.3390/microorganisms10091784 . . . . **147**

Anna Bonaterra, Esther Badosa, Núria Daranas, Jesús Francés, Gemma Roselló and Emilio Montesinos
Bacteria as Biological Control Agents of Plant Diseases
Reprinted from: *Microorganisms* **2022**, *10*, 1759, doi:10.3390/microorganisms10091759 . . . . **165**

About the Editors

Elvira Fiallo-Olivé

Elvira Fiallo-Olivé received her MSc and PhD degrees from the University of Havana (Cuba). Her expertise is in the characterization of whitefly-transmitted viruses that affect important vegetable crops and wild plants worldwide, with emphasis on begomoviruses and criniviruses, and their interactions with the insect vector. Her current research is focused on the molecular and biological characterization of deltasatellites, a novel class of DNA satellites associated with begomoviruses.

Soledad Sacristán

Soledad Sacristán received her PhD degree from the Universidad Politécnica de Madrid (Spain). She has worked on different aspects related to the evolution and adaptation to hosts of phytopathogenic viruses and fungi. She is now specially interested in the molecular and environmental factors that affect fungal interactions with plants, either beneficial or pathogenic.

Ana Palacio-Bielsa

Ana Palacio-Bielsa received her PhD degree from the University of Valencia (Spain). Her current research is focused on the study of phytopathogenic bacteria affecting both crops and forestry. The main aspects of her research include the development of detection, identification, epidemiology, and control methods. She is also involved in the study of plant–pathogen interaction mechanisms.

microorganisms

MDPI

Editorial

Plant Pathogenic Microorganisms: State-of-the-Art Research in Spain

Elvira Fiallo-Olivé [1,*], Ana Palacio-Bielsa [2] and Soledad Sacristán [3,4]

[1] Instituto de Hortofruticultura Subtropical y Mediterránea "La Mayora" (IHSM-UMA-CSIC), Consejo Superior de Investigaciones Científicas, 29750 Algarrobo-Costa, Spain

[2] Centro de Investigación y Tecnología Agroalimentaria de Aragón, Departamento de Sistemas Agrícolas, Forestales y Medio Ambiente, Instituto Agroalimentario de Aragón—IA2, CITA—Universidad de Zaragoza, 50059 Zaragoza, Spain

[3] Centro de Biotecnología y Genómica de Plantas, Instituto Nacional de Investigación y Tecnología Agraria y Alimentaria (INIA/CSIC), Universidad Politécnica de Madrid (UPM), Campus de Montegancedo UPM, 28223 Pozuelo de Alarcón, Spain

[4] Departamento de Biotecnología-Biología Vegetal, Escuela Técnica Superior de Ingeniería Agronómica, Alimentaría y de Biosistemas, Universidad Politécnica de Madrid (UPM), 28040 Madrid, Spain

* Correspondence: efiallo@eelm.csic.es

Citation: Fiallo-Olivé, E.; Palacio-Bielsa, A.; Sacristán, S. Plant Pathogenic Microorganisms: State-of-the-Art Research in Spain. *Microorganisms* **2023**, *11*, 816. https://doi.org/10.3390/microorganisms11030816

Received: 17 March 2023
Accepted: 20 March 2023
Published: 22 March 2023

Pathogenic microorganisms, including fungi, oomycetes, bacteria, viruses, and viroids, constitute a serious threat to agriculture worldwide. In Spain, one of the countries with the highest proportion of agricultural gross domestic product in Europe, the agri-food industry is the main manufacturing activity. Consequently, the presence and emergence of microorganisms causing serious plant diseases to economically important crops is especially relevant. In line with this, Spain has an important number of research groups interested in plant pathology, with scientists working on many aspects of pathogenic microorganism–plant interactions, from the basic aspects to more applied studies. In recent years, numerous important advancements have been achieved by scientists working in Spain in terms of the biological and molecular characterization of plant pathogenic microorganisms, in elucidating mechanisms of microbe pathogenesis, plant resistance to microbe infection, and plant–microbe–vector interactions. All these new achievements have provided essential knowledge for agricultural researchers worldwide.

The aim of this Special Issue was to provide a platform for Spanish researchers interested in plant pathogenic microorganisms to share their recent results related to microbe–plant host interactions, microbe–vector interactions, microbe–microbe interactions, evolution, ecology, and control strategies. A total of 11 papers have been contributed by 47 authors to the Issue, including 8 research articles and 3 reviews.

The availability of sensitive and accurate pathogen detection methods is crucial for plant pathologists and disease management. In this Special Issue, the paper by Quintana et al. [1] focused on this topic. The authors assessed different DNA extraction methods for the detection of the uncultured bacterium '*Candidatus* Liberibacter solanacearum' (CaLsol) by qPCR in the psyllid vector. They also evaluated the influence on the detection of CaLsol by qPCR in *Bactericera trigonica* of four specimen preparations (entire body, ground, cut off head, and punctured abdomen) and seven DNA extraction methods. Although optimum results were obtained through grinding, destructive procedures were not essential in order to detect CaLsol. The HotSHOT method was accurate, fast, simple, and sufficiently sensitive to detect the bacterium within the vector. This work provides a valuable guide when choosing a method to detect CaLsol in vectors according to the purpose of the study.

The knowledge of the presence and distribution of plant pathogens is a key aspect in the study of their epidemiology. In this Special Issue, three papers cover aspects related to the detection, molecular and biological characterization, and distribution of plant pathogens in Spain.

Fiallo-Olivé et al. [2] detected the mastrevirus sweet potato symptomless virus 1 in sweet potato plants for the first time in Spain (southern continental Spain and the Canary Islands) and Europe. Sequence analysis of the full-length genomes of isolates from Spain showed novel molecular features. Additionally, in this work, the first agroinfectious clone for the virus was developed, and infectivity assays showed that it was able to asymptomatically infect *Nicotiana benthamiana*, *Ipomoea nil*, *I. setosa*, and sweet potato.

In their contribution to this Special Issue, Fernández-Sanz et al. [3] studied the biochemical diversity and pathogenicity of the bacterium *Pseudomonas viridiflava* from common bean and weeds in Northern Spain and carried out a phylogenetic analysis. Regardless of their origin, the isolates displayed biochemical diversity. Phylogenetic analysis revealed two clusters of strains containing different pathogenicity islands, with no correlation observed for the plant host, biochemical profile, or pathogenicity. Ten new weed genera/species were identified as the new host of the bacterium, and more than half of the weed isolates were pathogenic in common bean. This work supports the role of weeds as reservoirs of *P. viridiflava* and a source of inoculum for common bean infection, further highlighting the role of weeds on the epidemiology of the disease.

The work by Hernández et al. [4] is focused on the fungal pathogens associated with aerial symptoms of avocado in Tenerife (Canary Islands). Avocado is one of the most important crops in this region, and the production area has been continuously increasing in recent years, with the rise of diseases associated with symptoms such as dieback, the external necrosis of branches and inflorescences, cankers on branches and trunks, or the stem-end rot of fruits. Hernández et al. obtained 297 isolates from 158 vegetal samples collected from 2018 to 2022 that were identified using morphological and molecular methods. Most of the isolates were the Botryosphaeriaceae species, and the authors have reported the first occurrence of *Lasiodiplodia brasiliensis* as an avocado dieback-producing pathogen and *N. cryptoaustrale/stellenboschiana* as a cause of dieback and stem-end rot symptoms in avocado.

The ability of plant pathogens to adapt to environmental conditions is of special concern since it has direct consequences on their survival and capacity to expand to new regions. The work by Álvarez et al. [5] is focused in *Ralstonia solanacearum*, a bacterial phytopathogen, originally from tropical and subtropical areas, whose ability to survive in temperate environments is of concern due to global warming. In this study, two strains from either cold or warm habitats were stressed by a simultaneous exposure to natural oligotrophy at low, temperate, or warm temperatures in environmental water. In their study, *R. solanacearum* adapted through different survival responses, irrespective of their cold or warm origin. In addition, starved, cold-induced viable but no culturable state, and/or resuscitated cells maintained virulence *in planta*. This work first describes the natural nutrient availability of environmental water favoring *R. solanacearum* survival, adaptations, and resuscitation in conditions that can be found in natural settings.

Understanding the mechanisms that pathogens use to sense and colonize the host has upmost importance to design control strategies that interfere with this process, as well as to evaluate the risk of pathogen adaptation to new hosts. Sena-Vélez et al. [6] have studied the chemotactic responses of bacteria from the genus *Xanthomonas* with different host ranges. The authors identified different chemotactic responses for carbon sources and apoplastic fluids depending on the *Xanthomonas* strain and the host plant from which the apoplastic fluid was derived. These differential chemotactic responses suggest that *Xanthomonas* strains sense host specific signals that facilitate their location and entry of stomatal openings or wounds.

Plant pathogens developed different strategies for their multiplication in the hosts and transmission between plants, ensuring their evolution. In this Special Issue, Martín-Hernández and Pagán [7] conducted comparative genomic approaches using the genus *Begomovirus* of plant viruses as a model. These analyses showed that terminal gene overlapping decreases the rate of virus evolution, which is associated with the lower frequency of both synonymous and nonsynonymous mutations. In contrast, terminal overlapping has little effect on the pace of virus evolution. The analyses carried out in this work support a role for gene overlapping in the evolution of begomoviruses and provide novel information on the factors that shape their genetic diversity.

Losses in crop yields due to disease need to be reduced in order to meet increasing global food demands associated with the growth of the human population. There is a well-recognized need to reduce the use of chemical pesticides and, therefore, new environmentally friendly control strategies to combat crop disease should be developed and implemented. Focused on this theme, four papers have been published in this Special Issue.

Cabrefiga et al. [8] described an integrated approach for the control of *Alternaria* spp., the causal agent of apple leaf blotch and fruit spot disease, which recently appearance in Spain, where it causes important losses. This disease is difficult to control and requires a lot of treatments during the season. In addition, treatments must be carried out until the end of the season with the implications on fruit residues. For these reasons, an environmentally friendly strategy should be implemented. The results obtained in the present work indicate that the reduction in primary inoculum production, through the removal of winter fallen leaves and also with the treatment of leaves with the biological agent *Trichoderma asperellum*, is an important key to increase the control efficiency and to help in the reduction in phytosanitary products.

Quetglas et al. [9] analyzed the control action plan implemented in the Balearic Islands after the detection, in October 2016, of *Xylella fastidiosa*, a quarantine bacterium in the European Union. The pathogen is already naturalized in the territory, so the application of eradication or containment measures are not substantiated in epidemiological and evolutionary terms. In this review, the authors describe which control measures may or may not work for the epidemiological situations of *X. fastidiosa* on each island and how the perception of control measures has been changing as knowledge of the epidemiological situation has increased.

Badosa et al. [10] have reviewed their twenty-year experience in the field of plant diseases control, in which they have improved the methods and technologies for the in vitro or *in planta* screening of a large number of linear and cyclic peptides as well as peptide conjugates. These screening methods include the assessment of antibacterial, hemolytic, and phytotoxicity activities, and also for their plant defense elicitor capacity. As a result, the authors have been able to identify sequences that can be considered promising candidates to develop effective phytosanitary plant protection products for their use in agriculture.

Bonaterra et al. [11] have reviewed the role of biological control as a sustainable alternative or complement to conventional plant protection products for the management of fungal and bacterial plant diseases. Some of the most intensively studied biological control agents are bacteria that use multiple mechanisms involved in limiting the development of plant disease. Although several bacteria-based plant protection products have already been registered and commercialized, efforts are still required to increase the number of products available on the market. This review shows some relevant examples of known bacterial biocontrol agents. The importance of the selection process and key steps in the development of such agents is highlighted, and some improvement approaches and future trends are considered.

Author Contributions: E.F.-O., A.P.-B. and S.S. contributed equally to this Special Issue and all aspects of this Editorial. All authors have read and agreed to the published version of the manuscript.

Acknowledgments: We would like to thank all the authors who contributed their papers to this Special Issue and the reviewers for their helpful recommendations. We are also grateful to Marta Colomer, Josephine Xue, and members of the *Microorganisms'* Editorial Office for their continuous support in managing and organizing this Special Issue. In addition, we would like to acknowledge the academic editors James F. White, Mohamed Hijri, and Michael J. Bidochka who made decisions for certain papers.

Conflicts of Interest: The authors declare no conflict of interest.

References

1. Quintana, M.; de-León, L.; Cubero, J.; Siverio, F. Assessment of Psyllid Handling and DNA Extraction Methods in the Detection of '*Candidatus* Liberibacter Solanacearum' by qPCR. *Microorganisms* **2022**, *10*, 1104. [CrossRef] [PubMed]
2. Fiallo-Olivé, E.; García-Merenciano, A.C.; Navas-Castillo, J. Sweet Potato Symptomless Virus 1: First Detection in Europe and Generation of an Infectious Clone. *Microorganisms* **2022**, *10*, 1736. [CrossRef] [PubMed]
3. Fernández-Sanz, A.M.; Rodicio, M.R.; González, A.J. Biochemical Diversity, Pathogenicity and Phylogenetic Analysis of *Pseudomonas viridiflava* from Bean and Weeds in Northern Spain. *Microorganisms* **2022**, *10*, 1542. [CrossRef] [PubMed]
4. Hernández, D.; García-Pérez, O.; Perera, S.; González-Carracedo, M.A.; Rodríguez-Pérez, A.; Siverio, F. Fungal Pathogens Associated with Aerial Aymptoms of Avocado (*Persea americana* Mill) in Tenerife (Canary Islands, Spain) Focused on Species of the Family *Botryosphaeriaceae*. *Microorganisms* **2022**, *11*, 585. [CrossRef]
5. Álvarez, B.; López, M.M.; Biosca, E.G. *Ralstonia solanacearum* Facing Spread-Determining Climatic Temperatures, Sustained Starvation, and Naturally Induced Resuscitation of Viable but Non-Culturable Cells in Environmental Water. *Microorganisms* **2022**, *10*, 2503. [CrossRef] [PubMed]
6. Sena-Vélez, M.; Ferragud, E.; Redondo, C.; Graham, J.H.; Cubero, J. Chemotactic Responses of *Xanthomonas* with Different Host Ranges. *Microorganisms* **2023**, *11*, 43. [CrossRef] [PubMed]
7. Martín-Hernández, I.; Pagán, I. Gene Overlapping as a Modulator of *Begomovirus* Evolution. *Microorganisms* **2022**, *10*, 366. [CrossRef] [PubMed]
8. Cabrefiga, J.; Salomon, M.V.; Vilardell, P. Improvement of Alternaria Leaf Blotch and Fruit Spot of Apple Control through the Management of Primary Inoculum. *Microorganisms* **2023**, *11*, 101. [CrossRef] [PubMed]
9. Quetglas, B.; Olmo, D.; Nieto, A.; Borràs, D.; Adrover, F.; Pedrosa, A.; Montesinos, M.; de Dios García, J.; López, M.; Juan, A.; et al. Evaluation of Control Strategies for *Xylella fastidiosa* in the Balearic Islands. *Microorganisms* **2022**, *10*, 2393. [CrossRef] [PubMed]
10. Badosa, E.; Planas, M.; Feliu, L.; Montesinos, L.; Bonaterra, A.; Montesinos, E. Synthetic Peptides against Plant Pathogenic Bacteria. *Microorganisms* **2022**, *10*, 1784. [CrossRef] [PubMed]
11. Bonaterra, A.; Badosa, E.; Daranas, N.; Francés, J.; Roselló, G.; Montesinos, E. Bacteria as Biological Control Agents of Plant Diseases. *Microorganisms* **2022**, *10*, 1759. [CrossRef] [PubMed]

MDPI

Article

Assessment of Psyllid Handling and DNA Extraction Methods in the Detection of 'Candidatus Liberibacter Solanacearum' by qPCR

María Quintana [1,*], Leandro de-León [2], Jaime Cubero [3] and Felipe Siverio [1,4]

[1] Unidad de Protección Vegetal, Instituto Canario de Investigaciones Agrarias, 38270 San Cristóbal de La Laguna, Spain; fsiverio@icia.es
[2] Departamento de Bioquímica, Microbiología, Biología Celular y Genética, Facultad de Farmacia, Universidad de La Laguna, 38200 San Cristóbal de La Laguna, Spain; lleongue@ull.es
[3] Centro Nacional Instituto de Investigación y Tecnología Agraria y Alimentaria (INIA/CSIC), 28040 Madrid, Spain; cubero@inia.es
[4] Sección de Laboratorio de Sanidad Vegetal, Consejería de Agricultura, Ganadería, Pesca y Aguas del Gobierno de Canarias, 38270 San Cristóbal de La Laguna, Spain
* Correspondence: mariaquingoncha@gmail.com

Abstract: '*Candidatus* Liberibacter solanacearum' (CaLsol) is an uncultured bacterium, transmitted by psyllids and associated with several diseases in *Solanaceae* and *Apiaceae* crops. CaLsol detection in psyllids often requires insect destruction, preventing a subsequent morphological identification. In this work, we have assessed the influence on the detection of CaLsol by PCR in *Bactericera trigonica* (Hemiptera: Psyllidae), of four specimen preparations (entire body, ground, cut-off head, and punctured abdomen) and seven DNA extraction methods (PBS suspension, squashing on membrane, CTAB, Chelex, TRIsure^{TM}, HotSHOT, and DNeasy®). DNA yield and purity ratios, time consumption, cost, and residues generated were also evaluated. Optimum results were obtained through grinding, but it is suggested that destructive procedures are not essential in order to detect CaLsol. Although CaLsol was detected by qPCR with DNA obtained by the different procedures, HotSHOT was the most sensitive method. In terms of time consumption and cost, squashed on membrane, HotSHOT, and PBS were the fastest, while HotSHOT and PBS were the cheapest. In summary, HotSHOT was accurate, fast, simple, and sufficiently sensitive to detect this bacterium within the vector. Additionally, cross-contamination with CaLsol was assessed in the ethanol solutions where *B. trigonica* specimens were usually collected and preserved. CaLsol-free psyllids were CaLsol-positive after incubation with CaLsol-positive specimens. This work provides a valuable guide when choosing a method to detect CaLsol in vectors according to the purpose of the study.

Keywords: vector; disease; bacterium; *Bactericera trigonica*; HLB; ethanol contamination

Citation: Quintana, M.; de-León, L.; Cubero, J.; Siverio, F. Assessment of Psyllid Handling and DNA Extraction Methods in the Detection of '*Candidatus* Liberibacter Solanacearum' by qPCR. *Microorganisms* **2022**, *10*, 1104. https://doi.org/10.3390/microorganisms10061104

Academic Editors: Elvira Fiallo-Olivé

Received: 5 May 2022
Accepted: 25 May 2022
Published: 26 May 2022

Publisher's Note: MDPI stays neutral with regard to jurisdictional claims in published maps and institutional affiliations.

1. Introduction

'*Candidatus* Liberibacter solanacearum' (CaLsol) is an uncultured, phloem-limited, Gram-negative bacterium of the alphaproteobacteria class belonging to the Rhizobiaceae family [1]. It is associated with Zebra Chip in potato and several vegetative disorders in *Solanaceae* and *Apiaceae* crops [2–7]. Eleven haplotypes of CaLsol were described according to their geographical distributions, host plant, and insect vector [7–12]. Haplotypes A and B are considered as regulated non-quarantine pests in the European legislation (Commission implementing regulation-EU 2019/2072). These haplotypes have been associated with Zebra Chip in solanaceous in North and Central America, whereas haplotype A is also found in New Zealand [13]. Haplotype C is present in the North of Europe [8,9,14–16] and haplotypes D and E were described in southern Europe, northern Africa, and in the Mediterranean basin [9,17–23]. The last three haplotypes are mainly detected in *Apiaceae*, but CaLsol C and CaLsol E were also found in potatoes in Finland and Spain, respectively [24–26].

Haplotype U has been described in *Urtica dioica* L. [10], haplotype F in a single potato tuber in the United States [11] and haplotype G was found in *Solanum umbelliferum* Eschsch. in California (USA) [27]. Two haplotypes designated as H were described, one in carrots and in two species of *Poligonaceae*: *Persicaria lapathifolia* L. and *Fallopia convolvulus* L. in Finland [12], and another in *Convolvulaceae* species in USA [28]. Finally, haplotypes Cra1 and Cra2 have been found in psyllids from the Aphalaridae family: *Craspedolepta nebulosa* (Zetterstedt, 1828) and *C. subpunctata* (Foerster, 1848) in United Kingdom [29].

Currently, six species of psyllids (Hemiptera: Triozidae) have been described as CaLsol vectors. *Bactericera trigonica* Hodkinson, 1981 lives in carrot-producing areas of southern Europe, northern Africa and the Near East, next to the Mediterranean basin and the nearby Atlantic Coast [30–32]. This species feeds primarily in carrot crops, but it can also be found in other species of *Apiaceae* transmitting CaLsol [7,18]. The potato psyllid, *B. cockerelli* (Šulc, 1909), which naturally affects potato and tomato, is the cited vector for haplotypes A and B [33]. *Bactericera nigricornis* Foerster, 1848, which naturally affects carrot and potato, may transmit haplotypes D and E [34,35]. *Trioza apicalis* Foerster, 1948, naturally affects carrots and was associated to haplotype C [8]. Finally, the psyllids *T. urticae* (Linné, 1758) and *T. anthrisci* Burckhardt, 1986, are vectors of haplotype U [10,36].

International trade of plants, vegetables, and fruits between different countries has contributed to the worldwide spread of harmful pests. The correct taxonomic identification of the pest is crucial in order to adopt the most appropriate and effective control measures. This identification may be achieved by classical morphological identification or by DNA-based molecular approaches. Usually, DNA extraction methods from arthropod samples involve the destruction of the specimen, preventing its subsequent morphological identification [37–44]. Although the preservation of the whole specimen structure is not always considered, some researchers have approached the use of non-destructive DNA extraction techniques to obtain both morphological and molecular identification [45–49]. Comparative studies of different DNA extraction methods have been performed using commercial extraction kits that offer standardized methods to ensure reliable results [39,41,50]. However, these kits are often expensive and their use becomes impractical to process a large number of samples or when funding resources are limited.

Insects not only produce direct harm to plants when they feed, but also indirect harm by transmitting diseases into the crops. Diseases transmitted by insects cause great economic losses in production areas and are one of the main concerns of the sector. Thus, in some cases, analysis of insect specimens for the detection of vectored plant pathogen is a first step to study the involvement of the insect in disease transmission. DNA extraction procedures used for the molecular identification of arthropods might not be appropriate to detect the pathogen they carry. Besides, the carried microorganism is generally unevenly distributed or at low concentrations in insect tissues and therefore not easily detected, as in the case of CaLsol [43].

In addition, water traps with detergents and preservatives—such as polyethylene glycol—are used for insect capture. The insects are usually collected and preserved altogether in ethanol solutions supplemented with glycerol. The use of these solutions to capture or preserve insect vectors could provide a source of cross-contamination with the target microorganism. This might lead to incorrect interpretation in studies of CaLsol prevalence in insect populations.

The overall objective of this work was to provide an appropriate method to detect CaLsol in its vectors, allowing their subsequent identification. Four specimen preparations and seven DNA extraction methods were assessed and compared in terms of DNA yield and purity ratio, results of conventional and real-time PCR (qPCR) analysis, time consumption, cost, and residues generated. Additionally, cross-contamination with CaLsol was also evaluated in the ethanol solutions where *B. trigonica* specimens are usually collected and preserved.

2. Materials and Methods

2.1. Source of Insects

Psyllids were collected from an experimental carrot field in the Instituto Canario de Investigaciones Agrarias (ICIA), located in Valle Guerra (Tenerife, Spain). They were captured using an entomological net and immediately taken to the laboratory for identification using a binocular microscope. *Bactericera trigonica* specimens were confined in entomological cages and fed on pesticide-free carrot plants (*Daucus carota* L. var. Bangor F1) infected with CaLsol to establish a positive colony. Under these conditions, psyllids were maintained for more than three months to ensure the acquisition of the bacteria. The CaLsol-free specimens were provided by the Instituto de Ciencias Agrarias of Consejo Superior de Investigaciones Científicas from Madrid (ICA-CSIC, Madrid, Spain). To guarantee the presence and absence of CaLsol in positive and negative colonies, respectively, adults from these colonies were randomly tested with two validated qPCR protocols: the CaLsol protocol [7] and the Lso protocol [51].

Specimens were collected from the colonies and killed by freezing at −20 °C without aqueous solution. The insects were then identified and sexed using a binocular microscope and processed individually according to each assay as described in subsequent sections.

2.2. Insect Preparations

All DNA extraction methods were evaluated on ground material. Each specimen was manually ground with a tapered pestle in 1.5 mL microtube with 5 µL of absolute ethanol. Subsequently, the ethanol was evaporated at room temperature for 15 min.

Three additional non-destructive preparations were also evaluated: (i) intact full insects without treatment or manipulation (hereinafter 'whole'); (ii) decapitated insects by splitting the head from the rest of the body ('cut off head'); and (iii) abdomen punctured once by using a sterile entomological needle ('punctured abdomen').

2.3. DNA Extraction Methods

The number and type of samples used in each DNA extraction method and specimen preparation are summarized in Table 1. Seven DNA extraction procedures were assessed: CTAB [52], Chelex (Bio-Rad Laboratories, Hercules, CA, USA) [53], TRIsure™ (Bioline, London, UK), squashed on membrane [54], HotSHOT [55,56], PBS suspension, and DNeasy Blood and Tissue kit® (QIAGEN, Hilden, Germany) (hereinafter 'DNeasy®') [49]. Fourteen individuals of *B. trigonica* (seven females and seven males) were used to evaluate each combination specimen preparation/method. All DNA extraction methods were assessed and compared from ground specimen preparation. In addition, the three non-destructive preparations were evaluated using Chelex, HotSHOT, PBS, and DNeasy® (14 specimens per method and preparation). Description and modifications of the DNA extraction methods are explained below. Extracted DNA was stored at −20 °C until use.

Table 1. Number and sex of *B. trigonica* specimens used to evaluate each DNA extraction method and preparation procedure.

Methods *	Specimen Preparation			
	Whole	Grinding	Cut-Off Head	Punctured Abdomen
CTAB	N/A	7 ♀ + 7 ♂	N/A	N/A
Chelex	7 ♀ + 7 ♂	7 ♀ + 7 ♂	7 ♀ + 7 ♂	7 ♀ + 7 ♂
TRIsure™	N/A	7 ♀ + 7 ♂	N/A	N/A
Squashed on membrane	N/A	7 ♀ + 7 ♂	N/A	N/A
HotSHOT	7 ♀ + 7 ♂	7 ♀ + 7 ♂	7 ♀ + 7 ♂	7 ♀ + 7 ♂
PBS	7 ♀ + 7 ♂	7 ♀ + 7 ♂	7 ♀ + 7 ♂	7 ♀ + 7 ♂
DNeasy®	7 ♀ + 7 ♂	7 ♀ + 7 ♂	7 ♀ + 7 ♂	7 ♀ + 7 ♂

* Methods: CTAB [52]; Chelex 100 (Biorad) [53]; TRIsure™; Squashed on membrane [54]; HotSHOT [55,56]; PBS, saline phosphate buffer + Tween 20 (5%); and DNeasy® Blood and Tissue kit (Qiagen) [49]. N/A: Not applicable. ♀: female; ♂: male.

CTAB method described by Murray and Thompson [52] was carried out with some modifications. First, 400 µL of CTAB buffer (2% (weight/volume, w/v) CTAB in 100 mM Tris HCl 1 M, 1.4 M NaCl, 20 mM EDTA 0.5 M, 1% (w/v) PVP, 0.1% (w/v) sodium bisulphite) were added to the ground specimen and heated at 65 °C for 15 min. The microtube was manually shaken, heated again at 65 °C for 15 min, and centrifuged at 850 rcf for 5 min. Supernatant (400 µL) was collected and carefully mixed with an equal volume of chloroform: isoamyl alcohol (24:1) and centrifuged at 18,500 rcf for 5 min. Two hundred microliters of the supernatant were mixed with 120 µL of isopropanol and kept at −20 °C for 30 min. The mixture was centrifuged at 18,500 rcf for 20 min, and the supernatant was discarded before the addition of 1 mL of 70% ethanol. The samples were mixed and centrifuged at 18,500 rcf for 10 min and the supernatant discarded, without disturbing the pellet. Finally, the pellet was dried for 2 h at room temperature and resuspended in 100 µL of RNAse-DNAse free water.

Chelex 100 (Biorad) was performed according to Casquet et al. [53]. Ten µL of proteinase k (10 mg·mL^{-1}) and 150 µL of Chelex (10% Chelex 100 Resin, 0.1 mg·mL^{-1}) were added to each sample, incubated at 55 °C for 24 h, and then cooled to room temperature.

TRIsure^{TM} (Bioline, Sydney, Australia) method was performed according to the manufacturer's instruction [57]: 1.11 mL of TRIsure^{TM} were added to the specimen preparation, mixed with a vortex and centrifuged at 12,000 rcf for 7 min. The supernatant was collected, mixed with 200 µL of chloroform, and shaken for 10–15 s before centrifugation at 12,000 rcf for 15 min. The aqueous phase was removed and 350 µL of absolute ethanol were added to organic phase, homogenized by mixing and centrifuged at 12,000 rcf for 7 min. The supernatant was discarded and the pellet was washed twice for 2 min with 1 mL of 0.1 M sodium citrate supplemented with 10% ethanol before centrifugation at 12,000 rcf for 7 min. The pellet was washed once with 1.5 mL of 75% ethanol, gently mixed for 30 s and centrifuged at 12,000 rcf for 7 min. The supernatant was discarded and the resultant pellet was dried out at 65 °C for 2 h. Finally, the pellet was resuspended in 50 µL of RNAse-DNAse free water.

For squashed on membrane method [54], a specimen was crushed with the bottom of a microtube on a positively charged nylon membrane (NYLM-Ro Roche, Merck KGaA, Darmstadt, Germany, ref. 11209299001). The piece of membrane with the squashed insect (~0.5 cm^2) was introduced to a microtube with 100 µL of RNAse–DNAse free water. The sample was vortexed and incubated at room temperature for 5 min before use.

For HotSHOT method [55,56], each specimen was heated at 100 °C for 15 min in 20 µL of 25 mM NaOH (pH = 12). The sample was cooled at 4 °C for 5 min and 20 µL of 40 mM Tris-HCl (pH = 5) was added to neutralize the reaction, at which point the sample was ready for analysis.

For PBS method, 100 µL of buffer (8.0 g NaCl, 2.4 g Na$_2$HPO$_4$·12H$_2$O, 0.2 g KH$_2$PO$_4$, 0.2 g KCl, 1 L H$_2$O, pH 7.2–7.4) supplemented with Tween 20 (0.05%) was added to the specimen preparation, mixed with an orbital shaker for 30 min, and pulse centrifuged.

For DNeasy$^®$ method, the commercial DNA extraction kit Blood and Tissue kit$^®$ from Qiagen was used according to Sjölund [49].

2.4. DNA Yield and Purity

Quantity (ng·µL^{-1}) and quality (A$_{260}$/A$_{280}$) of extracted DNA were measured with spectrophotometer ND100 (NanoDrop Technologies, Wilmington, DE, USA). To evaluate the efficiency of each extraction procedure, the DNA yield ratio was calculated considering the resuspension volume and body weight of the psyllid [41] according to the Formula (1):

$$\text{DNA yield ratio } \left(\text{ng·mg}^{-1}\right) = \text{DNA quantity (ng·µL)}/\text{volume of DNA resuspension (µL) mean } B.\ trigonica \text{ body weight ng} \quad (1)$$

Groups of 31 males and 18 females of *B. trigonica* were weighed, obtaining 9.6 mg (an average of 309 ng/specimen) and 8.3 mg (461 ng/specimen), respectively.

2.5. Amplification Conditions by Conventional and qPCR

Conventional PCR was used to detect CaLsol by using Lso TX 16/23 primers [58] and KAPA3G Plant PCR Kit (2×) (KAPA Biosystems, Cape Town, South Africa). PCRs were carried out using a SimpliAmp Thermal Cycler (Applied Biosystems, Forest City, CA, USA). The amplification products were visualized in 1.5% (w/v) agarose gel in 0.5 M TAE buffer (40 mM Tris base, 40 mM acetic acid, 1 mM EDTA) and stained with Good View™ (SBS Genentech Co., Ltd., Beijing, China). Gel images were recorded and processed with EL Logic 100 Image system and Kodak Molecular Imaging software v.4.0.5.

The qPCR analyses were performed in StepOne Plus thermal cycler (Applied Biosystems) with SensiFAST Probe Hi-ROX kit (Bioline) following two protocols previously mentioned [7,51]. The fluorescence threshold was established automatically by the software based on the background noise for the plate. The qPCR detections were repeated with the samples preserved at −80 °C for 130 weeks, to evaluate the storing effect on the results.

Table S1 (Supplementary Materials), shows the sequences and amplification conditions of the PCR and qPCR protocols used in the present study.

2.6. Morphological State of Psyllids after DNA Extraction

Specimens exposed to non-destructive treatments for DNA extraction (DNeasy®, PBS, HotSHOT and Chelex) were examined with a stereomicroscope (Nikon SMZ800, Tokyo, Japan). The psyllid body parts with taxonomic characters (forewings, head, metathoracic legs, male and female terminalia, etc.) of treated and non-treated psyllids were compared.

2.7. Time Consumed, Cost Estimation, and Residues Generated

The time to complete each DNA extraction method was estimated with a group of 10 samples that were processed at the same time, from the first step until the DNA was ready for PCR reaction. The generation of hazardous waste during DNA extraction was also recorded for the same 10 samples. The cost per sample for the entire process was estimated for each extraction method based on the current prices of consumables and reagents at the time of the evaluation.

2.8. Cross-Contamination Assays

To evaluate if cross-contamination occurs during capture and handling of psyllids, three different assays were performed with *B. trigonica*. For all assays, tubes with CaLsol-free and CaLsol-positive psyllids were used as negative and positive controls, respectively.

2.8.1. Assay 1

Twenty CaLsol-positive psyllids (10 males and 10 females), and 20 CaLsol-free psyllids (10 males and 10 females), were introduced in a microtube (one specimen per tube) with 50 µL of 70% ethanol and incubated at room temperature for 24 h (Figure 1). Next, 20 µL of ethanol from tubes with CaLsol-positive insects were taken and mixed with 20 µL of ethanol from tubes containing CaLsol-free specimens. Additionally, volumes of ethanol (20 µL) from tubes with positive and negative specimens were also individually analyzed. The ethanol of the tubes was dried at 65 °C for 2 h. Half of the tubes were then resuspended in 100 µL of RNAse-DNAse free H_2O and the other half in 100 µL of PBS. Volumes of 3 µL from H_2O and PBS solutions were used as a template for qPCR following Lso protocol [51].

2.8.2. Assay 2

Twenty same-sex couples of *B. trigonica*, one CaLsol-positive and one CaLsol-free, were introduced in a microtube with 50 µL of 70% ethanol (10 tubes with two males and 10 tubes with two females), and incubated at room temperature for 24 h (Figure 2). The tubes were dried at 65 °C for 2 h without removing specimens. Half of the tubes were resuspended in 100 µL of RNAse-DNAse free H_2O and the other half in 100 µL of PBS. Volumes of 3 µL from H_2O and PBS solutions were used as a template for qPCR following Lso protocol [51].

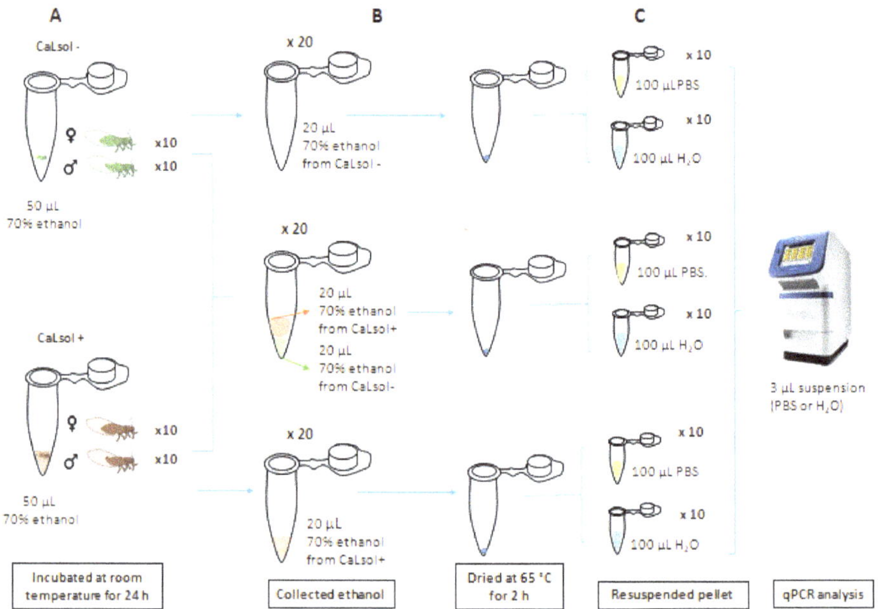

Figure 1. Cross-contamination study: Assay 1. CaLsol-positive and CaLsol-negative psyllids were individually incubated in Eppendorf tubes with 70% ethanol (**A**). Aliquots of ethanol were individually transferred to tubes or by mixing aliquots of positive and negative samples prior to drying (**B**). The samples were resuspended in water or PBS before direct qPCR analysis (**C**). ♀: female; ♂: male.

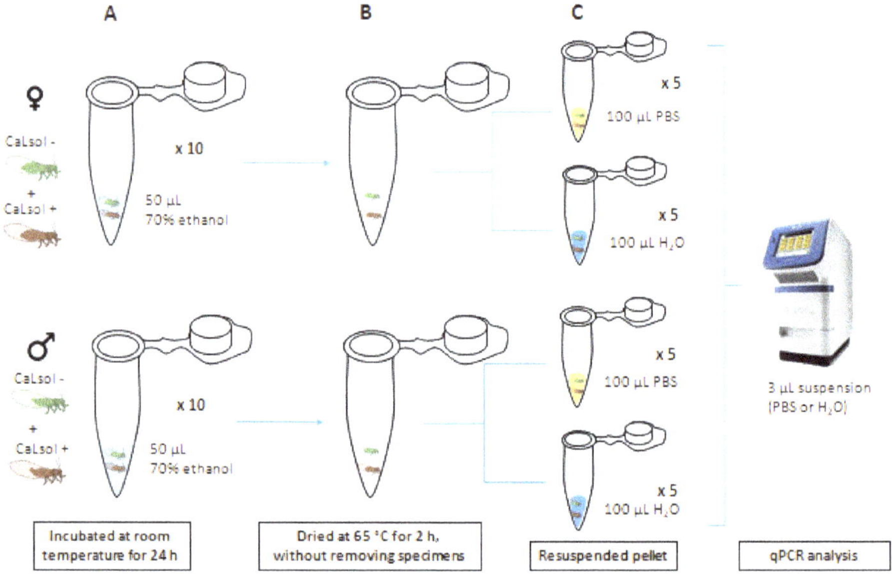

Figure 2. Cross-contamination study: Assay 2. Couples of same sex psyllids, one CaLsol-positive and one CaLsol-free, were incubated in ethanol (**A**). Ethanol was dried without removing the specimens (**B**). The samples were resuspended in water or PBS before direct qPCR analysis (**C**). ♀: female; ♂: male.

2.8.3. Assay 3

Twenty couples of *B. trigonica*, one adult male CaLsol-free and one adult female CaLsol-positive, were introduced in 50 µL of 70% ethanol and incubated 24 h at room temperature (Figure 3). Tubes were dried at 65 °C for 2 h without removing specimens. Then, half of the tubes were resuspended in 100 µL of RNAse-DNAse free H_2O and the other half in 100 µL of PBS. To detect whether there was cross-contamination from the positive female to the negative male insect, DNA extraction was performed from each specimen individually using the HotSHOT method.

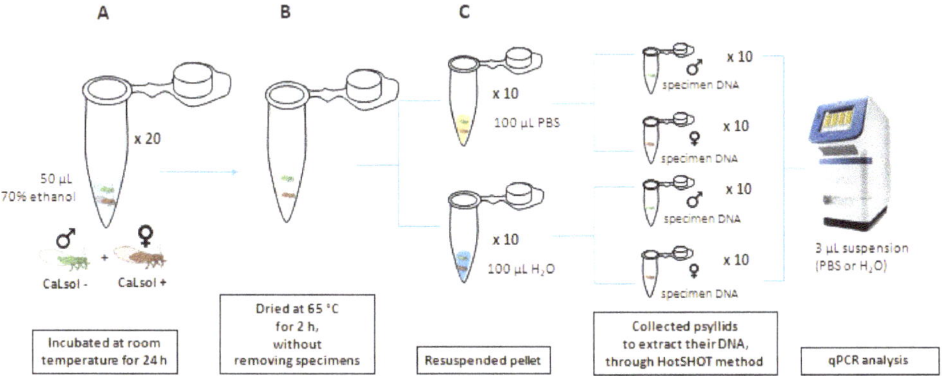

Figure 3. Cross-contamination study: Assay 3. Samples containing a pair of psyllids [one CaLsol-positive female (♀) and one CaLsol-free male (♂)] were incubated in ethanol (**A**). The ethanol was dried without removing the specimens (**B**). The samples were resuspended in water or PBS (**C**) and the psyllids were individually separated in different tubes before DNA extraction by the HotSHOT method.

2.9. Data Analysis

Statistical analysis was performed to elucidate significant differences among DNA extraction methods and specimen preparations according to the means of absorbance ratio value (A_{260}/A_{280}), DNA yield ratio ($ng \cdot mg^{-1}$) and cycle quantification (C_q) values obtained with the two protocols previously mentioned. The comparison among the specimen preparation procedures (grinding, whole, cut-off head, and punctured abdomen) were conducted with the following methods: Chelex, HotSHOT, PBS, and DNeasy®; whereas the comparison among all the DNA extraction methods was performed using ground specimen preparation. Results of qPCR of samples stored 130 weeks, where compared with previous C_q values obtained. SPSS Statistic version 22 software (IBM) was used for statistical analysis. All the data recorded were checked for normality using Shapiro–Wilk W test. Comparisons between treatments were made by ANOVA–Welch (Gaussian variables) or by Kruskal–Wallis test (for non-Gaussian variables). Post hoc testing of Gaussian variables was carried out using the Games–Howell procedure. *p*-values lower than 0.05 were considered statistically significant.

3. Results

3.1. DNA Yield and Purity

Results of DNA yield ($ng \cdot mg^{-1}$) and purity (A_{260}/A_{280}) of the seven DNA extraction methods and their comparison by using ground specimen preparation are shown in Table 2. Yield ratio of DNA ranged between 1757.1 ± 295.8 $ng \cdot mg^{-1}$ and $24,328 \pm 1638$ $ng \cdot mg^{-1}$ (mean $\pm$ SE) with significant differences among the methods evaluated (F = 43.438, df = 6, $p < 0.000$). Chelex, TRIsure™, and HotSHOT showed the highest values in DNA yield ratio followed by squashed on membrane, PBS, DNeasy®, and CTAB. Significant differences were also observed between methods when purity ratio was assessed (F = 51.530, df = 6,

$p < 0.000$), where CTAB, DNeasy® and TRIsure™ provided the highest values. By contrast, PBS and squashed on membrane provided the lowest DNA purities.

Table 2. Comparison among DNA yield ratios and DNA purity ratios (mean $\pm$ SE, $n = 14$) of seven DNA extraction methods with ground specimen preparation.

Methods	DNA Yield (ng·mg^{-1}) *	DNA Purity (A_{260}/A_{280})
CTAB	1757 $\pm$ 296 d	1.89 $\pm$ 0.11 a
Chelex	24,328 $\pm$ 1638 a	1.15 $\pm$ 0.03 b
TRIsure™	20,746 $\pm$ 4511 ab	1.59 $\pm$ 0.07 a
Squashed on membrane	6471 $\pm$ 713 cd	0.67 $\pm$ 0.05 c
HotSHOT	11,964 $\pm$ 1187 bc	1.03 $\pm$ 0.04 b
PBS	3850 $\pm$ 489 cd	0.62 $\pm$ 0.06 c
DNeasy®	3271 $\pm$ 380 d	1.85 $\pm$ 0.07 a

* DNA yield was calculated according to Chen et al. [41] based on DNA volume and average adult body weight of *B. trigonica*. Data and statistical analyses are presented in columns. The different letters mean significant differences among DNA extraction methods ($p < 0.05$).

Table 3 summarizes the comparison among the specimen preparations (grinding, whole, cut-off head, and punctured abdomen) according to their DNA yield. Chelex and HotSHOT provided the highest values in DNA yield with no significant differences among preparation procedures. However, significant differences in DNA yield within specimen preparation were observed with PBS (F = 18.741, df = 3, $p = 0.007$) and DNeasy (F = 2.96, df = 3, $p = 0.041$). Ground preparation and the punctured abdomen gave the highest NanoDrop concentration readings in PBS and DNeasy®, respectively, while the intact full insect preparation obtained the lowest values.

Table 3. Comparison of DNA yield ratios * expressed in ng·mg^{-1} (mean $\pm$ SE, $n = 14$) of the specimen preparation procedures in four DNA extraction methods.

Specimen Preparations	DNA Extraction Methods			
	Chelex	HotSHOT	PBS	DNeasy®
Whole	31,164 $\pm$ 3261 a	8078 $\pm$ 1271 a	1582 $\pm$ 211 b	2386 $\pm$ 256 b
Grinding	24,328 $\pm$ 1638 a	11,964 $\pm$ 1187 a	3850 $\pm$ 489 a	3271 $\pm$ 380 ab
Cut off head	32,936 $\pm$ 2816 a	8664 $\pm$ 1021 a	2750 $\pm$ 388 ab	3121 $\pm$ 320 ab
Punctured abdomen	31,814 $\pm$ 1956 a	10,643 $\pm$ 1523 a	2661 $\pm$ 357 ab	3786 $\pm$ 373 a

* DNA yield ratio calculated according to Chen et al. [41] based on DNA volume and average adult body weight of *B. trigonica*. Statistical analysis is shown in columns. Data followed by different letters mean significant differences among specimen preparations ($p < 0.05$) within the same method.

Comparison among specimen preparations and descriptive parameters in DNA purity (A_{260}/A_{280}) are summarized in Table 4. DNeasy® provided the best values in DNA purity, without significant differences among specimen preparation; meanwhile, all values for Chelex, HotSHOT, and PBS were below the range usually considered acceptable for molecular purposes.

Table 4. Comparison among DNA purity ratios A_{260}/A_{280} (mean $\pm$ SE, $n = 14$) of the specimen preparation procedures in four DNA extraction methods.

Specimen Preparations	DNA Extraction Methods			
	Chelex	HotSHOT	PBS	DNeasy®
Whole	1.17 $\pm$ 0.02 a	0.80 $\pm$ 0.06 b	1.38 $\pm$ 0.16 a	1.62 $\pm$ 0.09 a
Grinding	1.15 $\pm$ 0.03 a	1.03 $\pm$ 0.04 a	0.62 $\pm$ 0.06 b	1.85 $\pm$ 0.07 a
Cut off head	1.19 $\pm$ 0.02 a	0.98 $\pm$ 0.04 ab	1.15 $\pm$ 0.07 a	1.78 $\pm$ 0.06 a
Punctured abdomen	1.10 $\pm$ 0.03 a	0.79 $\pm$ 0.06 b	1.23 $\pm$ 0.11 a	1.85 $\pm$ 0.07 a

Statistical analysis is shown in columns. Data followed by different letters mean significant differences among specimen preparations ($p < 0.05$) within the same method.

3.2. Conventional PCR and qPCR

Results of conventional PCR for the detection of CaLsol are shown in Figure 4. Positive PCR results were consistently obtained with DNA extracted from all the non-destructive preparations: cut-off head, punctured abdomen, and whole. No PCR products were generated with DNAs from ground psyllids samples extracted with TRIsure™ and PBS. The brightest bands were observed in all specimen preparations in Chelex method, while whole specimen preparation in PBS, DNeasy, and HotSHOT provided the lowest intensity bands.

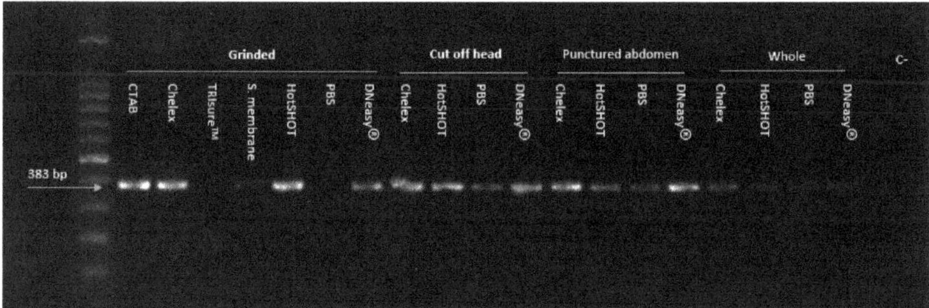

Figure 4. Agarose gel electrophoresis of products by Lso TX 16/23 primers, showing the 383 bp CaLsol-specific DNA fragment (arrow), obtained for seven DNA extraction methods and four specimen preparations. C-: negative control (right line). A 100 bp DNA ladder was included (left line).

Table 5 shows the results of the CaLsol detection of each DNA extraction method by using two different qPCR amplification protocols in ground samples. Data are expressed as positive result proportion (PP) and cycle quantification values (C_q). In general, the Lso protocol turned out to be more sensitive than CaLsol protocol to detect the bacteria from psyllid tissues (Z = -5.672; $p < 0.000$). According to the PP, CaLsol was detected in all samples regardless of the extraction method and qPCR protocol, with the exception of TRIsure™ and PBS. Significant differences in C_q values of the DNA extraction methods were obtained from both, CaLsol (F = 71.347, df = 6, $p < 0.000$) and Lso (F = 76.347, df = 6, $p < 0.000$) protocols. By means of CaLsol protocol, the HotSHOT method showed the lowest C_q values (mean ± SE, 18.15 ± 0.36) followed by Chelex, DNeasy®, and CTAB without significant differences. Similarly, with Lso protocol, the HotSHOT method showed the lowest C_q values (mean ± SE, 16.64 ± 0.23) without significant differences with Chelex and DNeasy®. The lowest sensitivity was obtained in both qPCR protocols with TRIsure™ (mean ± SE, CaLsol protocol: 29.80 ± 0.70; Lso protocol: 27.30 ± 0.88) and squashed on membrane (mean ± SE, CaLsol protocol: 29.85 ± 0.79; Lso protocol: 24.80 ± 1.03).

Table 5. CaLsol detection in ground specimen preparation with seven DNA extraction methods by two qPCR protocols.

DNA Extraction Methods	qPCR Protocols *			
	CaLsol		Lso	
	PP	C_q	PP	C_q
CTAB	100	21.9 ± 1.0 **ab**	100	22.4 ± 1.1 **c**
Chelex	100	20.1 ± 0.5 **ab**	100	18.7 ± 0.4 **ab**
TRIsure™	100	29.8 ± 0.7 **c**	92.8	27.3 ± 0.9 **c**
Squashed on membrane	100	29.8 ± 0.8 **c**	100	24.8 ± 1.0 **c**
HotSHOT	100	18.1 ± 0.4 **a**	100	16.6 ± 0.2 **a**
PBS	64.3	27.5 ± 1.2 **bc**	64.3	21.6 ± 0.7 **bc**
DNeasy®	100	20.2 ± 0.2 **ab**	100	18.1 ± 0.2 **ab**

* qPCR protocols: CaLsol [7] and Lso [51]. Results are expressed as positive results proportion. PP = 100·(No. of samples CaLso positives)/(no. of samples analyzed) and C_q mean ± SE (n = 14). Data and statistical analysis are shown in columns. Different letters mean significant differences among DNA extraction methods ($p < 0.05$).

Table 6 shows results of qPCR analysis by Li et al. [51] and Teresani et al. [7] protocols in four specimen preparations compared with the different DNA extraction methods evaluated. The Lso protocol showed a higher sensitivity compared to the CaLsol protocol as demonstrated by the lower C_q values obtained for all the samples analyzed regardless of the DNA extraction method used. In both protocols, the HotSHOT provided the lowest C_q values. Ground specimens showed the best results among other preparation methods except for the PBS procedure in CaLsol protocol. The combination of HotSHOT DNA extraction method and ground preparations provided the highest sensitivity. When Chelex was used, significant differences were shown between ground and non-destructive preparation procedures in CaLsol protocols (F = 17.274, df = 3. p = 0.001). Significant differences were also shown among specimen preparations when Lso protocol and DNeasy® extraction were used (F = 31.355, df = 3, p < 0.000). However, no differences were observed among preparations in PBS, HotSHOT, and DNeasy® with CaLsol protocol; or PBS and HotSHOT, with Lso protocol.

Table 6. CaLsol detection analysis in four specimen preparations and four DNA extraction methods by two qPCR protocols.

qPCR Protocols *	DNA Extraction Methods	Specimen Preparation			
		Whole	Ground	Cut-Off Head	Punctured Abdomen
CaLsol	Chelex	23.4 ± 0.8 b	20.1 ± 0.5 a	23.5 ± 1.1 b	23.9 ± 0.7 b
	HotSHOT	18.9 ± 0.7 a	18.1 ± 0.4 a	18.6 ± 0.6 a	20.1 ± 0.8 a
	PBS	27.4 ± 1.4 a	27.5 ± 1.2 a	26.3 ± 0.7 a	28.1 ± 1.1 a
	DNeasy®	25.4 ± 1.1 a	20.2 ± 0.2 a	22.4 ± 0.7 a	23.6 ± 1.0 a
Lso	Chelex	21.9 ± 0.7 b	18.7 ± 0.3 a	21.9 ± 0.9 b	21.6 ± 0.6 b
	HotSHOT	18.0 ± 0.5 a	16.6 ± 0.2 a	17.8 ± 0.5 a	18.4 ± 0.8 a
	PBS	23.9 ± 1.5 a	21.6 ± 0.7 a	21.7 ± 0.8 a	24.0 ± 1.2 a
	DNeasy®	25.1 ± 0.7 c	18.1 ± 0.1 a	20.9 ± 0.9 ab	21.5 ± 0.9 bc

* qPCR protocols: CaLsol [7] and Lso [51]. Results are expressed as C_q mean ± SE (n = 14). Statistical analysis is shown in rows. Data followed by different letters mean significant differences among specimen preparations (p < 0.05).

To check the usability of the isolated DNA after a long-term storage, extraction methods of ground specimen preparation with 100% of positive result proportion by qPCR were newly evaluated after 130 weeks of storage (Table 7). It was possible to detect CaLsol in all DNAs samples extracted by DNeasy®, Chelex, and CTAB by both qPCR protocols used. However, an increase in C_q values was observed after storage by CaLsol protocol (ranging from 1.8 to 5.7) and higher by Lso protocol (ranging from 4 to 7.9). In both qPCR protocols, DNeasy® and Chelex showed the lowest differences in C_q values when pre- and post-DNA storage were compared.

Table 7. Comparison of qPCR results by CaLsol and Lso protocols before (t = 0) and after 130-weeks of storage (t = 130) of ground samples in each method.

DNA Extraction Methods	qPCR Protocols *					
	CaLsol			Lso		
	PP	Cq (t = 0)	Cq (t = 130)	PP	Cq (t = 0)	Cq (t = 130)
CTAB	100	22.5 ± 1.7	28.2 ± 1.3	100	22.5 ± 1.3	27.4 ± 1.2
Chelex	100	21.0 ± 1.2	23.6 ± 0.6	100	19.4 ± 0.6	23.5 ± 0.6
Squashed on membrane	66.7	29.7 ± 1.3	32.4 ± 1.7	100	25.2 ± 1.5	31.2 ± 1.1
HotSHOT	100	18.0 ± 0.8	22.8 ± 0.4	83.3	17.4 ± 0.7	25.3 ± 1.5
DNeasy®	100	23.6 ± 2.4	25.4 ± 3.4	100	20.3 ± 1.9	24.3 ± 2.0

* qPCR protocols: CaLsol [7] and Lso [51]. Results are expressed as positive result proportion after 130 weeks of storage, PP = 100·(no. of samples CaLso positives)/(no. of samples analyzed) and C_q mean ± SE (n = 10).

3.3. Morphological State of Psyllids after DNA Extraction

After DNA extraction, all non-destructive specimen preparations tested (whole specimen, cut off head, and punctured abdomen) perfectly allowed a morphological identification of the psyllid to specific level (Figure 5). Under the conditions laid down in this work [49], DNA extraction with DNeasy® induced fragile structures in the psyllid that were easily broken when touched. This was inconvenient when handling the insect during preparation for its subsequent identification.

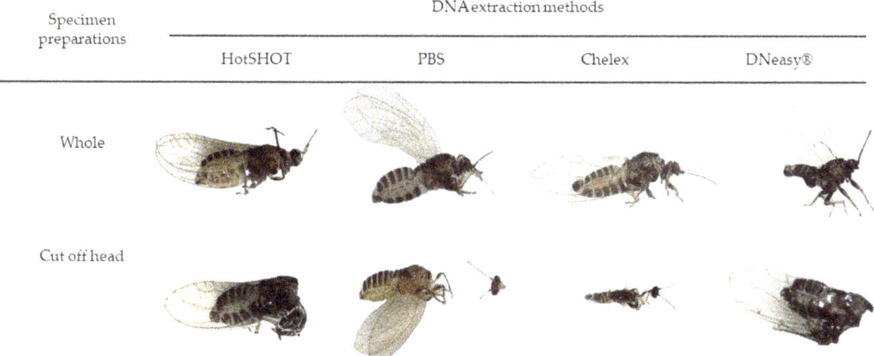

Figure 5. 'Whole' and 'Cut off head' individuals after DNA extraction with HotSHOT, PBS, Chelex, and DNeasy®.

3.4. Time Consumed, Cost Estimation, and Residues Generated

Table 8 summarizes the time consumed, cost of consumables, and reagent residues of each DNA extraction method evaluated.

Table 8. Summary of protocol according of time consumed, cost per sample, and reagent residues generated for each of the evaluated DNA extraction method.

DNA Extraction Methods	Time Consumed *	Cost Consumables (€/Sample)	Hazardous Reagent Residues *	Non-Hazardous Reagent Residues *
CTAB	2 h 20 min **	0.33	4 mL buffer CTAB + 4 mL chloroform:isoamyl alcohol (24:1)	1.2 mL isopropanol+ 10 mL 70% ethanol
Chelex	30 min ***	0.26	-	-
TRIsure[TM]	2 h **	1.16	10 mL TRIsure[TM] + 2 mL chloroform	10 mL 0.1 M sodium citrate with 10% ethanol + 3.5 mL 100% ethanol + 15 mL 75% ethanol
Squashed on membrane	15 min	0.24	-	-
HotSHOT	30 min	0.07	-	-
PBS	35 min	0.07	-	-
DNeasy®	30 min ***	2.56	-	0.2 mL proteinase K, 2 mL ethanol 100% and buffers: 1.8 mL ATL, 2 mL AL, 5 mL AW1 and 5 mL AW2.

* Estimated by processing 10 samples. ** It requires two extra hours of drying time (not included). *** It requires an overnight incubation period (not included).

Considering the time consumed during handling, the fastest methods of DNA extraction were squashed on membrane (15 min), followed by HotSHOT (30 min), and PBS (35 min). TRIsure[TM] (2 h) and CTAB (2 h 20 min), both including an extra period of 2 h for sample drying, required longer times to complete the procedure. Chelex and DNeasy® required an extra incubation time period of 24 h, although they only took 30 min of effective handling of the samples. Regarding chemical residues produced by the different protocols, Chelex, squashed on membrane, HotSHOT, and PBS did not generate waste, unlike DNeasy®, TRIsure [TM], and CTAB. In addition, these last two extraction methods produced hazardous residues such as CTAB buffer, chloroform: isoamyl alcohol (24:1), chloroform, and TRIsure[TM] lysis buffer.

The estimated cost of consumables per sample of CTAB, HotSHOT, Chelex, and PBS (ranging between 0.07 and 0.33 € per sample) was considerably lower than the cost of the commercial kits DNeasy® or TRIsure™ (2.56 € and 1.16 € per sample, respectively).

3.5. Cross Contamination Assay

The results of the cross-contamination assays during psyllid handling are shown in Table 9. The analysis of ethanol samples in which CaLsol-positive *B. trigonica* were incubated (assay 1) resulted in clear positive by qPCR from both PBS and water suspensions, with C_q values (mean ± SE) of 29.7 ± 0.6 and 30.4 ± 3.6, respectively. In the mixture of ethanol with CaLsol-positive psyllid and ethanol with CaLsol-negative psyllid, the bacterium was only detected in those samples resuspended in PBS (29.0 ± 0.6). When CaLsol-positive and CaLsol-free *B. trigonica* of the same sex were co-incubated (assay 2), the target was detected in water suspensions (31.8 ± 1.0) and PBS (28.9 ± 0.4). In both assays, the qPCR reactions ran with samples resuspended in PBS provided lower C_q values than those resuspended in water. The males of *B. trigonica* CaLsol-free co-incubated with CaLsol positive females (assay 3) gave positive results in all cases when they were individually analyzed by qPCR. Values of C_q for the male specimens were considerably higher (33.5 ± 0.4 in water and 30.9 ± 2.3 in PBS) than those obtained with the females (17.3 ± 0.8 in water and 20.2 ± 2.5 in PBS).

Table 9. Detection of CaLsol by qPCR analysis according to Li et al. (2009) in cross-contamination assays.

Assay No.	Description	Sample Analyzed	PP **	C_q, Mean ± SE, n = 10 ***
1	10 specimens CaLsol- (5♀and 5♂) (A)	H_2O	0%	nd
	10 specimens CaLsol+ (5♀and 5♂) (B)	H_2O	100%	30.4 ± 3.6
	10 mixtures of ethanol (A + B)	H_2O	0%	nd
	10 specimens CaLsol- (5♀and 5♂) (A)	PBS	0%	nd
	10 specimens CaLsol+ (5♀and 5♂) (B)	PBS	100%	29.7 ± 0.6
	10 ethanol mixture (A + B)	PBS	100%	29.0 ± 0.6
2	10 same-sex couples (5 ♀CaLsol+ ♀CaLsol- and 5 ♂CaLsol+ ♂CaLsol−)	H_2O	100%	31.8 ± 1.0
	10 same-sex couples (5 ♀CaLsol+ ♀CaLsol- and 5 ♂CaLsol + ♂CaLsol−)	PBS	100%	28.9 ± 0.4
3 *	10 ♂specimens CaLsol-	Insect	100%	33.5 ± 0.4
	10 ♀specimens CaLsol+	Insect	100%	17.2 ± 0.8
	10 ♂specimens CaLsol-	Insect	100%	30.9 ± 2.3
	10 ♀specimens CaLsol+	Insect	100%	20.2 ± 2.5

* The DNA analyzed was extracted from individual insects previously incubated in pairs (one positive female + one negative male). ** PP (Positive result proportion) = 100 (no. of samples CaLso positives)/ (no. of samples analyzed). *** Cycle quantification values of qPCR analysis by Lso protocol [51]. nd: not detected. ♀: female; ♂: male.

4. Discussion

In vector-borne diseases, such as those caused by Liberibacters that are transmitted by psyllids, it is usually necessary to detect the pathogen in the vector by molecular techniques such as PCR that require DNA extraction. Moreover, pathogen detection is sometimes a preliminary step before vector species identification, so keeping the insect structure is required. Moreover, a good DNA extraction method should be simple and fast to perform, as well as efficient in order to obtain sufficient DNA at a reasonable purity level. Comparison studies of different methods to extract DNA from insects have been carried out in previous works [41,48,50,59]. They highlight the importance of developing an accurate, fast, simple, cheap, and environmentally friendly method to facilitate and standardize work in laboratories.

This study evaluates four specimen preparations and seven DNA extraction methods, including commercial kits and in-house protocols, to detect CaLsol in the psyllid vector *B. trigonica*. The main goal was to find a non-destructive method to detect this endogenous bacterium by PCR, allowing for the subsequent morphological identification of the psyllid.

For each method, time consumed, monetary cost, and hazardous residues generated were determined. Cross-contamination with CaLsol was also assessed between psyllids handled in vials with ethanol as preservative solution.

Contrary to what might be initially expected, the crushing of the insects prior to DNA extraction did not result in a considerable increase of the total DNA yield when compared to non-destructive preparation methods. However, our study suggests that crushing specimens may improve the qPCR sensitivity for detecting CaLsol. In some ground samples, negative results were obtained by conventional PCR, which might indicate the presence of inhibitors that interfere in the reaction. In addition, our data confirm that an accurate qualitative detection of the target pathogen is also possible with four non-destructive methods, thereby allowing the subsequent morphological identification of the insects. The idea that destructive procedures are necessary to release the bacteria from the internal insect tissues, improving the detection of the target, is not fulfilled here. Obtaining enough DNA from a small insect such as *B. trigonica* (approximately 4 mm in length) is already a difficult task [60,61], and it is even more difficult to detect an endogenous bacterium that may be present in a very low concentration. However, the qPCR system provides sufficient sensitivity and specificity to allow for the detection of a few target molecules in an almost intact insect sample. In our case, CaLsol was detected in less than 0.05 µg of insect, and that level of detection was achieved through non-destructive preparations methods (unprocessed specimens, cut off head, or punctured abdomen).

DNA yield and purity were unrelated to the sample preparation procedure, but mainly to the DNA extraction protocol used. The nucleic acid purification methods showed differences among yield and purity in the DNA obtained. The best results were achieved by Chelex and TRIsure^{TM} that provided higher values in terms of DNA yield compared to the other methods. However, DNA purity obtained by these two methods was poor with an absorbance ratio (A_{260}/A_{280}) below 1.8, which is considered the minimum accepted value when determining the quality of DNA [62]. These observations could indicate the presence of impurities not removed during DNA extraction, which can lead to erroneous DNA concentration readings, obtaining abnormally high values. The Chelex method involves the addition of a chelating ion exchange resin and proteinase K that allow DNA release after a heating step [53]. Part of these products remains in the sample after DNA extraction as impurities, which may interfere with the correct determination of yield of the nucleic acids. However, it seems that there are not PCR inhibitors affecting the efficiency of the amplification by conventional PCR or qPCR. Chelex also allows valid amplifications for sequencing, genotyping, specific detection, or other applications and it is largely used to extract DNA from different matrix such as forensic materials, insects, plants, bacteria, etc. [63–66]. TRIsure^{TM}, which was developed to isolate both RNA and DNA [57], demonstrated poor sensitivity to detect CaLsol by PCR or qPCR. This low efficiency during the amplification could be caused by traces of chloroform or/and ethanol, which might remain in the extracted DNA, inhibiting the PCR reaction. The HotSHOT method rendered intermediate values in the DNA yield with poor quality, although, data from conventional PCR and qPCR revealed that it was one of the most sensitive in detecting CaLsol from psyllids.

The CTAB and the DNeasy$^®$ methods, widely used as DNA extraction procedures in diagnostic laboratories [13], provided A_{260}/A_{280} between 1.8 and 2.0, indicating high purity of the DNA [67], but with low DNA yields comparable to those obtained by PBS and squashed on membrane methods. Similar results were achieved for DNeasy$^®$ in western corn rootworm beetles [41] and in mealybugs [59], with a low DNA yield but high purity. Finally, the low-quality of the DNA obtained with PBS and squashed on membrane was not consistently and efficiently amplified by conventional PCR or qPCR. In both methods, the DNA was probably accompanied by PCR inhibitors causing a loss of efficiency or even in false negative results.

The analysis of samples which had been stored for 130 weeks at $-20\ °C$ showed a decrease in sensitivity for CaLsol detection by qPCR. Cycle quantification values were higher than those obtained previously, demonstrating a possible partial degradation of DNA, as it was noted in previous research [67–69]. However, the DNA extraction methods assessed in this study allowed the use of the isolated DNAs for more than 2 years with ground sample preparation for detection purposes. According to Hajibabbaei et al. [60] the storage potential of isolated DNA obtained through DNeasy® is moderate, being feasible one year at −20 °C, and similar results were obtained by Chakraborty et al. [69] in which DNA extracted with CTAB was stable, with a durability of about two years.

In terms of time consumption, TRIsure™ and CTAB were the most laborious methods, involving several steps during sample processing. Chelex and DNeasy® are two simple and easy to perform extraction procedures, but they include an overnight incubation step that extends the time to obtain the isolated DNA [49,53]. For DNeasy®, we followed the method optimized by Sjölund [49] for the detection of CaLsol in psyllids, although other studies have used this kit according to the manufacturer's recommendation or modifying the procedure by reducing the incubation periods [41,59,70]. Squashed on membrane and HotSHOT were the fastest procedures and excellent alternatives when a quick screening is required.

Comparing the cost of all the extraction methods, taking as a reference the cheapest ones (HotSHOT and PBS, 0.07 €/sample), DNeasy®, TRIsure™, CTAB, Chelex, and squashed on membrane were approximately 36, 17, 5, 4, and 3 times more expensive, respectively. At the same time, extracting DNA with a minimum of hazardous residues is always advisable and a factor to be taken into account when choosing the most appropriate procedure. TRIsure™ and CTAB generated hazardous reagents [57,71], which required the use of proper facilities—such as fume hoods, personal protective equipment—and a protocol to manage them safely, according to the European legislation (Regulation (CE) 1272/2008). Moreover, an additional disadvantage of using TRIsure™ is the risk of burns in contact with the skin, mucous membranes, or eyes [57]. Other DNA extraction procedures such as DNeasy® or CTAB require the use of storage containers and the costly provision of waste collection and treatment services. Therefore, the DNA extraction methods assessed for the detection of CaLsol in *B. trigonica*, offers a wide range of options regarding time consumption, price, and residues handling, thereby making it a process adaptable to the needs and circumstances of each laboratory.

In summary, HotSHOT and squashed on membrane were not only the fastest methods, but also the cheapest. Several other authors have also defined squashed on membrane and HotSHOT as fast methods to obtain DNA for diagnosis purposes [7,55,59]. Truett et al. [56] indicated that HotSHOT produced less nonspecific amplification than traditional methods and provided the cleanest PCR products. Results obtained in this work revealed that HotSHOT is the most sensitive method for detecting CaLsol by qPCR in psyllid samples. In addition, the simplicity of this method makes the procedure highly recommended for the analysis of psyllid samples in the detection of CaLsol, because it enables the performance in a single tube, keeps the structure of the specimen intact, requires less than 30 min, and can be carried out without generating contaminating residues.

During the development of this study, it was possible to detect CaLsol in psyllid samples without specimen preparation using the PBS method. This observation might indicate that, once collected, the insect could release the endogenous CaLsol into the surrounding environment. This is especially important during field prospections since many insects are usually collected and mixed in the same vials containing ethanol or other preservative solutions. Our results demonstrate that CaLsol-positive psyllids were able to cross-contaminate CaLsol-negative psyllids collected in the same tubes, resulting in positive reactions by qPCR from specimens not initially carrying the bacteria. These results must be considered in studies which seek to determine the prevalence of this endogenous organism in order to avoid an overestimation of its proportion in the psyllid population. Thus, it is advisable to collect the insects with CO_2 or by freezing at −20 °C, and to store the specimens in individual vials with the desired ethanol solution.

International trade of plants, vegetables, and fruits between different countries has caused the worldwide spread of harmful pests. The correct taxonomic identification of pests is crucial in order to adopt the most appropriate and effective control measures. Nowadays, the use of barcoding molecular techniques for this aim is more frequent and, consequently, the available information of specimen's sequences in databases is increasing [72]. When recognition and identification of arthropods by morphological approaches is required, keeping the arthropod exoskeleton after DNA extraction is mandatory for proper identification by trained specialists. Our work, which improves the knowledge on different specimen preparations and DNA extraction methods, was aimed at finding the best method to detect CaLsol in the vector. However, it is suggested that it could also be useful for the identification of insects using molecular barcoding techniques.

Moreover, information shown in this work could be a valuable guide for the detection of other species of these bacteria in their vectors, with a special focus on the species '*Ca.* L. africanus' (CaLaf), '*Ca.* L. americanus', and '*Ca.* L. asiaticus' (CaLas); associated with Huanglongbing (HLB), the most severe disease of citrus plants (*Citrus* spp.) [1,73]; and transmitted by *Diaphorina citri* [6,73,74] and *Trioza erytreae* [6,75]. These vectors could be handled as *B. trigonica* in this study as an alternative to the current protocols used [76,77]. This work provides a wide range of options to detect CaLsol in its vectors according to the purpose of the study. Among the extraction methods evaluated, HotSHOT was the fastest, cheapest, safest, and did not require destructive preparation to consistently detect CaLsol in psyllid vectors by qPCR.

Supplementary Materials: The following supporting information can be downloaded at: https://www.mdpi.com/article/10.3390/microorganisms10061104/s1, Table S1: References, sequences and amplification conditions of the PCR and qPCR protocols used in the present study.

Author Contributions: Conceptualization, M.Q., L.d.-L., J.C. and F.S.; Methodology, M.Q. and L.d.-L.; Formal analysis, M.Q.; Investigation, M.Q.; Resources, F.S.; Data curation, M.Q.; Writing—original draft preparation, M.Q.; Writing—review and editing, L.d.-L., J.C. and F.S.; Funding acquisition, F.S. All authors have read and agreed to the published version of the manuscript.

Funding: This research was funded by Instituto Nacional de Investigación y Tecnología Agraria y Alimentaria, INIA E-RTA2014-00008-C4-01. María Quintana-González de Chaves was recipient of a 2017–2021 PhD grant from INIA.

Acknowledgments: We thank Aránzazu Moreno (Instituto de Ciencias Agrarias of Consejo Superior de Investigaciones Científicas from Madrid, ICA-CSIC) for providing us of specimens of *B. trigonica* CaLsol free, Estíbaliz Padilla for her contribution in the assay development, Carlos Álvarez (Instituto Canario de Investigaciones Agrarias, ICIA) for helping us with the statistical analysis, and Servicio de Sanidad Vegetal de la Dirección General de Agricultura del Gobierno de Canarias for allowing the use of its laboratory equipment.

Conflicts of Interest: The authors declare no conflict of interest. The funders had no role in the design of the study; in the collection, analyses, or interpretation of data; in the writing of the manuscript, or in the decision to publish the results.

References

1. Jagoueix, S.; Bové, J.M.; Garnier, M. The phloem-limited bacterium of greening disease of citrus is a member of the subdivision of the proteobacteria. *Int. J. Syst. Bacteriol.* **1994**, *44*, 379–386. [CrossRef] [PubMed]
2. Hansen, A.K.; Trumble, J.T.; Stouthamer, R.; Paine, T.D. A new huanglongbing species, '*Candidatus* Liberibacter psyllaurous', found to infect tomato and potato, is vectored by the psyllid *Bactericera cockerelli* (Sulc). *Appl. Environ. Microbiol.* **2008**, *74*, 5862–5865. [CrossRef] [PubMed]
3. Liefting, L.W.; Perez-Egusquiza, Z.C.; Clover, G.R.G. A New '*Candidatus* Liberibacter species' in *Solanum tuberosum* in New Zealand. *Plant Dis.* **2009**, *92*, 1474. [CrossRef]
4. Liefting, L.W.; Weir, B.S.; Pennycook, S.R.; Clover, G.R.G. '*Candidatus* Liberibacter solanacearum', associated with plants in the family Solanaceae. *Int. J. Syst. Evol. Microbiol.* **2009**, *59*, 2274–2276. [CrossRef] [PubMed]
5. Munyaneza, J.E. Zebra Chip Disease of potato: Biology, epidemiology and Management. *Am. J. Potato Res.* **2012**, *89*, 329–350. [CrossRef]

6.   Haapalainen, M. Biology and epidemics of '*Candidatus* Liberibacter specie', psyllid-transmitted plant-pathogenic bacteria. *Ann. Appl. Biol.* **2014**, *165*, 172–198. [CrossRef]

7.   Teresani, G.R.; Bertolini, E.; Alfaro-Fernández, A.; Martínez, C.; Tanaka, F.A.O.; Kitajima, E.W.; Roselló, M.; Sanjuan, S.; Ferrándiz, J.C.; López, M.M.; et al. Association of '*Candidatus* Liberibacter solanacearum' with a vegetative disorder of celery in Spain and development of a real-time PCR method for its detection. *Phytopathology* **2014**, *104*, 804–811. [CrossRef]

8.   Nelson, W.R.; Fisher, T.W.; Munyaneza, J.E. Haplotypes of '*Candidatus* Liberibacter solanacearum' suggest long-standing separation. *Eur. J. Plant Pathol.* **2011**, *130*, 5–12. [CrossRef]

9.   Nelson, W.R.; Sengoda, V.G.; Alfaro-Fernández, A.O.; Font, M.I.; Crosslin, J.M.; Munyaneza, J.E. A new haplotype of '*Candidatus* Liberibacter solanacearum' identified in the Mediterranean region. *Eur. J. Plant Pathol.* **2013**, *135*, 633–639. [CrossRef]

10.   Haapalainen, M.L.; Wang, J.; Latvala, S.; Lehtonen, M.T.; Pirhonen, M.; Nissinen, A. Genetic variation of '*Candidatus* Liberibacter solanacearum' haplotype C and identification of a novel haplotype from *Trioza urticae* and stinging nettle. *Phytopathology* **2018**, *108*, 925–934. [CrossRef]

11.   Swisher-Grimm, K.D.; Garczynski, S.F. Identification of a new haplotype of '*Candidatus* Liberibacter solanacearum' in Solanum tuberosum. *Plant Dis.* **2019**, *103*, 468–474. [CrossRef] [PubMed]

12.   Haapalainen, M.; Latvala, S.; Wickström, A.; Wang, J.; Pirhonen, M.; Nissinen, A.I. A novel haplotype of '*Candidatus* Liberibacter solanacearum' found in Apiaceae and Polygonaceae family plants. *Eur. J. Plant Pathol.* **2020**, *156*, 413–423. [CrossRef]

13.   EPPO, European and Mediterranean Plant Protection. PM 7/143 (1) '*Candidatus* Liberibacter solanacearum'. *EPPO Bull.* **2020**, *50*, 49–68. [CrossRef]

14.   Munyaneza, J.E.; Fisher, T.W.; Sengoda, V.G.; Garczynski, S.F.; Nissinen, A.; Lemmetty, A. Association of '*Candidatus* Liberibacter solanacearum' with the psyllid, *Trioza apicalis* (Hemiptera: Triozidae) in Europe. *J. Econ. Entomol.* **2010**, *103*, 1060–1070. [CrossRef] [PubMed]

15.   Munyaneza, J.E.; Lemmetty, A.; Nissinen, A.; Sengoda, V.G.; Fisher, T.W. Molecular detection of aster yellows phytoplasma and '*Candidatus* Liberibacter solanacearum' in carrots affected by the psyllid *Trioza apicalis* (Hemiptera: Triozidae) in Finland. *J. Plant Pathol.* **2011**, *93*, 697–700.

16.   Munyaneza, J.E. Zebra chip disease, '*Candidatus* Liberibacter', and potato psyllid: A global threat to the potato industry. *Am. J. Potato Res.* **2015**, *92*, 230–235. [CrossRef]

17.   Alfaro-Fernández, A.; Siverio, F.; Cebrián, M.C.; Villaescusa, F.J.; Font, M.I. '*Candidatus* Liberibacter solanacearum' associated with *Bactericera trigonica*-affected carrots in the Canary Islands. *Plant Dis.* **2012**, *96*, 581. [CrossRef]

18.   Alfaro-Fernández, A.; Cebrián, M.C.; Villaescusa, F.J.; Hermoso de Mendoza, A.; Ferrándiz, J.C.; Sanjuán, S.; Font, M.I. First report of '*Candidatus* Liberibacter solanacearum' in carrots in mainland Spain. *Plant Dis.* **2012**, *96*, 582. [CrossRef]

19.   Loiseau, M.; Garnier, S.; Boirin, V.; Merieae, M.; Leguay, A.; Renaudin, I.; Renvoisé, J.P.; Gentit, P. First report of '*Candidatus* Liberibacter solanacearum' in carrot in France. *Plant Dis.* **2014**, *98*, 839. [CrossRef]

20.   Tahzima, R.; Maes, M.; Achbani, E.H.; Swisher, K.D.; Munyaneza, J.E.; De Jonghe, K. First Report of '*Candidatus* Liberibacter solanacearum' on carrot in Africa. *Plant Dis.* **2014**, *98*, 1426. [CrossRef]

21.   Holeva, M.C.; Glynos, P.E.; Karafla, C.D. First report of '*Candidatus* Liberibacter solanacearum' on carrot in Greece. *Plant Dis.* **2017**, *101*, 1654. [CrossRef]

22.   Ben Othmen, S.; Morán, F.E.; Navarro, I.; Barbé, S.; Martínez, C.; Marco-Noales, E.; Chermiti, B.; López, M.M. '*Candidatus* Liberibacter solanacearum' haplotypes D and E in carrot plants and seeds in Tunisia. *J. Plant Pathol.* **2018**, *100*, 197–207. [CrossRef]

23.   Mawassi, M.; Dror, O.; Bar-Joseph, M.; Piasezky, A.; Sjölund, J.M.; Levitzky, N.; Shoshana, N.; Meslenin, L.; Haviv, S.; Porat, C.; et al. '*Candidatus* Liberibacter solanacearum' is tightly associated with carrot yellows symptoms in Israel and transmitted by the prevalent psyllid vector *Bactericera trigonica*. *Phytopathology* **2018**, *108*, 1056–1066. [CrossRef] [PubMed]

24.   Haapalainen, M.; Latvala, S.; Rastas, M.; Wang, J.; Hannukkala, A.; Pirhonen, M.; Nissinen, A.I. A Carrot Pathogen '*Candidatus* Liberibacter solanacearum' Haplotype C detected in symptomless potato plants in Finland. *Potato Res.* **2018**, *61*, 1–20. [CrossRef]

25.   MAPA. *Programa Para La Aplicación De La Normativa Fitosanitaria De La Patata*; Ministerio de Agricultura, Pesca y Alimentación: Madrid, Spain, 2019; Available online: https://www.mapa.gob.es/es/agricultura/temas/sanidad-vegetal/manualpatata2019_tcm30-135971.pdf (accessed on 26 July 2021).

26.   Ruiz-Padilla, A.; Redondo, C.; Asensio, A.; Garita-Cambronero, J.; Martínez, C.; Pérez-Padilla, V.; Marquínez, R.; Collar, J.; García-Méndez, E.; Alfaro-Fernández, A.; et al. Assessment of multilocus sequence analysis (MLSA) for identification of '*Candidatus* Liberibacter solanacearum' from different host plants in Spain. *Microorganisms* **2020**, *8*, 1446. [CrossRef] [PubMed]

27.   Mauck, K.E.; Sun, P.; Meduri, V.; Hansen, A.K. New *Ca*. Liberibacter psyllaurous' haplotype resurrected from a 49-year-old specimen of *Solanum umbelliferum*: A native host of the psyllid vector. *Sci. Rep.* **2019**, *9*, 9530. [CrossRef]

28.   Contreras-Rendón, A.; Sánchez-Pale, J.R.; Fuentes-Aragón, D.; Alanís-Martínez, I.; Silva-Rojas, H.V. Conventional and qPCR reveals the presence of '*Candidatus* Liberibacter solanacearum' haplotypes A, and B in *Physalis philadelphica* plant, seed, and *Bactericera cockerelli* psyllids, with the assignment of a new haplotype H in Convolvulaceae. *Antonie Leeuwenhoek Int. J. Gen. Mol. Microbiol.* **2020**, *113*, 533–551. [CrossRef]

29.   Sumner-Kalkun, J.C.; Highet, F.; Arnsdorf, Y.M.; Back, E.; Carnegie, M.; Madden, S.; Carboni, S.; Billaud, W.; Lawrence, Z.; Kenyon, D. '*Candidatus* Liberibacter solanacearum' distribution and diversity in Scotland and the characterisation of novel haplotypes from *Craspedolepta* spp. (Psyllidae: Aphalaridae). *Sci. Rep.* **2020**, *10*, 1–11. [CrossRef]

30. Antolínez, C.A.; Fereres, A.; Moreno, A. Risk assessment of '*Candidatus* Liberibacter solanacearum' transmission by the psyllids *Bactericera trigonica* and *B. tremblayi* from Apiaceae crops to potato. *Sci. Rep.* **2017**, *7*, 45534. [CrossRef]
31. Teresani, G.R.; Hernández, E.; Bertolini, E.; Siverio, F.; Moreno, A.; Fereres, A.; Cambra, M. Transmission of '*Candidatus* Liberibacter solanacearum' by *Bactericera trigonica* Hodkinson to vegetable hosts. *Span. J. Agric. Res.* **2017**, *15*, e1011. [CrossRef]
32. EPPO, European and Mediterranean Plant Protection. EPPO Global Database: Bactericera Trigonica. Available online: https://gd.eppo.int/taxon/BCTCTR (accessed on 6 October 2020).
33. Mustafa, T.; Horton, D.R.; Cooper, W.R.; Swisher, K.D.; Zack, S.; Pappu, H.R.; Munyaneza, J.E. Use of electrical penetration graph technology to examine transmission of '*Candidatus* Liberibacter solanacearum' to potato by three haplotypes of potato psyllid (*Bactericera cockerelli*; Hemiptera: Triozidae). *PLoS ONE* **2015**, *10*, e0138946. [CrossRef] [PubMed]
34. Teresani, G.; Hernández, E.; Bertolini, E.; Siverio, F.; Marroquín, C.; Molina, J.; de Mendoza, A.H.; Cambra, M. Search for potential vectors of '*Candidatus* Liberibacter solanacearum': Population dynamics in host crops. *Span. J. Agric. Res.* **2015**, *13*, e1002. [CrossRef]
35. Antolínez, C.A.; Moreno, A.; Ontiveros, I.; Pla, S.; Plaza, M. Seasonal abundance of psyllid species associated with carrots and potato fields in Spain. *Insects* **2019**, *10*, 287. [CrossRef] [PubMed]
36. Sjölund, M.J.; Clark, M.; Carnegie, M.; Greenslade, A.F.C.; Ouvrard, D.; Highet, F.; Sigvald, R.; Bell, J.R.; Arnsdorf, Y.M.; Cairns, R.; et al. First report of '*Candidatus* Liberibacter solanacearum' in the United Kingdom in the psyllid *Trioza anthrisci*. *New Dis. Rep.* **2017**, *36*, 4. [CrossRef]
37. Tian, E.W.; Yu, H. A Simple and rapid DNA extraction protocol of small insects for PCR amplification. *Entomol. News.* **2014**, *123*, 303. [CrossRef]
38. Zeh, D.W.; May, C.A.; Coffroth, M.A.; Bermingham, E. MboI and Macrohaltica—Quality of DNA fingerprints is strongly enzyme-dependent in an insect (Coleoptera). *Mol. Ecol.* **1993**, *2*, 61–63. [CrossRef]
39. Ball, S.L.; Armstrong, K.F. Rapid, One-Step DNA Extraction for insect pest identification by using DNA barcodes. *J. Econ. Entomol.* **2008**, *101*, 523–532. [CrossRef]
40. Calderón-Cortés, N.; Quesada, M.; Cano-Camacho, H.; Zavala-Páramo, G. A simple and rapid method for DNA isolation from xylophagous insects. *Int. J. Mol. Sci.* **2010**, *11*, 5056–5064. [CrossRef]
41. Chen, H.; Rangasamy, M.; Tan, S.Y.; Wang, H.; Siegfried, B.D. Evaluation of five methods for total DNA extraction from western corn rootworm beetles. *PLoS ONE* **2010**, *5*, e11963. [CrossRef]
42. Oliveira, N.; Luís, M.I.; Revers, F. Developing a rapid, efficient and low cost method for rapid DNA extraction from arthropods. *Ciênc. Rural* **2011**, *419419*, 1563–1570.
43. Musapa, M.; Kumwenda, T.; Mkulama, M.; Chishimba, S.; Norris, D.E.; Thuma, P.E.; Mharakurwa, S. A Simple Chelex Protocol for DNA Extraction from *Anopheles* spp. *J. Vis. Exp.* **2013**, *71*, 1–6. [CrossRef] [PubMed]
44. Freitas, M.T.S.; Gomes-Júnior, P.P.; Batista, M.V.A.; Leal-Balbino, T.C.; Araujo, A.L.; Balbino, V.Q. Novel DNA extraction assay for molecular idenification of *Aedes* spp. eggs. *Genet. Mol. Res.* **2014**, *13*, 8776–8782. [CrossRef] [PubMed]
45. Phillips, A.J.; Simon, C. Protocol for Arthropods. *Ann. Entomol. Soc. Am.* **1995**, *88*, 281–283. [CrossRef]
46. Gilbert, M.T.P.; Moore, W.; Melchior, L.; Worebey, M. DNA extraction from dry museum beetles without conferring external morphological damage. *PLoS ONE* **2007**, *2*, e272. [CrossRef]
47. Castalanelli, M.A.; Severtson, D.L.; Brumley, C.J.; Szito, A.; Foottit, R.G.; Grimm, M.; Munyard, K.; Groth, D.M. A rapid non-destructive DNA extraction method for insects and other arthropods. *J. Asia Pac. Entomol.* **2010**, *13*, 243–248. [CrossRef]
48. Miura, K.; Higashiura, Y.; Maeto, K. Evaluation of easy, non-destructive methods of DNA extraction from minute insects. *Appl. Entomol. Zool.* **2017**, *52*, 349–352. [CrossRef]
49. Sjölund, M.J. Non-destructive DNA extraction from Psyllids. POnTE, Pest Organisms Threatening Europe. 2017. Available online: https://www.ponteproject.eu/protocols-calsol/non-destructive-dna-extraction-psyllids/ (accessed on 5 March 2021).
50. Asghar, U.; Malik, M.F.; Anwar, F.; Javed, A.; Raza, A. DNA extraction from insects by using different techniques: A review. *Adv. Entomol.* **2015**, *3*, 132–138. [CrossRef]
51. Li, W.; Abad, J.A.; French-Monar, R.D.; Rascoe, J.; Wen, A.; Gudmestad, N.C.; Secor, G.A.; Lee, M.; Duan, Y.; Levy, L. Multiplex real-time PCR for detection, identification and quantification of '*Candidatus* Liberibacter solanacearum' in potato plants with zebra chip. *J. Microbiol. Methods.* **2009**, *78*, 59–65. [CrossRef]
52. Murray, M.G.; Thompson, W.F. Rapid isolation of high molecular weight plant DNA. *Nucleic Acids Res.* **1980**, *8*, 4321–4326. [CrossRef]
53. Casquet, J.; Thebaud, C.; Gillespie, R.G. Chelex without boiling, a rapid and easy technique to obtain stable amplifiable DNA from small amounts of ethanol-stored spiders. *Mol. Ecol. Resour.* **2012**, *12*, 136–141. [CrossRef]
54. Bertolini, E.; Felipe, R.T.A.; Sauer, A.V.; Lopes, S.A.; Arilla, A.; Vidal, E.; Mourão Filho, F.D.A.A.; Nunes, W.M.C.; Bové, J.M.; López, M.M.; et al. Tissue-print and squash real-time PCR for direct detection of '*Candidatus* Liberibacter' species in citrus plants and psyllid vectors. *Plant Pathol.* **2014**, *63*, 1149–1158. [CrossRef]
55. Collado-Romero, M.; Mercado-Blanco, J.; Olivares-García, C.; Valverde-Corredor, A.; Jiménez-Díaz, R.M. Molecular variability within and among *Verticillium dahliae* vegetative compatibility groups determined by fluorescent amplified fragment length polymorphism and polymerase chain reaction markers. *Phytopathology* **2006**, *96*, 485–495. [CrossRef] [PubMed]
56. Truett, G.E.; Heeger, P.; Mynatt, R.L.; Walker, J.A.; Warman, M.L. Myocardial atrophy and chronic mechanical unloading of the failing human heart: Implications for cardiac assist device-induced myocardial recovery. *J. Am. Coll. Cardiol.* **2014**, *64*, 1602–1612.

57. Bioline. Meridian Bioscience. TRIsure™. 2020. Available online: https://www.bioline.com/trisure.html (accessed on 20 November 2020).
58. Ravindran, A.; Levy, J.; Pierson, E.; Gross, D.C. Development of primers for improved PCR detection of the potato zebra chip pathogen, '*Candidatus* Liberibacter solanacearum'. *Plant Dis.* **2011**, *95*, 1542–1546. [CrossRef] [PubMed]
59. Wang, Y.S.; Dai, T.M.; Tian, H.; Wan, F.H.; Zhang, G.F. Comparative analysis of eight DNA extraction methods for molecular research in mealybugs. *PLoS ONE* **2019**, *14*, e0226818. [CrossRef] [PubMed]
60. Hajibabaei, M.; DeWaard, J.R.; Ivanova, N.V.; Ratnasingham, S.; Dooh, R.T.; Kirk, S.L.; Mackie, P.M.; Hebert, P.D. Critical factors for assembling a high volume of DNA barcodes. *Philos. Trans. R. Soc. B Biol. Sci.* **2005**, *360*, 1959–1967. [CrossRef]
61. Dittrich-Schröder, G.; Wingfield, M.J.; Klein, H.; Slippers, B. DNA extraction techniques for DNA barcoding of minute gall-inhabiting wasps. *Mol. Ecol. Resour.* **2012**, *12*, 109–115. [CrossRef]
62. Olson, N.D.; Morrow, J.B. DNA extract characterization process for microbial detection methods development and validation. *BMC Res. Notes* **2012**, *5*, 668. [CrossRef]
63. Konakandla, B.; Park, Y.; Margolies, D. Whole genome amplification of Chelex-extracted DNA from a single mite: A method for studying genetics of the predatory mite *Phytoseiulus persimilis*. *Exp. Appl. Acarol.* **2006**, *40*, 241–247. [CrossRef]
64. Martín-Platero, A.M.; Peralta-Sánchez, J.M.; Soler, J.J.; Martínez-Bueno, M. Chelex-based DNA isolation procedure for the identification of microbial communities of eggshell surfaces. *Anal. Biochem.* **2010**, *397*, 253–255. [CrossRef]
65. Stephen, C.Y.; Lin, S.; Lai, K. An evaluation of the performance of five extraction methods: Chelex® 100, QIAamp® DNA Blood Mini Kit, QIAamp® DNA Investigator Kit, QIAsymphony® DNA Investigator® Kit and DNA IQ™. *Sci. Justice* **2015**, *55*, 200–208.
66. Turan, C.; Nanni, I.M.; Brunelli, A.; Collina, M. New rapid DNA extraction method with Chelex from *Venturia inaequalis* spores. *J. Microbiol. Methods.* **2015**, *115*, 139–143. [CrossRef] [PubMed]
67. Sambrook, J. *The Condensed Protocols from Molecular Cloning: A Laboratory Manual*; Cold Spring Harbor Laboratory Press: New York, NY, USA, 2006.
68. Fang, G.; Hammar, S.; Grumet, R. A quick and inexpensive method for removing polysaccharides from plant genomic DNA. *Biotechniques* **1992**, *13*, 52–54.
69. Chakraborty, S.; Saha, A.; Neelavar Ananthram, A. Comparison of DNA extraction methods for non-marine molluscs: Is modified CTAB DNA extraction method more efficient than DNA extraction kits? *3 Biotech* **2020**, *10*, 1–6. [CrossRef] [PubMed]
70. Qiagen. Supplementary Protocol: Purification of Total DNA from Insects Using the DNeasy®Blood &Tissue Kit. Available online: https://www.qiagen.com/br/resources/faq?id=58ddc6c7-d1f4-45d8-9156-8174fed15bf3&lang=en (accessed on 6 May 2020).
71. Merck. Safety Data Sheet: Hexadecyltrimethylammonium Bromide Version 6.0. Available online: https://www.sigmaaldrich.com/MSDS/MSDS/DisplayMSDSPage.do?country=ES&language=es&productNumber=H9151&brand=SIGMA&PageToGoToURL=https%3A%2F%2Fwww.sigmaaldrich.com%2Fcatalog%2Fproduct%2Fsigma%2Fh9151%3Flang%3Des (accessed on 6 May 2020).
72. Ashfaq, M.; Hebert, P.D.N. DNA barcodes for bio-surveillance: Regulated and economically important arthropod plant pests. *Genome* **2016**, *59*, 933–945. [CrossRef] [PubMed]
73. Teixeira, D.C.; Saillard, C.; Eveillard, S.; Danet, J.L.; Ayres, A.J.; Bové, J.M. '*Candidatus* Liberibacter americanus', associated with citrus Huanglongbing (greening disease) in São Paulo State, Brazil. *Int. J. Syst. Evol. Microbiol.* **2005**, *55*, 1857–1862. [CrossRef] [PubMed]
74. Capoor, D.P.; Rao, D.G.; Viawanath, S.M. *Diaphoria citri* Kuway., a vector of the greening disease of citrus in India. *Indian Agric. Res. Inst.* **1967**, *376*, 572–579.
75. McClean, A.P.D.; Oberholzer, P.C.J. Citrus psylla, a vector of the greening disease of Sweet Orange. *S. Afr. J. Agric. Sci.* **1965**, *81*, 297–298.
76. Manjunath, K.L.; Halbert, S.E.; Ramadugu, C.; Webb, S.; Lee, R.F. Detection of '*Candidatus* Liberibacter asiaticus' in *Diaphorina citri* and its importance in the management of citrus huanglongbing in Florida. *Phytopathology* **2008**, *98*, 387–396. [CrossRef]
77. Tabachnick, W.J. *Diaphorina citri* (Hemiptera: Liviidae) Vector Competence for the Citrus Greening Pathogen '*Candidatus* Liberibacter Asiaticus'. *J. Econ. Entomol.* **2015**, *108*, 839–848. [CrossRef]

Article

Sweet Potato Symptomless Virus 1: First Detection in Europe and Generation of an Infectious Clone

Elvira Fiallo-Olivé *, Ana Cristina García-Merenciano and Jesús Navas-Castillo

Instituto de Hortofruticultura Subtropical y Mediterránea "La Mayora" (IHSM-UMA-CSIC), Consejo Superior de Investigaciones Científicas, Avenida Dr. Wienberg s/n, 29750 Algarrobo-Costa, Málaga, Spain
* Correspondence: efiallo@eelm.csic.es

Abstract: Sweet potato (*Ipomoea batatas*), a staple food for people in many of the least developed countries, is affected by many viral diseases. In 2017, complete genome sequences of sweet potato symptomless virus 1 (SPSMV-1, genus *Mastrevirus*, family *Geminiviridae*) isolates were reported, although a partial SPSMV-1 genome sequence had previously been identified by deep sequencing. To assess the presence of this virus in Spain, sweet potato leaf samples collected in Málaga (southern continental Spain) and the Spanish Canary Islands of Tenerife and Gran Canaria were analyzed. SPSMV-1 was detected in samples from all the geographical areas studied, as well as in plants of several entries obtained from a germplasm collection supposed to be virus-free. Sequence analysis of full-length genomes of isolates from Spain showed novel molecular features, i.e., a novel nonanucleotide in the intergenic region, TCTTATTAC, and a 24-nucleotide deletion in the V2 open reading frame. Additionally, an agroinfectious clone was developed and infectivity assays showed that the virus was able to asymptomatically infect *Nicotiana benthamiana, Ipomoea nil, I. setosa*, and sweet potato, thus confirming previous suggestions derived from observational studies. To our knowledge, this is the first report of the presence of SPSMV-1 in Spain and Europe and the first agroinfectious clone developed for this virus.

Keywords: sweet potato; mastrevirus; *Geminiviridae*; infectious clone; agroinoculation

Citation: Fiallo-Olivé, E.; García-Merenciano, A.C.; Navas-Castillo, J. Sweet Potato Symptomless Virus 1: First Detection in Europe and Generation of an Infectious Clone. *Microorganisms* **2022**, *10*, 1736. https://doi.org/10.3390/microorganisms10091736

Academic Editor: Mohamed Hijri

Received: 15 August 2022
Accepted: 26 August 2022
Published: 28 August 2022

Publisher's Note: MDPI stays neutral with regard to jurisdictional claims in published maps and institutional affiliations.

1. Introduction

Sweet potato (*Ipomoea batatas*) is one of the most consumed crops in the world and is a staple food for people in many of the least developed countries. This crop is the sixth most important food crop worldwide after rice, wheat, potato, maize, and cassava. Sweet potato crops produce more edible energy per hectare per day than other important crops, such as wheat, rice, and cassava, and are an important source of carbohydrates, beta-carotene, and vitamins B, C, and E. Sweet potato is vegetatively propagated by vine cuttings, which cause the accumulation of pathogens, mainly viruses. Consequently, this crop is particularly affected by viral diseases that substantially decrease the quality and yield of the storage roots [1].

In Europe, the demand for sweet potatoes has greatly increased in recent years, leading to an increase in imports and production in southern European countries, including Spain. Although extensive field studies of the losses caused by sweet potato viruses have not been carried out in Spain, several viruses have been reported to infect this crop. They are the crinivirus (genus *Crinivirus*, family *Closteroviridae*) sweet potato chlorotic stunt virus [2]; the potyviruses (genus *Potyvirus*, family *Potyviridae*) sweet potato feathery mottle virus, sweet potato virus G, and sweet potato virus 2 [2,3]; and the begomoviruses (genus *Begomovirus*, family *Geminiviridae*) sweet potato leaf curl virus and sweet potato leaf curl Canary virus [4,5]. In addition, the presence of deltasatellites (genus *Deltasatellite*, family *Tolecusatellitidae*) associated with the begomoviruses infecting sweet potato has also been reported in this country [6,7].

The genus *Mastrevirus* is one of 14 genera in the family *Geminiviridae* [8]. Mastreviruses infect dicot and monocot plant species and are transmitted by leafhoppers. Like all members of the family *Geminiviridae*, mastrevirus genomes are encapsidated in unique twinned (geminate) icosahedral particles [8]. Mastrevirus genomes have a single component of circular single-stranded DNA of 2.6–2.8 kb containing four open reading frames (ORFs) encoded on the virion-sense (V1, coat protein and V2, movement protein) or the complementary-sense strands (C1 and C1:C2, proteins involved in replication). An intron occurs between ORFs C1 and C2 of all mastrevirus genomes. A short and a long intergenic region are also present; the latter contains sequence elements necessary for viral replication and transcription, including a stem-loop motif containing the canonical nonanucleotide sequence TAATATTAC [8].

In 2009, the partial genome sequence of a mastrevirus was identified in sweet potato by the deep sequencing of small RNAs in Peru [9]. However, it was not until 2017 that the complete genome sequence of the virus was obtained [10]. Until now, the presence of the virus, named sweet potato symptomless virus 1 (SPSMV-1), has been reported to infect sweet potato plants from Brazil, China, Ecuador (Galapagos Islands), Kenya, Korea, Peru, Taiwan, Tanzania, Uruguay, and the USA [10–13].

In this study, the presence of SPSMV-1 was assessed for the first time in Spain, with the virus found in samples from all of the geographical areas studied. This is the first record of the virus in Europe. SPSMV-1 was also detected in several samples of pathogen-tested in vitro plants obtained from a germplasm collection. In addition, the first agroinfectious clone of the virus was developed, and infectivity assays showed that it was able to infect *Nicotiana benthamiana*, *Ipomoea nil*, *I. setosa*, and sweet potato with no noticeable symptoms.

2. Materials and Methods

2.1. Plant Samples

Leaf samples were collected from 95 plants from Málaga province (southern continental Spain) and the Spanish Canary Islands of Tenerife and Gran Canaria (Table 1 and Table S1). Sampled plants included cultivated sweet potato (*I. batatas*) (*n* = 71) from Málaga, Tenerife, and Gran Canaria and ornamental or naturalized *I. indica* (*n* = 24) from Tenerife and Gran Canaria. Most of the plants did not show any conspicuous symptoms.

Table 1. Sweet potato and *Ipomoea indica* samples analyzed in this study. Additional details are given in Table S1.

Location	No. of Infected Sweet Potato Plants/Total No. of Sweet Potato Plants (Percentage)	No. of Infected *I. indica* Plants/Total No. of *I. indica* Plants
Tenerife (Canary Islands)	16/34 (47.05%)	0/12
Gran Canaria (Canary Islands)	14/22 (63.63%)	0/12
Málaga (southern continental Spain)	6/15 (40%)	-
Total	36/71 (50.70%)	0/12

2.2. DNA Extraction, PCR Amplification, and Cloning

Total DNA was extracted from leaf samples using a cetyltrimethylammonium bromide (CTAB)-based purification method [14]. PCR to detect sweet potato symptomless virus 1 was carried out using primers Detect-1F (5′-CCTAAGTCGTCGTCCGATAG-3′)/Detect-1R (5′-TTGAGTCCAGGTAAACTGAGC-3′) and Full-4F (5′-TGGATATTAGTAAACCGGGTCA-3′)/Full-4R (5′-CACCATTCGACGTCACAA-3′) [10]. PCR was carried out with BIO-TAQ DNA polymerase (Bioline, London, UK). For primers Detect-1F/Detect-1R, the first PCR step was denaturation for 3 min at 95 °C, followed by 34 cycles of denaturation for 45 s at 95 °C, hybridization for 45 s at 52 °C, and extension for 45 s at 72 °C, followed by a final extension step of 5 min at 72 °C. For primers Full-4F/Full-4R, a denaturation for 3 min at 95 °C was used, followed by 34 cycles of denaturation for 1 min at 95 °C, hybridization for 1 min at 53 °C, and extension for 3 min at 72 °C, followed by a final extension step of 5 min at 72 °C. In addition, nested PCR was designed to amplify 329 nt of the coat protein gene of the virus. The first PCR was carried out with primers MA2924 (5′-CTACCTGGGATGATGTGGCTAGAC-3′)/MA2925 (5′-CCATTCGACGTCACAATCGTCTTC-3′) (first denaturation step of 95 °C for 3 min, followed by 34 cycles of 30 s at 95 °C, 30 s at 61.7 °C, and 30 s at 72 °C, and a final step of 5 min at 72 °C). The second PCR was carried out with primers MA2926 (5′-CTACGAGATCGACCGAGTCTGCAG-3′)/MA2927 (5′-GCAACAGTCCACGTATTTGG GAAG-3′) (first denaturation step of 95 °C for 3 min, followed by 34 cycles of 30 s at 95 °C, 30 s at 64.2 °C, and 30 s at 72 °C, and a final step of 5 min at 72 °C). DNA fragments obtained with primers Detect-1F/Detect-1R and MA2926/MA2927 were directly sequenced. DNA fragments obtained with Full-4F/Full-4R and MA2782/MA2784 were cloned into the pGEM-T-Easy Vector (Promega, Madison, WI, USA), and inserts of selected clones were sequenced. All sequence reactions were carried out at Macrogen Inc. (Seoul, Korea).

2.3. Sequence Analysis

Sequences were analyzed with the Lasergene sequence analysis package (DNAStar Inc., Madison, WI, USA). The BLASTn algorithm was used to determine sequence similarity in the GenBank database. The BLASTn program [15] (http://www.ncbi.nlm.nih.gov/ Blast.cgi (accessed on 1 May 2022)) was used to perform the sequence similarity search. Pairwise identity scores were calculated with the Sequence Demarcation Tool (SDT) [16] after sequence alignment with MUSCLE [17]. The best-fit model of nucleotide substitution was determined based on the corrected Akaike and Bayesian information criteria, as implemented in MEGA7 [18].

2.4. Development of an Agroinfectious Clone and Plant Agroinoculation

Primers MA2782 (GGTACCGTGTATTTGATGACGATGTAC)/MA27784 (GGTACCCC CTGGGTTGAACACAAC) were designed to amplify the full-length genome of an isolate of SPSMV-1 present in sample B2 from Málaga (GenBank accession numbers ON526997). These primers included the sequence of the *Kpn*I restriction site (underlined) naturally present in the viral genome sequence. PCR was carried out as follows: denaturation for 3 min at 95 °C, 34 cycles of denaturation for 1 min at 95 °C, hybridization for 1 min at 66 °C, and extension for 3 min at 72 °C, followed by a final extension step of 5 min at 72 °C. The infectious dimeric SPSMV-1 clone was constructed essentially as described by Ferreira et al. [19]. The full-length SPSMV-1 genome was released with *Kpn*I from the pGEM-T-Easy Vector, religated, and used as a template for rolling circle amplification (RCA) (TempliPhi DNA Amplification Kit, GE Healthcare, Little Chalfont, UK). The RCA product was partially digested with *Kpn*I to produce dimeric molecules that were cloned in the pUC18 vector. The insert of the dimeric clone was excised with *Bam*HI/*Eco*53KI and subcloned in the *Bam*HI/*Sma*I sites of the pCAMBIA0380 binary vector.

An *Agrobacterium tumefaciens* C58C1 culture harboring the dimeric clone of SPSMV-1 was grown for 2 days at 28 °C in yeast extract peptone (YEP) liquid medium containing kanamycin (50 µg/mL) and rifampicin (50 µg/mL). For agroinoculation assays, *A. tumefaciens* cultures were used to inoculate plants by stem puncture, as previously described [20,21]. Plants inoculated with *A. tumefaciens* C58C1 cultures containing the empty pCAMBIA0380 vector were used as negative controls (mock). *Nicotiana benthamiana* was inoculated at the four-leaf stage. *Ipomoea nil* and *I. setosa* were inoculated at the two-leaf stage and cuttings of sweet potato cv. 'Tanzania' and 'Camote Morado' were inoculated. Two independent experiments were carried out for each plant species. Plants were maintained in an insect-free growth chamber (25/18 °C day/night, 70% relative humidity, and a 16 h photoperiod at 250 µmol s^{-1} m^{-2} of photosynthetically active radiation) until analyzed.

3. Results

3.1. Sweet Potato Symptomless Virus 1 Naturally Infects Sweet Potato in Continental Spain and the Canary Islands

The DNA extracted from the leaf samples of sweet potato and *I. indica* plants from Málaga (southern continental Spain) and the Spanish Canary Islands of Tenerife and Gran Canaria was analyzed for the presence of SPSMV-1. PCR with primers Detect-1F/Detect-1R showed the expected DNA fragment of 418 bp in sweet potato samples from all locations (Tenerife, Gran Canaria, and Málaga) (Tables 1 and S1; Figure 1A). Overall, 36 of 71 sweet potato samples were infected with SPSMV-1 (16/34 from Tenerife, 14/22 from Gran Canaria, and 6/15 from Málaga). In contrast, none of the *I. indica* samples showed the expected DNA fragments (Tables 1 and S1).

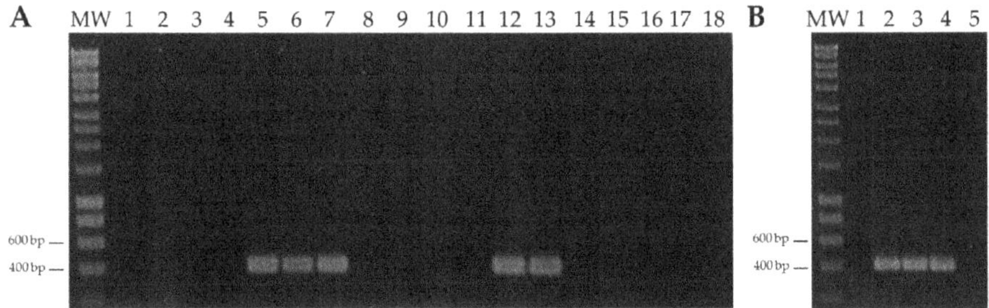

Figure 1. Agarose gel electrophoresis of PCR products of sweet potato symptomless virus 1 using primers Detect-1F/1R. MW, HyperLadder 1kb (Bioline). (**A**) Lanes 1–18, sweet potato samples CI62–CI79 from the Canary Islands. (**B**) Samples from a sweet potato germplasm collection. Lanes 1–5, cv. 'Tanzania,' 'Blanca,' 'Beauregard,' 'Promesa', and 'Camote Morado,' respectively.

3.2. Sweet Potato Symptomless Virus 1 Is Present in Pathogen-Tested In Vitro Plants Obtained from a Germplasm Collection

Sweet potato plants of cultivars 'Tanzania', 'Blanca,' 'Beauregard,' 'Promesa', and 'Camote Morado' obtained from the USDA-ARS Plant Genetic Resources Conservation Unit (Griffin, GA, USA) and maintained by cuttings in insect-proof chambers for more than 10 years were screened with primers Detect-1F/Detect-1R to detect SPSMV-1. Surprisingly, the virus was detected in cultivars 'Blanca', 'Beauregard', and 'Promesa' (Figure 1B). To ensure that the virus was not present in sweet potato cultivars 'Tanzania' and 'Camote Morado' thus allowing the use of these cultivars in agroinoculation experiments, 10 leaf samples of each cultivar were used to individually detect the presence of the virus using nested PCR with primers MA2924/MA2925 and MA2926/MA2927; none of the 20 samples amplified the expected DNA fragment of 329 nt (Figure S1).

3.3. Sweet Potato Symptomless Virus 1 Isolates from the Canary Islands and Málaga Are Closely Related with Isolates from All over the World but Contain Novel Molecular Features

To molecularly characterize SPSMV-1 isolates from Spain and establish the phylogenetic relationships with isolates from other countries, the full-length viral genome was amplified with primers Full-4F/Full-4R from samples CI24 from Tenerife; CI61 from Gran Canaria; and B2, B3, and B14 from Málaga. The expected DNA fragments corresponding to the full-length SPSMV-1 genomes were cloned into the pGEM-T-Easy Vector. Two clones were sequenced per sample and deposited in the GenBank database under accession numbers ON526993–ON527002.

Sequences obtained in this work showed a similarity of 99.0–99.8% and at least 96.9% similarity with all SPSMV-1 isolates available in GenBank (Figure 2). The unusual nonanucleotide TAAGATTCC present in most Spanish SPSMV-1 sequences (9 out of 10, (ON526993–ON526997, ON526999–ON527002)) is coincident with the nonanucleotide previously reported for the virus [10]. However, one of the clones obtained from sample B2 contained a different nonanucleotide sequence, TCTTATTAC (ON526998). The full-length SPSMV-1 sequences obtained in this work ranged from 2578 to 2602 nt. This difference in size occurred because several isolates (ON526994, ON526999, ON527000, and ON527002) showed a deletion of 24 nucleotides near the 5′ end of V2 ORF. The movement protein encoded by the V2 ORF in the mentioned isolates lacked the corresponding eight amino acids present in all previously sequenced SPSMV-1 isolates deposited in GenBank.

Phylogenetic analysis of all SPSMV-1 isolates available, including those obtained in this work, showed no correlation between the isolates and their geographic origin (Figure 3).

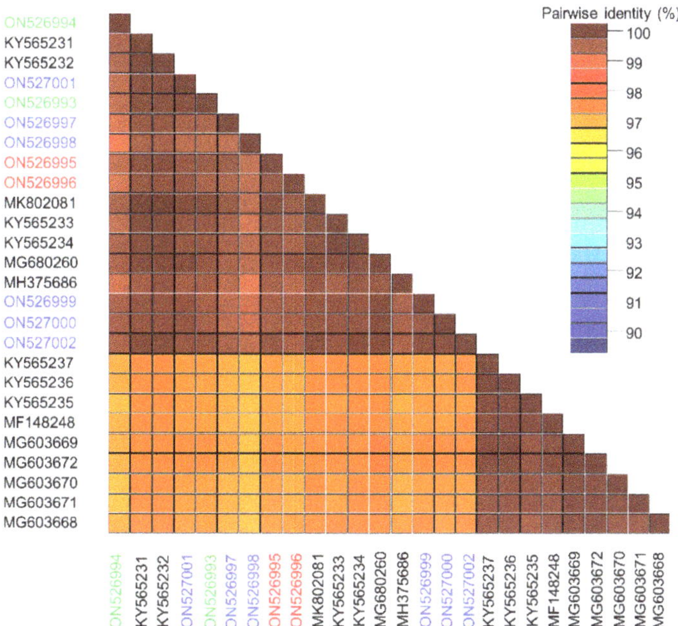

Figure 2. Color-coded matrix of pairwise sequence identity scores generated by the alignment of the full-length genomes of sweet potato symptomless virus 1 (SPSMV-1) obtained in this work. Samples from the Canary Islands of Tenerife and Gran Canaria are in green and red, respectively, and those from Málaga are in blue. All other full-length genomes of SPSMV-1 available from GenBank (in black) are included in the analysis.

3.4. A Dimeric Sweet Potato Symptomless Virus 1 Clone Is Infectious in N. benthamiana, I. nil, I. setosa, and Sweet Potato

A dimeric clone constructed for an SPSMV-1 isolate from Málaga was used to agroinoculate a range of plant species; the experimental hosts *N. benthamiana*, *I. nil*, and *I. setosa*; and the natural host sweet potato (cv. 'Tanzania' and 'Camote Morado'). Plants inoculated in two independent experiments and analyzed by nested PCR with primers MA2924/MA2925 and MA2926/MA2927 at 5 weeks post-inoculation showed that the virus was able to infect all of the plant species assayed (Table 2, Figure S2). The proportion of infected plants ranged from 12.5% for *I. nil* to 45.83% for *N. benthamiana*. No evident symptoms were observed in any of the inoculated plants, suggesting that the Spanish SPSMV-1 isolate does not cause symptoms in these hosts (Figure 4).

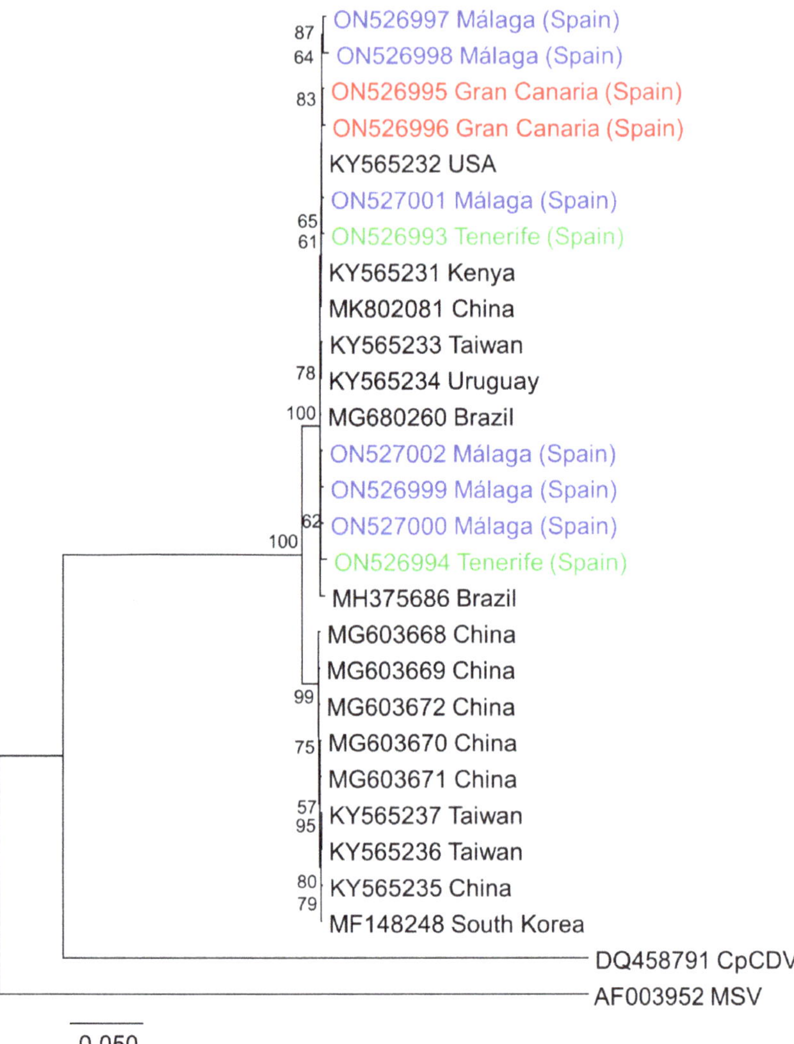

Figure 3. Phylogenetic tree illustrating the relationships of the sweet potato symptomless virus 1 (SPSMV-1) genomes obtained in this work with SPSMV-1 isolates previously reported and one representative isolate each of the dicot-infecting mastrevirus chickpea chlorotic dwarf virus (CpCDV) and the monocot-infecting mastrevirus maize streak virus (MSV). Samples were obtained from Málaga (southern continental Spain) (blue) and the Canary Islands of Tenerife (green) and Gran Canaria (red). The tree was constructed using the maximum likelihood method with the MEGA 7 program using the best fit model. HKY and bootstrap values (1000 replicates) are shown for supported branches (>50%). The bar below the tree indicates 0.050 nucleotide substitutions per site.

Figure 4. Plant and leaf details of *Nicotiana benthamiana* (**A,F**), *Ipomoea setosa* (**B,G**), *Ipomoea nil* (**C,H**), sweet potato cv. 'Tanzania' (**D,I**) and 'Camote Morado' (**E,J**) infected with the sweet potato symptomless virus 1 infectious clone. In each panel, mock-inoculated plants are shown on the left and infected plants on the right. Photographs of representative plants were taken 40 days post-inoculation.

Table 2. Infectivity of sweet potato symptomless virus 1 in *Nicotiana benthamiana*, *Ipomoea nil*, *I. setosa*, and sweet potato plants. Data from two agroinoculation experiments and mock-inoculated plants were included.

	No. of Infected Plants/No. of Agroinoculated Plants		
Plant Species	**Experiment 1**	**Experiment 2**	**Total (%)**
Nicotiana benthamiana	9/12 0/12 (mock)	2/12 0/12 (mock)	45.83
Ipomoea nil	3/24 0/12 (mock)	3/24 0/12 (mock)	12.5
Ipomoea setosa	6/24 0/12 (mock)	9/24 0/12 (mock)	31.25
Sweet potato cv. 'Tanzania'	2/18 0/5 (mock)	4/10 0/3 (mock)	21.42
Sweet potato cv. 'Camote Morado'	2/13 0/5 (mock)	2/14 0/3 (mock)	14.81

4. Discussion

Mastrevirus SPSMV-1 has been reported to infect sweet potato plants in several countries around the world [10–13]. In this work, the presence of the virus was assessed for the first time in Spain. SPSMV-1 was detected in approximately half of the sweet potato plants analyzed (50.70%) and was present in all sampled areas, Málaga (40%) in southern continental Spain and the Spanish Canary Islands of Tenerife (47.05%) and Gran Canaria (63.63%). To our knowledge, this is the first report of the presence of SPSMV-1 in Spain and Europe. Unexpectedly, the virus was also found infecting three sweet potato cultivars ('Blanca', 'Beauregard', and 'Promesa') that were supposed to be virus-free, as they were obtained as pathogen-tested in vitro plants maintained in a germplasm collection. This highlights the importance of developing methods to detect viruses present in germplasm collections that would otherwise remain undetected.

Sequence analysis of the full-length genomes of the SPSMV-1 isolates reported here showed a high sequence identity and a close phylogenetic relationship between them and the isolates available from GenBank. As with all SPSMV-1 isolates previously characterized [10], most of the isolates sequenced in this work showed the presence of the unusual nonanucleotide TAAGATTCC. However, a unique novel nonanucleotide, TCTTATTAC, was found in an isolate from Málaga. Another singularity of some of the SPSMV-1 isolates reported here is the presence of a shorter-than-usual V2 ORF with a 24-nt deletion close to the 5' end, thus encoding a shorter putative movement protein. Interestingly, these deletion mutants, not previously described, were found in both the Canary Island of Tenerife and in Málaga.

In this study, we developed for the first time an agroinfectious clone for SPSMV-1. Our results showed that the clone was able to infect the model plant *N. benthamiana*, two species used as bioindicators for sweet potato viruses, *I. nil* and *I. setosa*, and two cultivars of sweet potato ('Tanzania' and 'Camote Morado'). No evident symptoms of the disease were observed in any of the infected plants. Thus, our experimental results support previous observations that SPSMV-1 could not be associated with any conspicuous symptoms in sweet potato plants [9,10]. However, although this virus does not seem to play an important pathological role, at least in the sweet potato cultivars assayed, it may have an effect in the presence of other viruses. The occurrence of synergistic interactions between viruses infecting sweet potatoes has been widely reported [22,23]. The presence of symptomless/latent viruses is not uncommon in crops and sweet potato is not an exception. Examples of asymptomatic viruses infecting sweet potatoes include, in addition to SPSMV-1, the potyvirus sweet potato latent virus and the badnavirus sweet potato pakakuy virus [22,24].

In summary, SPSMV-1 was confirmed as an asymptomatic mastrevirus infecting sweet potato crops in an increasing number of countries including Spain. Several biological aspects remain to be elucidated about this virus, including the possible interactions with other viruses in mixed infections and the mode of transmission. The putative insect vector is likely a leafhopper, similar to other mastreviruses [8].

Supplementary Materials: The following supporting information can be downloaded at: https://www.mdpi.com/article/10.3390/microorganisms10091736/s1, Figure S1: Agarose gel electrophoresis of nested PCR products of sweet potato symptomless virus 1 using primers MA2924/MA2925 and MA2926/MA2927; Figure S2: Agarose gel electrophoresis of nested PCR products from plants agroinoculated with sweet potato symptomless virus 1 using primers MA2924/MA2925 and MA2926/MA2927; Table S1: Information on the sweet potato and *Ipomoea indica* samples used in this study.

Author Contributions: Conceptualization, E.F.-O. and J.N.-C.; investigation, E.F.-O. and A.C.G.-M.; writing—original draft preparation, E.F.-O.; writing—review and editing, J.N.-C. and E.F.-O.; funding acquisition, J.N.-C. All authors have read and agreed to the published version of the manuscript.

Funding: This work was supported by the Ministerio de Ciencia e Innovación (MICINN, Spain) (grant PID2019-105734RB-I00, MCIN/AEI/10.13039/501100011033) and co-financed by the European

Regional Development Fund (ERDF). E.F.-O. was the recipient of a "Ramón y Cajal" (RYC2019-028486-I/AEI/10.13039/501100011033) postdoctoral contract from MICINN.

Data Availability Statement: Nucleotide sequences obtained in this study have been deposited in GenBank under accession numbers ON526993-ON527002.

Acknowledgments: The authors are thankful to Remedios Tovar and José M. Cid for their excellent technical assistance and Remedios Tovar, Ana Espino, Cristo Medina, Puri Benito, and Jesús Ariza for their help with the field sampling.

Conflicts of Interest: The authors declare no conflict of interest. The funders had no role in the design of the study; in the collection, analyses, or interpretation of data; in the writing of the manuscript, or in the decision to publish the results.

References

1. Valverde, R.A.; Sim, J.; Lotrakul, P. Whitefly transmission of sweet potato viruses. *Virus Res.* **2004**, *100*, 123–128. [CrossRef] [PubMed]
2. Valverde, R.A.; Lozano, G.; Navas-Castillo, J.; Ramos, A.; Valdés, F. First report of *Sweet potato chlorotic stunt virus* and *Sweet potato feathery mottle virus* infecting sweet potato in Spain. *Plant Dis.* **2004**, *88*, 428. [CrossRef]
3. Trenado, H.P.; Lozano, G.; Valverde, R.A.; Navas-Castillo, J. First report of *Sweet potato virus G* and Sweet potato virus 2 infecting sweet potato in Spain. *Plant Dis.* **2007**, *9*, 1687. [CrossRef] [PubMed]
4. Banks, G.K.; Bedford, I.D.; Beitia, F.J.; Rodriguez-Cerezo, E.; Markham, P.G. A novel geminivirus of *Ipomoea indica* (*Convolvulacae*) from Southern Spain. *Plant Dis.* **1999**, *83*, 486. [CrossRef]
5. Lozano, G.; Trenado, H.P.; Valverde, R.A.; Navas-Castillo, J. Novel begomovirus species of recombinant nature in sweet potato (*Ipomoea batatas*) and *Ipomoea indica*: Taxonomic and phylogenetic implications. *J. Gen. Virol.* **2009**, *90*, 2550–2562. [CrossRef]
6. Lozano, G.; Trenado, H.P.; Fiallo-Olivé, E.; Chirinos, D.; Geraud-Pouey, F.; Briddon, R.W.; Navas-Castillo, J. Characterization of non-coding DNA satellites associated with sweepoviruses (genus *Begomovirus*, *Geminiviridae*)—definition of a distinct class of begomovirus-associated satellites. *Front. Microbiol.* **2016**, *7*, 162. [CrossRef] [PubMed]
7. Hassan, I.; Orílio, A.F.; Fiallo-Olivé, E.; Briddon, R.W.; Navas-Castillo, J. Infectivity, effects on helper viruses and whitefly transmission of the deltasatellites associated with sweepoviruses (genus *Begomovirus*, family *Geminiviridae*). *Sci. Rep.* **2016**, *6*, 30204. [CrossRef] [PubMed]
8. Fiallo-Olivé, E.; Lett, J.M.; Martin, D.P.; Roumagnac, P.; Varsani, A.; Zerbini, F.M.; Navas-Castillo, J.; ICTV Report Consortium. ICTV Virus Taxonomy Profile: *Geminiviridae* 2021. *J. Gen. Virol.* **2021**, *102*, 001696. [CrossRef]
9. Kreuze, J.F.; Perez, A.; Untiveros, M.; Quispe, D.; Fuentes, S.; Barker, I.; Simon, R. Complete viral genome sequence and discovery of novel viruses by deep sequencing of small RNAs: A generic method for diagnosis, discovery and sequencing of viruses. *Virology* **2009**, *388*, 1–7. [CrossRef]
10. Cao, M.; Lan, P.; Li, F.; Abad, J.; Zhou, C.; Li, R. Genome characterization of sweet potato symptomless virus 1: A mastrevirus with an unusual nonanucleotide sequence. *Arch. Virol.* **2017**, *162*, 2881–2884. [CrossRef]
11. Kwak, H.-R.; Kim, M.-K.; Shin, J.-C.; Lee, Y.-J.; Seo, J.-K.; Lee, H.-U.; Jung, M.-N.; Kim, S.-H.; Choi, H.-S. The current incidence of viral disease in Korean sweet potatoes and development of multiplex RT-PCR assays for simultaneous detection of eight sweet potato viruses. *Plant Pathol. J.* **2014**, *30*, 416–424. [CrossRef] [PubMed]
12. Mbanzibwa, D.R.; Tugume, A.K.; Chiunga, E.; Mark, D.; Tairo, F.D. Small RNA deep sequencing-based detection and further evidence of DNA viruses infecting sweetpotato plants in Tanzania. *Ann. Appl. Biol.* **2014**, *165*, 329–339. [CrossRef]
13. Wang, Y.-J.; Zhang, D.-S.; Zhang, Z.-C.; Wang, S.; Qiao, Q.; Qin, Y.-H.; Tian, Y.-T. First report on sweetpotato symptomless virus 1 (genus *Mastrevirus*, family *Geminiviridae*) in sweetpotato in China. *Plant Dis.* **2015**, *99*, 1042. [CrossRef]
14. Haible, D.; Kober, S.; Jeske, H. Rolling circle amplification revolutionizes diagnosis and genomics of geminiviruses. *J. Virol. Methods* **2006**, *135*, 9–16. [CrossRef] [PubMed]
15. Altschul, S.F.; Gish, W.; Miller, W.; Myers, E.W.; Lipman, D.J. Basic local alignment search tool. *J. Mol. Biol.* **1990**, *215*, 403–410. [CrossRef]
16. Muhire, B.; Varsani, A.; Martin, D.P. SDT: A virus classification tool based on pairwise sequence alignment and identity calculation. *PLoS ONE* **2014**, *9*, e108277. [CrossRef]
17. Edgar, R.C. MUSCLE: A multiple sequence alignment method with reduced time and space complexity. *BMC Bioinform.* **2004**, *5*, 113. [CrossRef]
18. Kumar, S.; Stecher, G.; Tamura, K. MEGA7: Molecular Evolutionary Genetics Analysis version 7.0 for bigger datasets. *Mol. Biol. Evol.* **2016**, *33*, 1870–1874. [CrossRef]
19. Ferreira, P.T.; Lemos, T.O.; Nagata, T.; Inoue-Nagata, A.K. One-step cloning approach for construction of agroinfectious begomovirus clones. *J. Virol. Methods* **2008**, *147*, 351–354. [CrossRef]
20. Ferro, C.G.; Zerbini, F.M.; Navas-Castillo, J.; Fiallo-Olivé, E. Revealing the complexity of sweepovirus-deltasatellite-plant host interactions: Expanded natural and experimental helper virus range and effect dependence on virus-host combination. *Microorganisms* **2021**, *9*, 1018. [CrossRef]

21. Fiallo-Olivé, E.; Tovar, R.; Navas-Castillo, J. Deciphering the biology of deltasatellites from the New World: Maintenance by New World begomoviruses and whitefly transmission. *New Phytol.* **2016**, *212*, 680–692. [CrossRef] [PubMed]
22. Untiveros, M.; Fuentes, S.; Salazar, L.F. Synergistic interaction of *Sweet potato chlorotic stunt virus* (*Crinivirus*) with carla-, cucumo-, ipomo-, and potyviruses infecting sweet potato. *Plant Dis.* **2007**, *91*, 669–676. [CrossRef] [PubMed]
23. Cuellar, W.J.; Galvez, M.; Fuentes, S.; Tugume, J.; Kreuze, J. Synergistic interactions of begomoviruses with *Sweet potato chlorotic stunt virus* (genus *Crinivirus*) in sweet potato (*Ipomoea batatas* L.). *Mol. Plant Pathol.* **2015**, *16*, 459–471. [CrossRef] [PubMed]
24. Kreuze, J.F.; Perez, A.; Gargurevich, M.G.; Cuellar, W.J. Badnaviruses of sweet potato: Symptomless coinhabitants on a global scale. *Front. Plant Sci.* **2020**, *11*, 313. [CrossRef]

 *microorganisms*

Article

Biochemical Diversity, Pathogenicity and Phylogenetic Analysis of *Pseudomonas viridiflava* from Bean and Weeds in Northern Spain

Ana M. Fernández-Sanz [1], M. Rosario Rodicio [2] and Ana J. González [1,*]

[1] Programa de Patología Vegetal, Servicio Regional de Investigación y Desarrollo Agroalimentario (SERIDA), Ctra AS-267, PK 19, 33300 Villaviciosa, Spain; anafsanz@gmail.com
[2] Área de Microbiología, Departamento de Biología Funcional, Universidad de Oviedo, Julián Claveria 6, 33006 Oviedo, Spain; rrodicio@uniovi.es
* Correspondence: anagf@serida.org

Abstract: *Pseudomonas viridiflava* was originally reported as a bean pathogen, and subsequently as a wide-host range pathogen affecting numerous plants species. In addition, several authors have reported the epiphytic presence of this bacterium in "non-host plants", which may act as reservoir of *P. viridiflava* and source of inoculum for crops. A new biotype of this bacterium, showing an atypical LOPAT profile, was found in Asturias, a Northern region of Spain, causing significant damage in beans, kiwifruit, lettuce, and *Hebe*. In order to investigate the involvement of weeds in bean disease, samples were collected from beans and weeds growing in the same fields. A total of 48 isolates of *P. viridiflava* were obtained, 39 from weeds and 9 from beans. 48% and 52% of them showed typical (L− O− P+ A− T+) and atypical (L+ O− P v A− T+) LOPAT profiles, and they displayed high biochemical diversity. Regarding virulence factors, the T-PAI and S-PAI pathogenicity islands were found in 29% and 70.8% of the isolates, 81.2% displayed pectinolytic activity on potato slices, and 59% of the weed isolates produced symptoms after inoculation on bean pods. A phylogenetic tree based on concatenated *rpoD*, *gyrB*, and *gltA* sequences separated the strains carrying S-PAI and T-PAI into different clusters, both containing isolates from beans and weeds, and pathogenic as well as non-pathogenic strains. Closely related strains were found in the two hosts, and more than half of the weed isolates proved to be pathogenic in beans. This is consistent with the role of weeds as a reservoir and source of inoculum for bean infection. Detection of *P. viridiflava* in weeds throughout the year further supports these roles.

Keywords: *Pseudomonas viridiflava*; *Phaseolus vulgaris*; LOPAT; pathogenicity islands; pectinolytic activity; phylogenetic analysis

Citation: Fernández-Sanz, A.M.; Rodicio, M.R.; González, A.J. Biochemical Diversity, Pathogenicity and Phylogenetic Analysis of *Pseudomonas viridiflava* from Bean and Weeds in Northern Spain. *Microorganisms* **2022**, *10*, 1542. https://doi.org/10.3390/microorganisms10081542

Academic Editor: Ana Palacio-Bielsa

Received: 5 July 2022
Accepted: 27 July 2022
Published: 29 July 2022

Publisher's Note: MDPI stays neutral with regard to jurisdictional claims in published maps and institutional affiliations.

1. Introduction

Pseudomonas viridiflava (Burkholder) Dowson was first described by Burkholder in 1930 as a bean (*Phaseolus vulgaris* L.) pathogen and subsequently reported to cause disease in many other crop plants. These bacteria produce several symptoms, including necrosis, spots, and rots, affecting different parts of the plant (stems, leaves, blossoms, and roots). For instance, it has been shown to cause bacterial blight in kiwifruit, necrosis in tomato, bacterial blight in pea, rots in alfalfa, carrot and *Cucumis sativus*, bacterial shoot blight in sweet crab apple, bacterial canker in stone fruit trees, and bacterial leaf spot in lettuce [1–10]. According to this, *P. viridiflava* is regarded as a generalist pathogen, able to attack multiple host species [11,12]. In Asturias, a Northern region of Spain, a new emerging biotype of *P. viridiflava* (termed BT2) was described [13]. This biotype differs from the typical biotype by the production of yellowish mucoid material in the sucrose medium used for the levan test, and by a variable pectinolytic activity on different potato varieties. First detected in 1999, BT2 has caused significant damage in several crops such as beans, in which symptoms ranged from red spots mainly in the petioles and pods; to plant death due to systemic infection associated with the destruction of the medulla; kiwifruit, showing dark brown spots in floral buds that developed into extensive rot, leading to the collapse of floral buds or the production of small or distorted fruits; and lettuce, generating soft rot of intermediate leaves, which could also progress and result in plant death [13]. Subsequently, this biotype was also shown to cause defoliation in *Hebe* [14], and has been isolated from rapeseed in South Korea [15].

P. viridiflava is a close relative of *P. syringae*. Based on phylogenetic analysis, it is located within the *P. syringae* species complex [16–18]. However, unlike *P. syringae*, the knowledge about the mechanisms of virulence in *P. viridiflava* is still limited. In this species, two alternative pathogenicity islands, T-PAI and S-PAI, have been detected [19]. Some isolates harbor a tripartite PAI, equivalent to that found in *P. syringae* (T-PAI), while in others the PAI has a single component (S-PAI). T-PAI and S-PAI occupy different chromosomal locations, but only one of them is present in each individual isolate [19]. T-PAI consists of a central *hrp/hrc* (hypersensitivity reaction and pathogenicity/hypersensitivity reaction and conserved) cluster, flanked by CEL (conserved effector locus) and EEL (exchangeable effector locus) regions. The *hrp/hrc* cluster encodes a Type III secretion system (TTSS) responsible for the formation of a syringe-like structure used to translocate effectors into the host cell. These effectors are encoded by the CEL and EEL loci. S-PAI is also composed by a *hrp/hrc* cluster, but it is not flanked by the CEL and EEL regions. Instead, the cluster is interrupted by an insertion that harbors genes encoding effector proteins [19]. Other virulence factors detected in *P. viridiflava* are the enzyme pectate lyase and extracellular proteases related to plant tissue maceration [20,21].

Apart from acting as a crop pathogen, *P. viridiflava* also exists as an epiphyte on weeds. Gitaitis et al. [22] found this bacterium on weeds associated with onion crops in the USA and observed that weed control is necessary to avoid plant disease. The weed species from which *P. viridiflava* was isolated included *Oenothera laciniata* Hill, *Taraxacum officinale* Weber, *Fumaria officinalis* L., *Gnaphalium purpureum* L., *Sonchus asper* (L.) Hill, *Lepidium virginicum* L., and *Raphanus raphanistrum* L. Similarly, Basavand & Khodaygan [23] detected *P. viridiflava* in *Alisma plantago-aquatica*, a perennial weed in rice fields in Northern Iran.

In the present study, we performed a comparative analysis of *P. viridiflava* isolated from the common bean type "granja asturiana", and from weeds growing in the bean fields. "Granja asturiana" beans are one of the most popular food products in our region, have great culinary value within the Asturian gastronomy, and are an important economic resource for small farmers. Therefore, the possible role of weeds as a reservoir of *P. viridiflava* and their involvement in transmission of the pathogen to bean crops are worthy of investigation. In this study, strains isolated from the two hosts were characterized, their phylogenetic relationships established, and weed isolates were tested for pathogenicity on bean.

2. Materials and Methods

2.1. Bacterial Isolates, Culture Conditions and Phenotypic Characterization

Forty-eight isolates of *P. viridiflava* from weeds and beans were recovered (39 and 9, respectively) in 10 fields located in four councils of Asturias: Navia (Anleo), Valdés (Busto 1 and 2, Constancios, Ronda and Pontigón), Tineo (Carbajal, Bárcena and Yerbo), and Siero (Argüelles).

Bacteria were identified by means of the KOH test, to determine their Gram stain [24], and by standard biochemical tests. These included oxidation-fermentation of glucose [25], the LOPAT determinative tests [26], utilization of mannitol, m-inositol, erythritol, and sorbitol in Hellmers broth [27], utilization of homoserine, D-tartrate, sucrose, L-lactate, trigonelline, quinate, betaine and adonitol in Ayer's solid medium [28], and hydrolysis of esculin, gelatin [29], casein, and Tween 80 [30]. Isolates showing identical features and collected from the same field site and host at the same period of the year, were considered as a single strain. In this way, a total of 39 strains were identified, 30 and 9 from weeds and bean, respectively.

2.2. 16S rDNA Sequencing and ARDRA (Amplified Ribosomal DNA Restriction Analysis

16S rDNA was amplified by PCR, with primers described by Edwards et al. [31], and ARDRA carried out with the *Sac*I and *Hinf*I restriction enzymes [13].

2.3. Detection of Pathogenicity Islands

Presence of T-PAI was detected with primers hoppsyAr1 (CYGGCTATGATTGATAAACG-CATCG) and shcAf1 (GGCGCACTTAACCCTCTGKTCAA TGA) [19], while the presence of S-PAI and absence of T-PAI was revealed with primers R1-orf41f1 (GCCTTGCCTCT-GATCTCATTC) and R1-dsorf78r1 (GTAGCAT TCGGCATATCCC) (M. San José, personal communication). The fragments expected are of 1113 bp for T-PAI and 889 bp for S-PAI.

2.4. Pathogenicity Assays

Pathogenicity was tested by inoculation of isolates grown on King B medium on pods of beans cv. *Helda* with a sterile toothpick [32]. The assays were repeated with three replicates each time.

2.5. Phylogenetic Analysis

The phylogenetic analysis was performed for the 39 detected strains (see above), using the *rpoD* (RNA polymerase sigma D factor), *gltA* (citrate synthase), and *gyrB* (DNA gyrase subunit B) genes [33]. Amplification was conducted as in Hwang et al. [34]. The obtained fragments were sequenced by Secugen S.L. (Spain) or Eurofins (Germany). Sequences were submitted to GenBank (accession numbers MT683625-MT683672, MT709110-MT709148, and ON838894-ON838932, for *gyrB*, *rpoD* and *gltA*, respectively). Concatenated sequences of the three genes were aligned using Clustal W [35], and phylogenetic trees were constructed using Maximum-Likelihood with the Tamura–Nei model [36]. Their topological robustness was evaluated by bootstrap analysis based on 1000 replicates using Mega 6 software [37]. Sequences from *P. viridiflava* DSM 6694^T, *P. asturiensis* LPPA 221^T and *P. protegens* Chao^T were included as references.

In addition, the number of segregating sites (S) and the mean of the nucleotide diversity (π), defined as the average number of nucleotide differences by site between sequences of the whole population [38], were calculated both for the individual genes and the concatenated sequences, also using the Kimura two parameters model [37].

3. Results and Discussion

3.1. Identification and Biochemical Characterization of the Isolates

Forty eight isolates identified as *P. viridiflava* by ARDRA, and recovered during the period 2007–2009, were included in the present study. Thirty nine of them were isolated from weeds and nine from beans (Table 1). All isolates shared the ARDRA profile characteristically associated with *P. viridiflava* (not shown) and were Gram-negative bacilli.

Table 1. Origin and general features of *Pseudomonas viridiflava* isolates used in this study.

Year	Site	Isolate	Host	BT	BP	PAI	PP	P
2007	Carbajal	LPPA 511	*Phaseolus vulgaris*	2	29	S	+	−
		LPPA 574	*Stellaria media*	1	2	S	+	+
2007	Bárcena	LPPA 513	*P. vulgaris*	2	25	S	+	−
2007	Pontigon	LPPA 1598 [a]	n.i.	1	13	S	−	−
		LPPA 1600 [a]	n.i.	1	2	S	+	+
		LPPA 1604	n.i.	1	12	T	−	+
2008	Carbajal	LPPA 593	*P. vulgaris*	1	3	S	+	+
		LPPA 842	*P. vulgaris*	1	3	S	+	+
2008	Anleo	LPPA 599	*P. vulgaris*	2	30	S	+	−

Table 1. *Cont.*

Year	Site	Isolate	Host	BT	BP	PAI	PP	P
2008	Busto 1	LPPA 820	*Cyperus rotundus*	1	5	T	+	−
2008	Constancios	LPPA 806	*Fumaria* sp.	1	8	T	+	+
		LPPA 824	*Senecio vulgaris*	1	4	T	+	+
2008	Ronda	LPPA 811	*Capsella bursa-pastoris*	1	10	T	+	−
		LPPA 813	*Sonchus oleraceus*	1	4	S	+	−
		LPPA 814	*Fumaria* sp.	1	12	S	−	+
2008	Argüelles	LPPA 827	*P. vulgaris*	2	23	S	+	+
2008	Yerbo	LPPA 846	*P. vulgaris*	1	11	S	+	−
2009	Busto 1	LPPA 1420	*P. vulgaris*	2	33	S	−	+
		LPPA 888	*Malva sylvestris*	1	6	T	+	+
		LPPA 891	*S. oleraceus*	1	2	T	+	−
		LPPA 894	*S. oleraceus*	1	1	T	+	−
		LPPA 896 [b]	*Fumaria* sp.	2	24	S	+	+
		LPPA 897 [b]	*Fumaria* sp.	2	17	S	+	+
		LPPA 1674	*Fumaria* sp.	2	27	S	+	+
		LPPA 1676	*Fumaria* sp.	2	15	S	+	+
		LPPA 1679	*Fumaria* sp.	1	7	S	+	−
		LPPA 934 [c]	*Hypochaeris radicata*	1	1	S	+	−
		LPPA 935 [c]	*H. radicata*	1	2	S	+	−
		LPPA 937 [c]	*H. radicata*	1	1	S	+	−
		LPPA 1421	*Galinsoga parviflora*	2	19	T	+	+
		LPPA1665 [d]	*C. rotundus*	2	28	S	+	−
		LPPA 1666 [d]	*C. rotundus*	2	15	S	+	+
		LPPA 1671	*C. rotundus*	1	9	S	+	−
		LPPA 1417	*Solanum nigrum*	2	14	S	+	+
		LPPA 1680	*S. nigrum*	2	15	T	+	−
		LPPA 1682	*S. nigrum*	2	14	T	+	−
		LPPA 939 [e]	n.i.	1	1	S	+	+
		LPPA 941 [e]	n.i.	1	1	S	+	+
2009	Busto 2	LPPA 1385	*Fumaria* sp.	2	16	S	−	+
		LPPA 1391	*Chenopodium album*	2	31	S	+	v
		LPPA 1393 [f]	*C. bursa-pastoris*	2	32	S	−	+
		LPPA 1394 [f]	*C. bursa-pastoris*	2	16	S	−	+
2009	Yerbo	LPPA 1432	*P. vulgaris*	2	26	T	+	+
2009	Ronda	LPPA 1446	*S. nigrum*	2	18	S	+	+
		LPPA 1451	*Fumaria* sp.	2	21	T	+	+
		LPPA 1452 [g]	*Trifolium* sp.	2	22	S	+	+
		LPPA 1454 [g]	*Trifolium* sp.	2	14	S	+	−
2009	Constancios	LPPA 1467	*S. oleraceus*	2	20	S	+	−

LPPA, Laboratory of Phytopathology of the Principality of Asturias; [a–g], isolated from the same sample; ni, not identified, but all were different; BT, biotype; 2, atypical profile; 1, typical profile; BP, biochemical profile according to Table S1; PAI, pathogenicity island; T, T-PAI; S, S-PAI; PP, pectinolysis on potato; P, pathogenicity; +, positive; −, negative; v, variable.

Regarding the LOPAT scheme, presence of the two previously described biotypes was revealed, with 48% of the isolates (20 from weeds and 3 from beans) showing the typical (L−, O−, P+, A−, T+), and 52% (19 from weeds and 6 from beans) showing the atypical

(L+, O−, Pv, A−, T+) biotype. The latter profile has persisted in beans in Asturias, at least since 2003 when it was first reported. In addition, it has also been described in South Korea in rapeseed [15]. Other biochemical features of the isolates are compiled in Table 2.

Table 2. Biochemical features of the isolates under study.

Test	Total (N = 48)	BT1 (N = 23)	BT2 (N = 25)
Levan	58.3	0	100
Oxidase	0	0	0
Potato rot	81.2	87	76
Arginine	0	0	0
Tobacco	100	100	100
Oxidative	100	100	100
Esculin	100	100	100
Sucrose	0	0	0
Casein	93.75	87	100
Tween80	50	60.8	40
Gelatin	91.6	82.6	100
Mannitol	97.9	100	96
Erythritol	89.5	91.3	88
Sorbitol	97.9	100	96
M-inositol	95.8	100	92
Adonitol	2	4.3	0
D-Tartrate	29.1	21.7	36
L-Lactate	79.1	82.6	76
Trigonelline	97.9	95.6	100
Betaine	87.5	95.6	80
Homoserine	2	4.3	0
Quinate	100	100	100
Xylose	100	100	100
Lactose	0	0	0

The numbers correspond to the percentage of isolates positive for a given test.

Results of the biochemical tests were highly variable, distributing the 48 isolates into 33 biochemical profiles (BP1 to BP33; Table 1 and Table S1). A correlation between profile and host plant or sample site was not found. Thus, several profiles were associated with the same host or field, and the same profile was shared by isolates from different hosts and sites. This wide variability makes phenotypic identification rather difficult. Consistent results were only obtained for the sucrose and lactose tests, both negative, and for the tobacco, esculin, quinate, and xylose tests, all positive, in 100% of the isolates.

Following the proposal of Billing [39], Wilkie et al. [40] found that the use of sucrose and tartrate could help in the initial identification of members of the species. Our results coincided in the case of sucrose utilization but not of D-tartrate, a test in which 29% of the isolates were positive. Nor do they agree with the results of Sarris et al. [41] who studied 18 isolates of *P. viridiflava* obtained from different hosts in Crete (Greece) and found no variability in the biochemical tests performed, except for the L-tartrate test. Regarding the latter, isolates obtained from tomato were positive, while the type strain was negative, as well as isolates from other hosts.

3.2. Occurrence of the Bacterium in Weeds and Bean Samples

In this study, *P. viridiflava* was isolated from twelve genera/species of weeds: *Capsella bursa-pastoris*, *Chenopodium album*, *Cyperus rotundus*, *Fumaria* sp, *Galinsoga parviflora*, *Hypochaeris radicata*, *Malva sylvestris*, *Senecio vulgaris*, *Solanum nigrum*, *Sonchus oleraceus*, *Stellaria media*, and *Trifolium* sp., and from three unidentified weeds which were different to each other (Table 1). The bacterium was most frequently found in *Fumaria* sp. (nine isolates from eight samples), followed by *S. olereaceus* and *S. nigrum* (four isolates from four samples), and *C. rotundus* (four isolates from three samples). This further expands the already wide host range of *P. viridiflava* which, as far as we know, has not been previously reported in 10 out of the 12 weed species/genera mentioned before. However, the bacterium has already been described by Gitaitis et al. [22] in *Sonchus* sp and *Fumaria* sp. associated with an onion crop. It is important to note that none of the weeds showed disease symptoms, consistent with an epiphytic existence of *P. viridiflava* in weeds.

Unlike Gitaitis et al. [22], who only isolated the bacteria from weeds during the onion growing season, we verified their presence throughout the year, i.e., before, during, and after the crop season. The survival of *P. viridiflava* in five species of weeds had already been described by Aysan and Uygur [42] before and after the tomato crop, and by Mariano and McCarter [43] also on tomato. In the latter study, persistence of the bacterium on the surface of the leaves of two weed species was observed by electron microscopy, for at least 16 weeks.

Pseudomonas viridiflava was less frequently detected in bean samples than in weeds. This species was isolated as a sole pathogen in 8% of the bean samples tested and together with *P. syringae* pv. *phaseolicola* in 1.1%. In contrast, a previous study in our region revealed *P. viridiflava* as the only pathogen in 28% of the samples, and together with *P. syringae* pv. *phaseolicola* and pv. *syringae* in a small percentage [13]. These differences could be due to the fact that the bean fields sampled in the present study had a significant presence of halo blight caused by *P. syringae* pv. *phaseolicola*, which could have displaced the less aggressive *P. viridiflava*. In any case, the simultaneous presence of *P. viridiflava* with other pathogens like *P. syringae* pv. *syringae* or pv. *phaseolicola* in the same sample highlights the epiphytic and opportunistic nature of the former species. Moreover, the relatively frequent detection of *P. viridiflava* in weeds, suggests that they could be an important reservoir and source of inoculum for crops. This is particularly true in Asturias, where the climatic conditions: mild temperatures, frequent rainfall, and high relative humidity values, are favorable both for growing of the weeds and for the development of the disease.

3.3. PAI Distribution, Pectinolysis Activity, and Pathogenicity Tests

Each isolate carried one of the pathogenicity islands (T-PAI or S-PAI) previously reported in *P. viridiflava*, thus confirming the polymorphism in terms of the presence/absence of these islands [19]. S-PAI was the most frequent, found in 26 and 8 of the isolates from weeds and beans, respectively (Table 1).

An important virulence factor in *P. viridiflava* is the enzyme pectate lyase, which causes maceration of plant tissues [20; 44]. In our study, 81.2% of the strains produced pectinolysis on potato slices, and 58.3% on bean pods (Table 1). Pectinolysis on potato slices was observed for isolates carrying both T-PAI (92.8%) and S-PAI (79.4%), although Jakob et al. [44] have shown that isolates with S-PAI had higher enzyme activity than those carrying T-PAI.

When pathogenicity tests were performed, different kinds of symptoms were observed (Figure 1). Some isolates caused only a small brown spot in the pods, around the inoculation point. Others produced a reddish or ferrous halo 24–48 h post-inoculation, which could be followed or not by maceration of the tissues, observed after 48–72 h. This coincides with results reported by Wilkie et al. [40] who, using the same method of inoculation, found different responses depending on the inoculated strain.

Twenty-seven strains of the 48 studied were pathogenic on bean pods, and one gave a variable response. Five of the pathogenic isolates came from bean samples, while the remaining 23 were from weeds belonging to the species *C. bursa-pastoris*, *C. rotundus*, *C. album*, *Fumaria* sp., *G. parviflora*, *M. sylvestris*, *S. vulgaris*, *S. nigrum*, *S. media*, and *Trifolium* sp. In total, 59% of the weed isolates were pathogenic to bean, which is highly relevant with respect to the epidemiology of the disease. This coincides with previous studies performed on tomato and onion, where *P. viridiflava* cause serious diseases [22,42,43] and a high percentage of weed isolates were pathogenic.

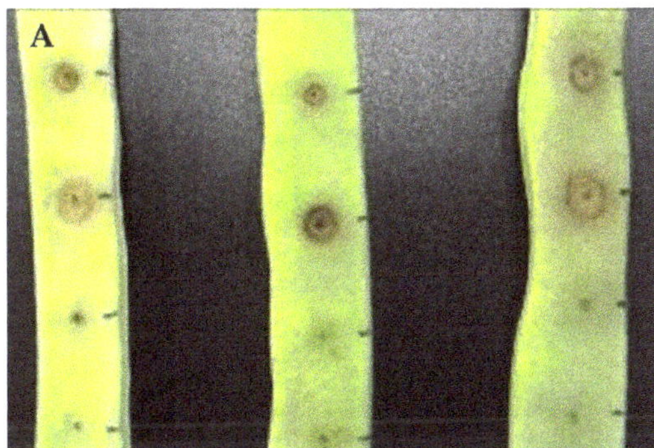
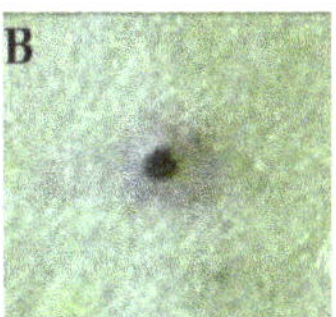
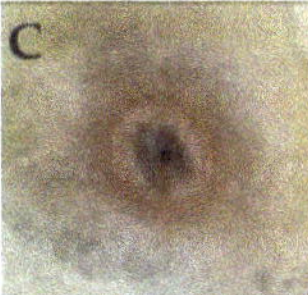

Figure 1. Inoculation of bean pods cv. *Helda* with representative isolates of *Pseudomonas viridiflava* (**A**) and enlarged details without (**B**) and with symptoms (**C**).

It is finally of note that the ability to produce maceration in bean pods did not correlate with biotype (since it was observed for 47.8% and 68% of the isolates with BT1 and BT2, respectively), nor with the type of PAI (soft rot was produce by 57% and 58.8% of the isolates with T-PAI and S-PAI, respectively). The latter observation is in line with results obtained by Bartoli et al. [45], who also found that the presence of S-PAI and T-PAI was not correlated with the ability to produce soft rot and with pathogenicity.

3.4. Phylogenetic Analysis

To establish the phylogenetic relationships of the isolates under study, the *gyrB*, *rpoD* and *gltA* genes from the 39 identified strains were sequenced. The 16S rDNA was not included because, being a highly conserved gene, it does not provide intraspecies variability [12]. The *gyrB* and *rpoD* genes were used for the investigation of populations of *P. viridiflava* in two previous studies [12,41].

The concatenated sequences of the three loci had a total length of 2450 bp (610 bp *gyrB*, 882 bp *rpoD*, and 958 bp *gltA*). By means of the Tajima's test of neutrality, we have been able to verify that the concatenated sequence had 168 segregating sites and a nucleotide diversity (π) of 0.019. Tajima's D-statistic test distinguishes between DNA sequences that evolve randomly ("neutrally") from those that evolve under a non-random process. In our case, the D value was >0 so there are more haplotypes than number of segregating sites (Table 3). The gene that most contributed to nucleotide diversity was *gyrB*, a result already obtained by Yin et al. [46]. However, *rpoD* was the gene that provided nucleotide diversity (0.019617) closer to that obtained with the three concatenated genes (0.019432).

Table 3. Results from Tajima's neutrality test.

Gene	m	n	S	π	D
gyrB	40	610	54	0.025542	0.812835
rpoD	40	882	63	0.019617	0.605882
gltA	40	958	51	0.015371	0.813580
gyrB + *rpoD* + *gltA*	40	2450	168	0.019432	0.760686

m = number of sequences, n = number of positions, S = number of segregating sites, π = nucleotide diversity, D = Tajima Test statistic.

Figure 2 shows the phylogenetic tree based on the three concatenated *gyrB*, *rpoD*, and *gltA* genes of the 39 strains. They were distributed into two clades, A and B, which contain the S-PAI- and T-PAI-positive strains, respectively. In contrast, the strains were not separated according to biotype, soft rot activity, and pathogenicity, since isolates with these properties appear in the two clusters. Strains from weeds and beans were represented in both clusters, and closely related strains were obtained from the two hosts.

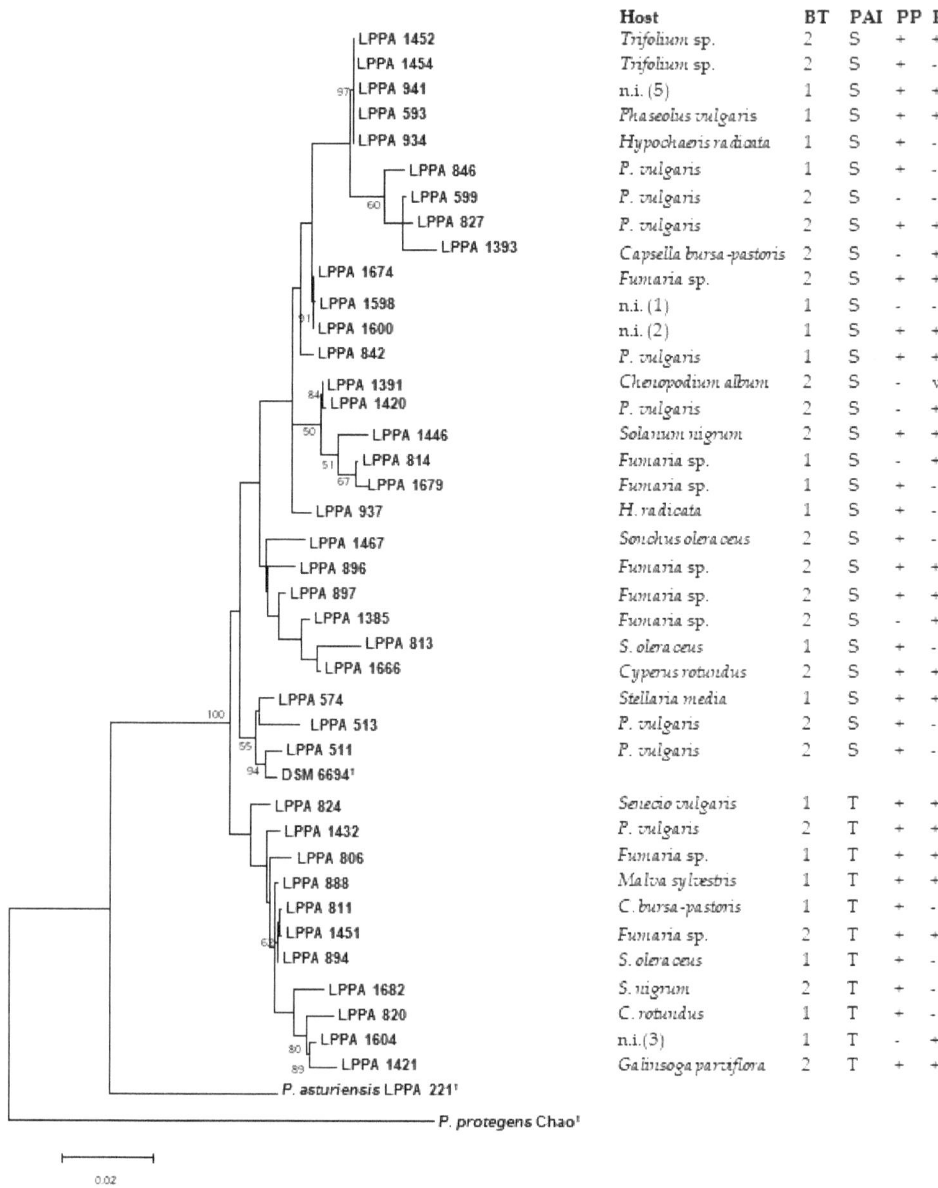

Figure 2. Phylogenetic tree based on concatenated partial sequences of the *gyrB*, *rpoD*, and *gltA* genes, inferred with the Maximum Likelihood method. The evolutionary distances were computed by the Tamura–Nei model. Bootstrap values $\geq$50% (based on 1000 replicates) are indicated at branch points. *P. viridiflava* DSM 6694[T] was used as control, *P. asturiensis* LPPA 221[T] as a member of the closest-related species, and *P. protegens* strain Chao[T] as outgroup. Bar scale, substitutions per site. Relevant features related to the strains are shown at the right of the figure. BT, Biotype; PAI, pathogenicity island; PP, pectinolysis on potato; P, pathogenicity on bean pods. Accession numbers of the sequences and the pairwise distance matrix used to construct the phylogenetic tree are shown in Tables S2 and S3, respectively.

4. Conclusions

P. viridiflava was isolated not only from beans but also from fifteen different weeds growing in the same fields. The bacterium is reported for the first time in ten of the twelve identified weed genera/species (with the remaining three weeds, which could not be identified, but were all different than each other). Regardless of their origin, the isolates displayed wide biochemical diversity, hindering identification by traditional methods. Consistent results were only obtained for the sucrose and lactose tests (negative), and for the tobacco, esculin, quinate, and xylose tests (positive). Phylogenetic analysis with concatenated *gyrB*, *rpoD*, and *gltA* sequences separated the strains carrying S-PAI and T-PAI into two different clusters, with no correlation observed for other characteristics, such as plant host, LOPAT profile, pectinolytic activity, or pathogenicity. Detection of *P. viridiflava* before, during, and after the crop season shows survival of the bacteria in weeds throughout the year, hence supporting the role of weeds as reservoir of *P. viridiflava*, and as a source of inoculum for bean infection. The fact that 59% of the weed isolates behave as bean pathogens, and that some strains recovered from beans and weeds were closely related, further highlights the role of weeds on the epidemiology of the disease.

Supplementary Materials: The following supporting information can be downloaded at: https://www.mdpi.com/article/10.3390/microorganisms10081542/s1, Table S1: Biochemical profiles of the isolates under study; Table S2: Accession numbers of the *gyrB*, *rpoD* and *gltA* sequences used for phylogenetic analysis; Table S3: Pairwise distance matrix used to construct the phylogenetic tree.

Author Contributions: Conceptualization, A.J.G. and M.R.R.; methodology, A.M.F.-S. and A.J.G.; formal analysis, A.M.F.-S., M.R.R., and A.J.G.; investigation, A.M.F.-S. and A.J.G.; resources, A.J.G. and M.R.R.; data curation, A.M.F.-S.; writing—original draft preparation, A.J.G. and M.R.R.; writing—review and editing, A.M.F.-S., M.R.R., and A.J.G.; supervision, M.R.R. and A.J.G.; project administration, A.J.G.; funding acquisition, A.J.G. All authors have read and agreed to the published version of the manuscript.

Funding: This research was funded by Instituto Nacional de Investigación y Tecnología Agraria y Alimentaria, grant number RTA2005-00076 and RTA2008-00019, and Ana M Fernández-Sanz is recipient of a grant from the "Instituto Nacional de Investigación y Tecnología Agraria y Alimentaria", both co-funded by the European Social Fund. The APC was funded by Consejería de Medio Rural y Cohesión Territorial, Government of the Principality of Asturias.

Institutional Review Board Statement: Not applicable.

Informed Consent Statement: Not applicable.

Data Availability Statement: The *gyrB*, *rpoD* and *gltA* sequences have been deposited in GenBank under accession numbers shown in Table S2.

Conflicts of Interest: The authors declare no conflict of interest.

References

1. Young, J.M.; Cheesmur, G.J.; Welham, F.V.; Henshall, W.R. Bacterial blight of kiwifruit *Ann. Appl. Biol.* **1988**, *112*, 91 105. [CrossRef]
2. Yildiz, H.N.; Aysan, Y.; Sahin, F.; Cinar, O. Potential inoculum sources of tomato stem and pith necrosis caused by *Pseudomonas viridiflava* in the Eastern Mediterranean Region of Turkey. *Z. Pflanzenkrankh. Pflanzenschutz* **2004**, *111*, 380–387.
3. Popović, T.; Ivanović, Ž.; Ignjatov, M. First report of *Pseudomonas viridiflava* causing pith necrosis of tomato (*Solanum lycopersicum*) in Serbia. *Plant Dis.* **2015**, *99*, 1033. [CrossRef]
4. Martín-Sanz, A.; Palomo, J.L.; Pérez de la Vega, M.; Caminero, C. First report of bacterial blight caused by *Pseudomonas viridiflava* on pea in Spain. *Plant Dis.* **2010**, *94*, 128. [CrossRef]
5. Heydari, A.; Khodakaramian, G.; Zafari, D. Characterization of *Pseudomonas viridiflava* causing alfalfa root rot disease in Hamedan Province of Iran. *J. Plant Pathol. Microbiol.* **2012**, *3*, 135. [CrossRef]
6. Almeida, I.M.G.; Maciel, K.W.; Rodrigues Neto, J.; Beriam, L.O.S. *Pseudomonas viridiflava* in imported carrot seeds. *Australas. Plant Dis. Notes* **2013**, *8*, 17–19. [CrossRef]
7. Al-Karablieh, N.; Mutlak, I.; Al-Dokh, A. Isolation and identification of *Pseudomonas viridiflava*, the causal agent of fruit rotting of *Cucumis sativus*. *J. J. Agri. Sci.* **2017**, *13*, 79–91.

8. Choi, O.; Lee, Y.; Kang, B.; Kim, S.; Bae, J.; Kim, J. Bacterial shoot blight of sweet crab apple caused by *Pseudomonas viridiflava*. *For. Pathol.* **2020**, *50*, e12603. [CrossRef]

9. Bophela, K.N.; Petersen, Y.; Bull, C.T.; Coutinho, T.A. Identification of *Pseudomonas* isolates associated with bacterial canker of stone fruit trees in the Western Cape, South Africa. *Plant Dis.* **2020**, *104*, 882–892. [CrossRef]

10. Canik Orel, D. Biocontrol of bacterial diseases with beneficial bacteria in lettuce. *Intern. J. Agric. Nat. Sci.* **2020**, *13*, 108–117.

11. Goumans, D.E.; Chatzaki, A.K. Characterization and host range evaluation of *Pseudomonas viridiflava* from melon, blite, tomato, chrysanthemum and eggplant. *Eur. J. Plant Pathol.* **1998**, *104*, 181–188. [CrossRef]

12. Goss, E.M.; Kreitman, M.; Bergelson, J. Genetic diversity, recombination and cryptic clades in *Pseudomonas viridiflava* infecting natural populations of *Arabidopsis thaliana*. *Genetics* **2005**, *169*, 21–35. [CrossRef]

13. González, A.J.; Rodicio, M.R.; Mendoza, M.C. Identification of an emergent and atypical *Pseudomonas viridiflava* lineage causing bacteriosis in plants of agronomic importance in a Spanish region. *Appl. Environ. Microbiol.* **2003**, *69*, 2936–2941. [CrossRef]

14. González, A.J.; Rodicio, M.R. *Pseudomonas viridiflava* causing necrotic leaf spots and defoliation on *Hebe* spp. in Northern Spain. *Plant Dis.* **2006**, *90*, 830. [CrossRef]

15. Myung, I.S.; Lee, Y.K.; Lee, S.W.; Kim, W.G.; Shim, H.S.; Ra, D.S. A new disease, bacterial leaf spot of rape, caused by atypical *Pseudomonas viridiflava* in South Korea. *Plant Dis.* **2010**, *94*, 1164. [CrossRef]

16. Anzai, Y.; Kim, H.; Park, J.; Wakabayashi, H.; Oyaizu, H. Phylogenetic affiliation of the *Pseudomonads* based on 16S rRNA sequence. *Int. J. Syst. Evol. Microbiol.* **2000**, *50*, 1563–1589. [CrossRef]

17. Yamamoto, S.; Kasai, H.; Arnold, D.L.; Jackson, R.W.; Vivian, A.; Harayama, S. Phylogeny of the genus *Pseudomonas*: Intrageneric structure reconstructed from the nucleotide sequences of *gyrB* and *rpoD* genes. *Microbiology* **2000**, *146*, 2385–2394. [CrossRef]

18. Mulet, M.; Lalucat, J.; García-Valdés, E. DNA sequence-based analysis of the *Pseudomonas* species. *Environ. Microbiol.* **2010**, *12*, 1513–1530.

19. Araki, H.; Tian, D.; Goss, E.M.; Jakob, K.; Halldorsdottir, S.S.; Kreitman, M.; Bergelson, J. Presence/absence polymorphism for alternative pathogenicity islands in *Pseudomonas viridiflava*, a pathogen of *Arabidopsis*. *Proc. Natl. Acad. Sci. USA* **2006**, *103*, 5887–5892. [CrossRef]

20. Liao, C.H.; Hung, H.Y.; Chatterjee, A.K. An extracellular pectate lyase is the pathogenicity factor of the soft-rotting bacterium *Pseudomonas viridiflava*. *Mol. Plant-Microbe Interact.* **1988**, *1*, 199–206. [CrossRef]

21. Liao, C.H.; McCallus, D.E.; Fett, W.F. Molecular characterization of two gene loci required for production of the key pathogenicity factor pectate lyase in *Pseudomonas viridiflava*. *Mol. Plant-Microbe Interact.* **1994**, *7*, 391–400. [CrossRef]

22. Gitaitis, R.; Macdonald, G.; Torrance, R.; Hartley, R.; Sumner, D.R.; Gay, J.D.; Johnson, W.C., III. Bacterial streak and bulb rot of sweet onion: II. Epiphytic survival of *Pseudomonas viridiflava* in association with multiple weed hosts. *Plant Dis.* **1998**, *82*, 935–938. [CrossRef]

23. Basavand, E.; Khodaygan, P. Common water-plantain, a new host of *Pseudomonas viridiflava* in rice fields in Iran. *J. Plant Pathol.* **2020**, *102*, 913. [CrossRef]

24. Ryu, E. A simple method of differentiation between gram-positive and gram-negative organisms without staining. *Kitasato Arch. Exp. Med.* **1940**, *17*, 58–63.

25. Hugh, R.; Leifson, E. The taxonomic significance of fermentative versus oxidative metabolism of carbohydrates by various Gram negative bacteria. *J. Bacteriol.* **1953**, *66*, 24–26. [CrossRef]

26. Lelliott, R.A.; Billing, E.; Hayward, A.C. A determinative scheme for fluorescent plant pathogenic bacteria. *J. Appl. Bacteriol.* **1966**, *29*, 470–478. [CrossRef]

27. Jansing, H.; Rudolph, K. A sensitive and quick test for determination of bean seed infestation by *Pseudomonas syringae* pv. *phaseolicola*. *Z. Pflanzenkrankh. Pflanzenschutz* **1990**, *97*, 42–55.

28. Schaad, N.W.; Jones, J.B.; Chun, W. *Laboratory Guide for Identification of Plant-Pathogenic Bacteria*, 3rd ed.; CPL APS Press: St. Paul, MN, USA, 2001; p. 398.

29. Noval, C. Medios de cultivo y pruebas de diagnóstico. In *Manual de Laboratorio. Diagnóstico de Hongos, Bacterias y Nematodos Fitopatógenos*; MAPA: Madrid, Spain, 1991; pp. 379–410.

30. Goszczynska, T.; Serfortein, J. Milk-Tween agar, a semiselective medium for isolation and differentiation of *Pseudomonas syringae* pv. *syringae*, *Pseudomonas syringae* pv. *phaseolicola* and *Xanthomonas axonopodis* pv. *phaseoli*. *J. Microbiol. Methods* **1998**, *32*, 65–72. [CrossRef]

31. Edwards, U.; Rogall, T.; Blöcker, H.; Emde, M.; Böttger, E.C. Isolation and direct complete nucleotide determination of entire genes. Characterization of a gene coding for 16S ribosomal RNA. *Nucleic Acids Res.* **1989**, *17*, 7843–7853. [CrossRef] [PubMed]

32. Cheng, G.Y.; Legard, D.E.; Hunter, J.E.; Burr, T.J. Modified bean pod assay to detect strains of *Pseudomonas syringae* pv. *syringae* that cause bacterial brown spot of snap bean. *Plant Dis.* **1989**, *73*, 149–423. [CrossRef]

33. Sarkar, S.F.; Guttman, D.S. Evolution of the core genome of *Pseudomonas syringae*, a highly clonal, endemic plant pathogen. *Appl. Environ. Microbiol.* **2004**, *70*, 1999–2012. [CrossRef]

34. Hwang, M.S.H.; Morgan, R.L.; Sarkar, S.F.; Wang, P.W.; Guttman, D.S. Phylogenetic characterization of virulence and resistance phenotypes of *Pseudomonas syringae*. *Appl. Environ. Microbiol.* **2005**, *71*, 5182–5191. [CrossRef]

35. Thompson, J.D.; Higgins, D.G.; Gibson, T.J. CLUSTAL W: Improving the sensitivity of progressive multiple sequence alignment through sequence weighting, position-specific gap penalties and weight matrix choice. *Nucleic Acids Res.* **1994**, *22*, 4673–4680. [CrossRef]

36. Saitou, N.; Nei, M. The Neighbor-Joining method: A new method for reconstructing phylogenetic trees. *Mol. Biol. Evol.* **1987**, *4*, 406–425.
37. Tamura, K.; Stecher, G.; Peterson, D.; Filipski, A.; Kumar, S. MEGA6: Molecular Evolutionary Genetics Analysis version 6. *Mol. Biol. Evol.* **2013**, *30*, 2725–2729. [CrossRef]
38. Nei, M.; Kumar, S. *Molecular Evolution and Phylogenetics*; Oxford University Press: New York, NY, USA, 2000; p. 333.
39. Billing, E. *Pseudomonas viridiflava* (Burkholder, 1930; Clara, 1934). *J. Appl. Bacteriol.* **1970**, *33*, 492–500. [CrossRef]
40. Wilkie, J.P.; Dye, D.W.; Watson, D.R.W. Further hosts of *Pseudomonas viridiflava*. *N. Z. J. Agric. Res.* **1973**, *16*, 315–323. [CrossRef]
41. Sarris, P.F.; Trantas, E.A.; Mpalantinaki, E.; Ververidis, F.; Goumas, D.E. *Pseudomonas viridiflava*, a multi host plant pathogen with significant genetic variation at the molecular level. *PLoS ONE* **2012**, *7*, e36090. [CrossRef]
42. Aysan, Y.; Uygur, S. Epiphytic survival of *Pseudomonas viridiflava*, causal agent of pith necrosis of tomato, on weeds in Turkey. *J. Plant Pathol.* **2005**, *87*, 135–139.
43. Mariano, R.D.; McCarter, S.M. Epiphytic survival of *Pseudomonas viridiflava* on tomato and selected weed species. *Microb. Ecol.* **2004**, *26*, 47–58. [CrossRef]
44. Jakob, K.; Kniskern, J.M.; Bergelson, J. The role of pectate lyase and the jasmonic acid defense response in *Pseudomonas viridiflava* virulence. *Mol. Plant-Microbe Interact.* **2007**, *20*, 146–158. [CrossRef]
45. Bartoli, C.; Berge, O.; Monteil, C.L.; Guilbaud, C.; Balestra, G.M.; Varvaro, L.; Jones, C.; Dangl, J.L.; Baltrus, D.A.; Sands, D.C.; et al. The *Pseudomonas viridiflava* phylogroups in the *P. syringae* species complex are characterized by genetic variability and phenotypic plasticity of pathogenicity-related traits. *Environ. Microbiol.* **2014**, *16*, 2301–2315. [CrossRef]
46. Yin, H.; Cao, L.; Xie, M.; Chen, Q.; Qiu, G.; Zhou, J.; Wu, L.; Wang, D.; Liu, X. Bacterial diversity based on 16S rRNA and *gyrB* genes at Yinshan mine, China. *Syst. Appl. Microbiol.* **2008**, *31*, 302–311. [CrossRef]

 *microorganisms*

Article

Fungal Pathogens Associated with Aerial Symptoms of Avocado (*Persea americana* Mill.) in Tenerife (Canary Islands, Spain) Focused on Species of the Family Botryosphaeriaceae

David Hernández [1], Omar García-Pérez [1], Santiago Perera [2], Mario A. González-Carracedo [3,4], Ana Rodríguez-Pérez [4] and Felipe Siverio [1,5,*]

1   Unidad de Protección Vegetal, Instituto Canario de Investigaciones Agrarias, 38270 San Cristóbal de La Laguna, Spain
2   Servicio Técnico de Agricultura y Desarrollo Rural del Cabildo Insular de Tenerife, 38007 Santa Cruz de Tenerife, Spain
3   Instituto Universitario de Enfermedades Tropicales y Salud Pública de Canarias, 38200 San Cristóbal de La Laguna, Spain
4   Departamento de Bioquímica, Microbiología, Biología Celular y Genética, Universidad de La Laguna, 38200 San Cristóbal de La Laguna, Spain
5   Sección de Laboratorio de Sanidad Vegetal, Consejería de Agricultura, Ganadería, Pesca y Aguas del Gobierno de Canarias, 38270 San Cristóbal de La Laguna, Spain
*   Correspondence: fsiverio@icia.es

Citation: Hernández, D.; García-Pérez, O.; Perera, S.; González-Carracedo, M.A.; Rodríguez-Pérez, A.; Siverio, F. Fungal Pathogens Associated with Aerial Symptoms of Avocado (*Persea americana* Mill.) in Tenerife (Canary Islands, Spain) Focused on Species of the Family Botryosphaeriaceae. *Microorganisms* **2023**, *11*, 585. https://doi.org/10.3390/microorganisms11030585

Academic Editors: Soledad Sacristán

Received: 31 December 2022
Revised: 20 February 2023
Accepted: 23 February 2023
Published: 25 February 2023

Abstract: Fungi of the family Botryosphaeriaceae are considered responsible for various symptoms in avocado such as dieback, external necrosis of branches and inflorescences, cankers on branches and trunks, or stem-end rot of fruits. In recent years, these problems are becoming more frequent in avocado orchards in the Canary Islands (Spain). This work includes the characterization of fungal species involved in these diseases, which were isolated from avocado crops in Tenerife Island between 2018 and 2022. A total of 158 vegetal samples were collected, from which 297 fungal isolates were culture-isolated. Fifty-two of them were selected according to their morphological features as representative isolates of Botryosphaeriaceae, and their molecular characterization was carried out, sequencing the ITS1-2 region as well as the *β-tubulin* and the *elongation factor 1-alpha* genes. Five species of Botryosphaeriaceae were isolated, including *Neofusicoccum australe*, *N. cryptoaustrale/stellenboschiana*, *N. luteum*, *N. parvum*, and *Lasiodiplodia brasiliensis*. This is the first time that *L. brasiliensis* has been associated with avocado dieback and that *N. cryptoaustrale/stellenboschiana* has been cited in avocado causing symptoms of dieback and stem-end rot. However, it was not possible to assign our isolates unequivocally to *N. cryptoaustrale* or *N. stellenboschiana* even additionally using the rpb2 marker for their molecular characterization. Botryosphaeriaceae family seem to be involved in avocado dieback, in the premature fall of fruits during their development in the field and in post-harvest damage in Tenerife, but further studies are needed to clarify the fungal pathogens associated with symptoms in relation to phenological plant growth stages or less frequently observed.

Keywords: *Lasiodiplodia brasiliensis*; *Neofusicoccum australe*; *Neofusicoccum cryptoaustrale*; *Neofusicoccum stellenboschiana*; *Neofusicoccum luteum*; *Neofusicoccum parvum*; *Persea americana*; dieback; canker; stem-end rot

1. Introduction

Avocado (*Persea americana* Mill., 1768) is one of the most important crops in the Canary Islands, with the production area increasing at 5% per annum over the past decade. Plantations currently occupy about 2300 ha (more than 10% of its dedicated cultivation area in Spain) [1,2]. The avocado crops on the Canary Islands are affected by various diseases, with those caused by fungi of the Botryosphaeriaceae family more frequent over the past years but depending on the environmental conditions. In other avocado-producing regions, species of Botryosphaeriaceae have been associated with cankers and dieback on young and adult avocado plants, as well as with fruit rot (mainly stem-end rot) in post-harvest conditions (Table 1).

Symptoms of dieback are characterized by necrosis and death of twigs and branch-es in the tree canopy, with symptomatic plants exhibiting dry branches, external necrosis in twigs, and wilting leaves and inflorescences [3–7]. Cankers associated with Botryosphaeriaceae species (formerly *Dothiorella canker*) can occur on branches and trunks, where infections are initiated by spores entering through fresh wounds resulting from pruning, mechanical injury, sunburn or frost damage [8]. Cankers on avocado may exudate reddish-brown sap that turns to a whitish-yellowish powder when dry [8–11]. The bark appears darkened and friable and can be easily removed in old cankers. Underneath the bark, cankers are reddish-brown in colour and variable in shape, sometimes penetrating into the heartwood and affecting the xylem [9,10]; this may lead to dieback and collapse of branches or the entire tree [5,8,10]. In addition, fruit rot caused by Botryosphaeriaceae species is a common problem in avocado-producing regions (Table 1), representing an important threat to avocado production throughout the word. Symptoms of fruit rot are mostly developed after harvest when fruits begin to ripen. Then dark-coloured spots appear mainly in the insertion area of the peduncle (stem-end rot). As fruit ages, lesions gradually increase in size and become sunken, while decay spreads inside the fruit affecting the flesh that turns brown and watery [6,11–14]. Botryosphaeriaceae species associated with stem-end rot are present in cankers and living and dead branches and twigs collected from avocado trees [6,7,14], indicating that stem-end infections can be originated from endophytic fungal colonization. However, stem-end rot appears to be mostly originated from infections of peduncle that occur during fruit harvesting [7].

The Botryosphaeriaceae family (order *Botryosphaeriales, Ascomycota fungi*) includes numerous species. Some of them are morphologically similar, having only been differentiated and described in recent years by combining morphological characterization with phylogenetic analysis [15–17]. The species most frequently associated with avocado damage belong to the genera *Botryosphaeria, Diplodia, Dothiorella, Lasiodiplodia* and, especially, *Neofusicoccum* (Table 1). Fungal species of the Botryosphaeriaceae family are pathogenic to a wide range of woody hosts. They can also be found in nature as endophytes, behaving as latent or opportunistic pathogens that develop their pathogenicity when their hosts are subjected to stress conditions [18]. However, species of *Colletotrichum, Alternaria* and other genera, have also been described as causing similar symptoms, which sometimes could be erroneously attributed to fungal species of the Botryosphaeriaceae family [19].

In this work we show results of the studies carried out between 2018 and 2022 in Tenerife (Canary Islands, Spain), in an effort to obtain information about the symptoms attributed to Botryosphaeriaceae species in avocado orchards and post-harvest fruits. Morphological and molecular methods were used to characterize fungal isolates obtained from symptomatic samples, to determine the fungal agents associated with each symptomatology. Furthermore, pathogenicity of Botryosphaeriacerae species was tested on avocado seedlings.

Table 1. Fungal species isolated from avocado plants or fruits showing symptoms usually attributed to fungi of the Botryosphaeriaceae family.

| Country | Reference | Symptoms | | | | Botryosphaeriaceae Species | | | | | | | | | | | | | | | | | Non-Botryosphaeriaceae Species |
| | | Plant | | Fruit | | | | | | | | | | | | | | | | | | | |
Country	Reference	Branch or Trunk Canker	Dieback	Stem-End Rot	Black Spots	*Botryosphaeria dothidea*	*Diplodia mutila*	*Diplodia pseudoseriata*	*Diplodia seriata*	*Dothiorella aromatica*	*Dothiorella dominicana*	*Dothiorella iberica*	*Dothiorella sp.*	*Lasiodiplodia pseudotheobromae*	*Lasiodiplodia theobromae*	*Neofusiccocum australe*	*Neofusiccocum luteum*	*Neofusiccocum mangiferae*	*Neofusiccocum mediterraneum*	*Neofusiccocum nonquaesitum*	*Neofusiccocum parvum*	*Neofusiccocum sp.*	Non-Botryosphaeriaceae Species
Australia	[20]														X	X	X				X	X	
Australia	[12]			X	X	X				X	X												
Australia	[21]			X	X					X													Cg
Brasil	[22]			X	X	X															X		
Chile	[3]		X													X							
Chile	[23]			X												X							
Chile	[6]	X	X	X			X	X	X			X			X	X				X	X		
Chile	[24]			X	X												X						
China	[25]			X		X																	
Colombia	[26]	X	X												X								
Colombia	[27]			X	X									X	X								Cg, Phs, Pp, Ps
Ethiopia	[28]			X	X																		Af, An, Cg, Cls, Fs, Pes
Italy	[29]			X																	X		Cf, Cg,
Mexico	[30]				X																X		
New Zealand	[12]			X		X											X				X		Ca, Cg, Phs
New Zealand	[7]		X	X		X											X				X		Ca, Cg, Phs
Peru	[11]	X	X	X											X								As, Cos, Fs
South Africa	[13]			X	X					X													Cg, Cls, Np, Phs, Pms, Pp, Ps
Spain	[4]		X																		X		
Spain	[5]	X	X												X	X	X		X		X		Cg

Table 1. *Cont.*

| Country | Reference | Branch or Trunk Canker | Dieback | Stem-End Rot | Black Spots | *Botryosphaeria dothidea* | *Diplodia mutila* | *Diplodia pseudoseriata* | *Diplodia seriata* | *Dothiorella aromatica* | *Dothiorella dominicana* | *Dothiorella iberica* | *Dothiorella sp.* | *Lasiodiplodia pseudotheobromae* | *Lasiodiplodia theobromae* | *Neofusiccocum australe* | *Neofusiccocum luteum* | *Neofusiccocum mangiferae* | *Neofusiccocum mediterraneum* | *Neofusiccocum nonquaesitum* | *Neofusiccocum parvum* | *Neofusiccocum sp.* | Non-Botryosphaeriaceae Species |
		Plant	Plant	Fruit	Fruit																		
Thailandia	[31]	X		X	X									X									
Taiwan	[32]				X													X					
Taiwan	[33]			X	X	X										X			X			X		
USA	[9]	X				X											X	X				X		
USA	[10]	X				X						X				X	X				X	X	Cos, Fs, Phs
USA	[8]	X										X				X	X				X		Phs
USA	[14]			X													X						Cg, Phs

Abbreviations: As, *Alternaria* spp.; Af, *Aspergillus flavus*; An, *Aspergillus niger*; Cls, *Cladosporium* spp.; Ca, *Colletotrichum acutatum*; Cf, *Colletotrichum fructicola*; Cg, *Colletotrichum gloeosporioides*; Cos, *Colletotrichum* spp.; Fs, *Fusarium* spp.; Np, *Nectria pseudotrichia*; Pes, *Pestalotiopsis* spp.; Phs, *Phomosis* spp.; Pms, *Phoma* spp.; Pp, *Pseudocercospora purpurea*; Ps, *Pestalotia* spp.

2. Materials and Methods

2.1. Sampling and Fungal Isolation

Between 2018 and 2022, 40 avocado orchards located in Tenerife (Canary Island, Spain), were surveyed for the presence of symptoms in the aerial parts of the trees. Orchards were selected based on information provided by Agricultural Extension Services (Council of Tenerife), or by technical advisors from local agricultural cooperatives who noticed symptoms of dieback of inflorescences, shoots and branches, branch or trunk cankers, or fruit rots. A total of 123 samples of necrotic panicles, shoots and branches, wood samples cut from cankers, or rotten avocado fruits were collected in the surveyed orchards. In addition, necrotic branches (11 samples), which were not located on the distribution maps, were directly provided by several avocado growers. Moreover, samples of postharvest fruits were obtained from two avocado packaging facilities (15 fruits), and from four retail stores (9 fruits). Samples were kept refrigerated at 5 °C until they were processed.

Vegetal fragments were taken with sterile scalpels from the transition zone between healthy and diseased tissue, disinfected with 0.5% sodium hypochlorite for 10 min, placed in 70% ethanol for 30 s, rinsed with sterile distilled water, and left to dry on sterile filter paper in a laminar flow cabinet. Afterwards, small pieces of about 5 mm^3 were cut from the disease progression zones with a sterile scalpel, and fragments were plated in potato dextrose agar (PDA; Becton Dickinson, Sparks, MD, USA) plates, supplemented with 500–1000 mg/L Streptomycin (Sigma-Aldrich, St. Louis, MO, USA). Plates were incubated in the dark at 25 °C and examined for fungal growth over a 7-day period. Fungal colonies were isolated by transferring hyphal tips from the edge of the colonies to fresh PDA-Streptomycin (PDAS) plates (one isolate per plate). Isolates were preserved at −20 °C and −80 °C as mycelial plugs (5 mm diameter) in cryotubes containing 1 mL of 30% glycerol, as part of the fungal culture collection maintained at the Instituto Canario de Investigaciones Agrarias (ICIA, Canary Islands, Spain).

2.2. Morphological and Molecular Identification of Fungal Isolates

Isolates were grown on PDA plates and examined for colony colour and growth pattern, mycelial morphology, as well as for production and morphology of reproductive structures. Isolates were grouped according to their morphological characteristics, in order to obtain a first tentative identification at genus level. Based on this preliminary characterization, representative isolates of Botryosphaeriaceae fungi, as well as other fungal species, were selected for molecular identification. Fungal isolates were grown on PDA plates for 7 days at 25 °C, and about 20 mm^2 of mycelia was scraped off with a sterile scalpel. Recovered mycelia was grounded in an Eppendorf tube with a micropestle, and DNA was extracted with the EZNA Tissue DNA Purification Kit (Omega BIO-TEK, Norcross, GA, USA) or using a modification of the rapid HotSHOT extraction method of Truett et al. [34] as described in Collado-Romero et al. [35]. Genomic DNA extracts were stored at –20 °C until used.

Molecular identification of Botryosphaeriaceae species was carried out by PCR amplification and sequencing of the ribosomal DNA ITS1 5.8S-ITS2 region (ITS1-2), a region from the *translation elongation factor 1-alpha (tef1)* gene, and part of the *beta-tubulin (tub2)* gene, using primers pairs ITS-1/ITS-4 (5'-CTTGGTCATTTAGAGGAAGTAA-3'/5'-TCCTCCGCTTATTGATATGC-3') [36], EF1-688F/EF1-986R (5'-CGGTCACTTGATCTA CAAGTGC-3'/5'-TACTTGAAGGAACCCTTACC-3') [37], and Bt2a/Bt2b (5'-GGTAACCA AATCGGTGCTGCTTTC-3'/5'-ACCCTCAGTGTAGTGACCCTTGGC-3') [38], respectively. Additionally, and in order to contribute to the identification of the isolates obtained close to the species *N. cryptoaustrale* or *N. stellenboschiana*, PCR amplification and sequencing of their DNA extracts were carried out with the primers RPB2bot6F/RPB2bot7R (5' GGTAGCGACGTCACTCCC-3'/5'-GGATGGATCTCGCAATGCG-3') corresponding to the *rpb2* region [39]. Molecular identification of non-Botryosphaeriaceae fungal isolates was based only on the ITS1-2 region.

PCR conditions for all amplifications consisted of an initial denaturation step (95 °C, 3 min), followed by 35 cycles of denaturation (95 °C, 30 s), annealing (53 °C, 30 s), and extension (72 °C, 1 min), and one final extension step (72 °C, 10 min). Each 25 μL PCR mix included 2.5 μL of 10× NH$_4$ Buffer), 1.5 μL of 50 mM MgCl$_2$, 0.25 μL of 100 mM dNTPs, 1.0 μL of 10 μM of each primer, 0.2 μL of 5 U/μL Taq DNA polymerase (VWR Life Science, Radnor, PA, USA), and 2.0 μL of DNA extract (substituted by 2.0 μL of H$_2$O in negative control reactions). PCRs were incubated in a Mastercycler Gradient Thermocycler (Eppendorf AG, Hamburg, German). Amplicons were analysed in 1% agarose gels, prepared in 1× TAE (40 mM Acetate, 2 mM EDTA, 40 mM Tris-HCl, pH 8.0), and containing 1× RealSafe Nucleic Acid Staining (Durviz, Valencia, Spain). Gels were photographed under UV light using a UV transilluminator 2000 (Bio-Rad, Hercules, CA, USA), equipped with a Gel Logic 100 Imaging System (Kodak, Rochester, NY, USA). ExoSAP-IT PCR Product Cleanup (Applied Biosystems, Waltham, MA, USA) was used for enzymatic cleanup of PCR products, which were sequenced in both directions, with the same primers used for PCR amplification, at the *Servicio de Genómica* of the Universidad de La Laguna (Canary Islands, Spain).

DNA sequences were manually inspected using the MEGA 11 package [40], to confirm basecalling quality, and edited to trim the tails. Next, a unique high-quality contig sequence was generated for each marker, after the alignment of forward and reverse reads. The contig sequence for each marker was independently used as query for a nucleotide BLAST search [41], against the NCBI GenBank nucleotide collection database (nr/nt) [42]. Species-level assignment was carried out considering the ten most closely related NCBI results, ordered by preference according to the minimum BLAST E-value. For species-level assignment, a minimum of 98% coverage and 98.5% identity percentage with a reference sequence was established as threshold. When reference sequences associated with different species were found inside the species-level assignment range, priority was established in base to published works. For Botryosphaeriaceae isolates, only those isolates identified as the same species independently using the three markers were considered as correctly characterized at species-level.

Representative isolates of the different Botryosphaeriaceae species on the basis of BLAST searches were subjected to further morphological characterization. Isolates were cultured in PDAS medium, and mycelial plugs (5 mm diameter) were cut after 7 days, seeded in salt-cellulose medium enriched with sugarcane bagasse (SC) [43], and incubated for up to 42 days at 25 °C in the dark. This process was performed in Petri dishes sealed for 14 days with Parafilm M, and also in unsealed plates. Some of these isolates were also cultured in water agar medium amended with autoclaved pine needles (WA, 20 g/L Bacto^{TM} Agar; (Becton Dickinson, Sparks, MD, USA)). These plates were incubated at 22 °C, 70% relative humidity and 12-h photoperiod, for 28 days. Plates were checked for pycnidia after 3 days, and then every 7 days. The time that each isolate took to produce pycnidia and conidia was recorded, and photographs of the cultures were taken. Pycnidia production and conidia characteristics were studied under a light microscope (Nikon Eclipse 80i) according to Marques et al. [44], Crous et al. [45], Yang et al. [46] and Phillips et al. [16]. The pycnidia were extracted from the medium and prepared on slides with 60% lactic acid. If the presence of conidia was detected, appearance, length and width were recorded for a total of 50 conidia per isolate.

Phylogenetic trees were constructed independently for *Neofusiccocum* and *Lasiodiplodia* genera. ITS1-2, *tef1* and *tub2* gene sequences of representative isolates obtained in this work (Table 2) were compared with reference sequences downloaded from the NCBI database, including ex-type strains (Tables S1 and S2). Sequence alignments were obtained for each marker separately, using the ClustalW algorithm implemented in the MEGA 11 software [40]. Each alignment was manually inspected to confirm gap positions and trimmed from both ends to obtain a rectangular matrix dataset, prior to the concatenation of the three alignments.

Table 2. GenBank accessions, representative isolates and groups of isolates with same sequences of *Neofusicoccum* and *Lasiodiplodia* species treated in the phylogenies.

Species	N°	Isolates [1]	Genbank Accessions		
			ITS	tef1	tub2
N. australe	2	**B018**, B149	OP788200	OQ236719	OQ181384
N. cryptoaustrale/stellenboschiana	12	**B012**, B013, B026, B027, B050, B150, B151, B152, B154, B157, B158, B160	OQ176234	OQ236715	OQ181395
	7	B029, B030, B031, B032, **B055**, B148, B155	OP788375	OQ236717	OQ181394
	3	**B041**, B042, B043	OP788376	OQ236716	OQ181393
N. luteum	9	**B003**, B004, B017, B019, B028, B047, B054, B057, B153	OP788373	OQ236713	OQ181386
	1	**B024**	OP788372	OQ236714	OQ181387
N. parvum	5	**B020**, B021, B022, B023, B147	OP788194	OQ236708	OQ181388
	2	**B034**, B035	OP788193	OQ236709	OQ181389
	2	**B037**, B038	OP788192	OQ236710	OQ181390
	2	**B044**, B045	OP788176	OQ236711	OQ181391
	1	**B156**	OQ176235	OQ236712	OQ181392
L. brasiliensis	1	**B161**	OP788199	OQ236718	OQ181385

[1] Representative isolates submitted to GenBank are in bold.

The concatenated alignment for *Neofusicoccum* species included 29 reference ingroups and sequences from 46 strains isolated in the present study. *Botryosphaeria dothidea* CBS 115476 strain was included as outgroup. The three loci resulted in 1137 aligned sites. Among them, 43 positions corresponded with gaps, which were excluded from subsequent analysis. In the final dataset (1094 sites), a total of 165 positions were variable, being 78 of them parsimony-informative, and 87 singleton variants. The best substitution model was estimated for each marker independently, using IQ-Tree v.2.1.2 tool [47], at the CIPRES server [48], considering the Bayesian Information Criterion (BIC) value. For the three markers, the best substitution model was GTR + F [49], with BIC values of 2471.26 (ITS1-2), 1905.60 (*tef1*), and 1814.28 (*tub2*). For *Lasiodiplodia* species, the alignment included 24 reference ingroups, one strain from the present study, and *Neodeightonia phoenicum* CBS 122528 as outgroup. The three markers resulted in 1030 aligned sites. However, 61 positions were excluded since they corresponded with gaps. Therefore, the final dataset contains 969 sites, being 107 variables, 23 parsimony-informative, and 84 singletons variants. The best substitution model, estimated as described above, was GTR + F with BIC values of 1496.25, 1407.46, and 1480.50, for ITS1-2, *tef1* and *tub2*, respectively.

Maximum Likelihood (ML) and Bayesian Inference (BY) approaches were conducted for phylogenetic analyses, using the concatenated alignments. The ML and BY analyses were performed on the CIPRES server, using RAxML-HPC2 v.8.2.12 [50] and MrBayes v.3.2.7 [51], respectively. For the ML phylogeny, parameters were maintained as default, but 1000 bootstrap replicates were included. For BY analyses, the posterior probabilities (PP) were calculated by four Markov Chain Monte Carlo (MCMC) runs. Each chain included 5×10^6 generations, and data were sampled every 100 generations. The first 12,000 calculations were discarded as the burn-in phase for each chain. The ML and BY trees of both genera were plotted using FigTree v.1.4.4 software, and topology of ML and BY trees were manually compared for congruence checking. As both approaches generate almost identical topologies, the BY tree was selected and ML bootstrap values were also included over the branches.

2.3. Pathogenicity Tests

Pathogenicity of representative isolates of the different Botryosphaeriaceae species that were isolated in this study was tested by inoculating 6- to 12-month-old avocado seedlings (cv. Topa-topa or West Indian) grown at room temperature in pots (16.5 cm diameter × 36.5 cm height) with a volume of approximately 5 L, which were watered on demand. Avocado seedlings were inoculated with mycelial plugs (0.5 cm in diameter) of each isolate, previously grown on PDA for 7 days at 25 °C in the dark. Control plants were inoculated with sterile PDA discs without mycelium. Several pathogenicity tests were conducted throughout this study. Briefly, for plant inoculation with *L. brasiliensis* (isolate B161), *N. luteum* (isolate B153) and *N. parvum* (isolate B156), a wound of 0.5 cm in diameter was made with a sterile cork borer on the stem of each plant at 5 cm from the ground (6–8 plants per isolate), in which it was introduced a mycelial plug. The inoculated wounds were sealed with Parafilm M, which was removed after 15 days. Immediately after inoculation, half of the plants were bagged for 10 days with transparent plastic to increase the relative humidity. In subsequent tests, pathogenicity of representative isolates of *N. australe* (B018) and *N. crytoaustrale/stellenboschiana* (B043, B050), as well as of additional, representative isolates of *N. luteum* (B004, B047) and *N. parvum* (B034, B045, B113), was tested by pruning the seedlings at the top (2–8 plants per isolate), and then a mycelial plug was placed onto the pruning-cut wound and it was sealed with Parafilm M, which was removed after 4 days. Plants inoculated in this way were not bagged. At the end of the tests, pathogenicity was evaluated by examining the plants for external symptoms (dieback necrosis) and by cutting the stems longitudinally to assess necrotic inner lesions developed from the inoculation point. Koch's postulates were confirmed after the reisolation from pieces of stem cut from the lesion margins as described above for fungal isolation.

3. Results

3.1. Sampling and Fungal Isolation

A total of 158 symptomatic avocado samples were studied in this work, from which 297 fungal strains were isolated. The analysed samples included 123 samples collected from 40 surveyed orchards; 11 samples of necrotic branches that were directly provided by growers; and 24 avocado fruits obtained from packaging facilities or stores. Symptoms usually observed in the aerial part of avocado plants included dieback, necrosis of twigs and inflorescences, or cankers on branches and trunks (Figure 1A–D). In addition, fruit rot was observed in postharvest, and also in developing fruits (Figure 1E,F).

The most frequent symptom in the surveyed orchards was necrosis of branches (mainly diebacks), which was observed in 90.0% of them (Figure S1). Dieback represents a serious problem in new avocado plantations (Figure S1A,B), as necrosis spreads down both inside and outside of the small tender trunk from the apex, causing the necrosis above the graft, and subsequently the rootstock sprout out or die. Necrosis could continue to rot the branches downwards despite pruning them in healthy areas below the advance front of necrosis (Figure S1D–F), causing the collapse and death of the whole plant in the most severe cases (Figure S1C). Another symptom found in adult plantations was panicle dieback (25.0% of orchards), which results in necrosis of inflorescences (Figure S2A–D). Therefore, if necrosis affects a significant proportion of the panicles, avocado production could be greatly reduced. This necrosis also prevented the development of the shoots and could progress downwards, affecting the corresponding branch (Figure S2E–H) that become necrotic form the apex. Usually, the entire tree was not affected and therefore it remained productive in the healthy parts of the tree's canopy. Cankers on branches and trunks were less frequent in affected plantations of the Canary Islands, as cankers were only observed in 7.5% of surveyed orchards. Sugary exudates could be seen in them, which once dry turns into a whitish-yellowish powder (Figure S3A–C), although avocado plants could produce the exudates for other reasons. In turn, the bark of the cankers could be cracked, dark in colour, or slightly sunken (Figure S3C). Beneath the canker, the inner bark and wood were brown, sometimes with reddish hues, instead of the normal pale colour. When the branch was cut transversely at the level of the canker, a wedge-shaped necrosis could be seen extending into the xylem (Figure S3D). In addition, fruit rot was observed in 12.5% of orchards. Symptoms of rot in developing fruits could appear either constricting the peduncle or directly affecting the fruit, which could detach or remains mummified hanging from the dry peduncle (Figure S3E–H). Regarding fruit rot in harvested avocados, they usually showed symptoms of stem-end rot, but also necrotic spots in other locations on the fruit (Figure S4). In postharvest, dark-coloured spots appeared mainly in the insertion area of the peduncle (stem-end rot), from where they gradually increase in size and can cover the entire surface of the fruit. Generally, the fungus also invaded the pulp (Figure S4), which began to discolour and gave a characteristic unpleasant odour when the fruit was opened for consumption.

Figure 1. Symptoms usually observed in the aerial part of the avocado plants. (**A**), panicle blight; (**B**), branch dieback; (**C**), trunk canker; (**D**), collapse of entire young plant; (**E**), rots in developing fruits; (**F**), postharvest stem-end rot of fruit.

3.2. Morphological and Molecular Identification of Fungal Isolates

After a preliminary morphological characterization, 74 isolates showed morphological and growth features according to the Botryosphaeriaceae family, from which 52 were selected for their molecular identification based on the ITS1-2, *tef1* and *tub2* genetic markers. PCR amplification of ITS1-2, *tef1* and *tub2* markers gave products of 500–540 bp, 246–300 bp and 410 bp, respectively, and absence of contamination and non-specific PCR products were confirmed from agarose gels (not shown). Sequences obtained allowed the molecular identification of five different Botryosphaeriaceae species, based on independent BLAST searches with the three sequenced markers. One fungal isolate was identified as *Lasiodiplodia brasiliensis*, while four species of the *Neofusiccocum* genus were detected, including *N. australe* (2 isolates), *N. cryptoaustrale/stellenboschiana* (22 isolates), *N. luteum* (10 isolates), and *N. parvum* (12 isolates). The remaining five isolates were not identified to species level as they could not be recovered from storage. Twenty one out of the 22 isolates identified as *N. cryptoaustrale/stellenboschiana* with ITS1-2, *tef1* and *tub2* genomic regions showed for *rpb2* the same sequence (OQ401613, OQ401614 or Q401615), which matches all

GenBank sequences from *N. stellenboschiana* and all-but-one *N. cryptoaustrale*. One isolate (B043) showed a sequence (OQ401617) equivalent to KX464014.1 which corresponds to that of the reference strain CBS 122813 of *N. cryptoaustrale*. These results were confirmed by repeating the analyses by means of a new DNA extraction using an alternative second procedure, PCR amplification and sequencing.

Colony growth, pycnidia and conidia of representative isolates of the different species of Botryosphaeriaceae identified in this work are shown in Figures S5–S9. All *Neofusicoccum* isolates generated pycnidia on SC medium with and without Parafilm M, except the isolate B022 of *N. parvum*, which only produced pycnidia on WA medium (Table 3). No conidia were obtained from *N. australe*, represented by isolate B018, but isolates of *N. cryptoaustrale/stellenboschiana* (B026), *N. luteum* (B003, B017, B024) *and N. parvum* (B020, B022) produced conidia on SC medium without Parafilm M and/or WA medium. In addition, the isolate B161, identified as *L. brasiliensis*, produced pycnidia and conidia on PDAS. Mature conidia of *L. brasiliensis* were dark-walled, one-septate, striate (Figure S9B), whereas conidia of *Neofusicoccum* species were hyaline and aseptate (Figures S6H, S7H and S8G). The conidia measurements are shown in Table 4, as well as reference measurements previously reported by other authors.

Table 3. Production of pycnidia/conidia by *Neofusicoccum* isolates.

Species	Isolate	Apparition of Pycnidia/Conidia (Days)		
		SC-P	SC	WA
N. australe	B018	14/NP	21/NP	14/NP
N. cryptoaustrale/stellenboschiana	B012	14/NP	14/NP	ND
	B026	14/NP	14/NP	14/21
	B030	21/NP	14/NP	ND
N. luteum	B003	21/NP	21/42	14/21
	B017	14/NP	14/14	ND
	B024	14/NP	14/28	ND
N. parvum	B020	28/NP	28/42	14/NP
	B022	NP/NP	NP/NP	21/21

SC-P: salt-cellulose agar medium + sugarcane bagasse sealed first 14 days with Parafilm M; SC: salt-cellulose agar medium + sugarcane bagasse without Parafilm M; WA: water-agar medium + pine needles; ND, not determined; NP, not produced.

Table 4. Conidial dimensions of *Lasiodiplodia* and *Neofusicoccum* isolates from this study and comparison with previous reports.

Species	Isolates	Values Obtained in This Study			References	
		Length (L) (μm)	Width (W) (μm)	L/W Ratio	Length (L) (μm)	Width (W) (μm)
L. brasiliensis	B161	23.49 ± 1.42	14.95 ± 1.18	1.57	25.1–27.3 [1]	13.3–14.79 [1]
N. cryptoaustrale/ stellenboschiana	B026	20.44 ± 1.41	6.02 ± 0.51	3.4	(18–) 20.5–21 (–26.5) [2] (17–) 19–21 (–22) [3]	5–6 (–6.5) [2] (4.5–) 5.5–6 [3]
N. luteum	B003	18.60 ± 2.72	5.93 ± 0.60	3.14		
	B017	21.63 ± 1.55	6.34 ± 0.69	3.41	(15–) 16.5–22.5 (–24) [4]	4.5–6 (–7.5) [4]
	B024	18.34 ± 1.56	5.50 ± 0.38	3.33		
N. parvum	B020	14.97 ± 1.05	5.70 ± 0.65	2.63	(12–) 13.5–21 (–24) [4]	4–6 (–10) [4]

The sizes of the conidia were taken from: [1] *L. brasiliensis* from Marques et al. [44], [2] *N. cryptoaustrale* from Crous et al. [45], [3] *N. stellenboschiana* from Yang et al. [46] and [4] *N. luteum* and *N. parvum* from Phillips et al. [16].

For the multiple alignment of *Neofusicocum* species, the ML and BY phylogenetic analysis generated identical topologies, showing six statistically well supported clades where the isolates from this study were placed in four of them. The BY tree is presented with PP/ML values at the corresponding branch (Figure 2). The first clade grouped twelve isolates from this study, along with the nine *N. parvum* reference sequences selected for the analysis, therefore confirming their molecular identification as *N. parvum* strains. The second clade, which includes the four reference sequences of *N. mediterraneum*, did not contain any isolate from the present work, suggesting that this species seems to be absent from the avocados sampled in Tenerife island. The third clade includes 22 of the new isolates along with *N. cryptoaustrale* CBS 122813 reference strain, but also with *N. stellenboschiana* CBS 110866 and CBS 118839. Interestingly, 12 of the isolates were placed in the main branch with the three sequences mentioned above, but the 10 remaining isolates were ubicated in two sub-clades, the first with seven isolates, and the second with three. The fourth clade corresponded with the *N. luteum* reference strains where ten isolates from this work were ubicated, all of them in the same group, along with reference strains CBS 562.92, CBS 110299, CBS 118842 and CBS 133502. Neither of the isolates of this study appeared in the fifth clade, corresponding this with *N. rapaneae*. Finally, the sixth and last clade grouped the two remaining isolates, along with *N. australe* reference strains.

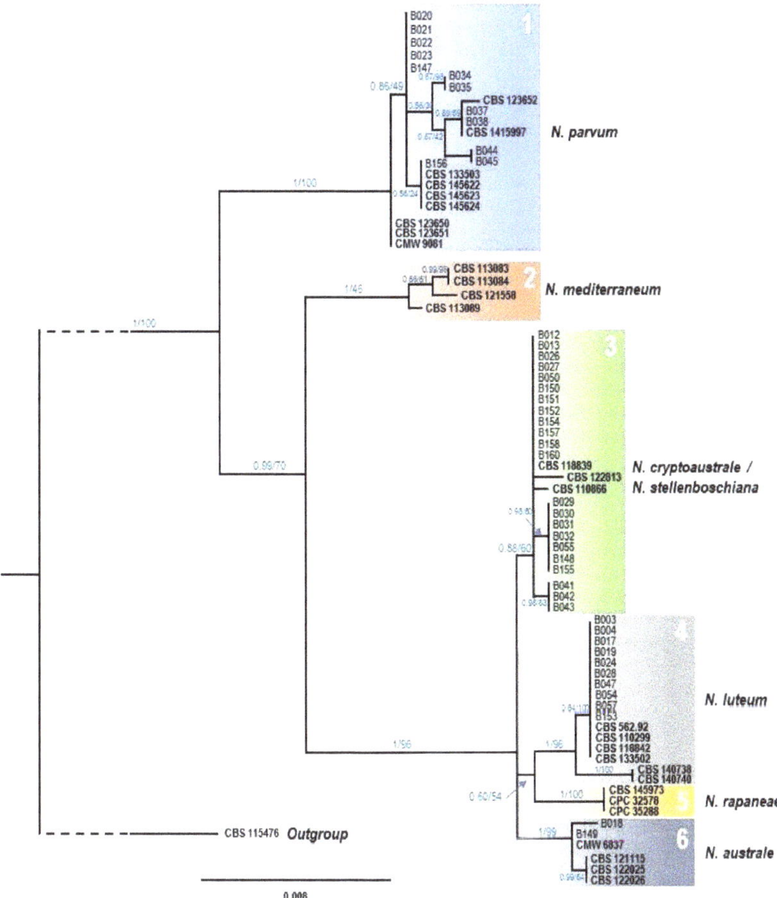

Figure 2. Phylogenetic tree of *Neofusicoccum* species resulting from a Bayesian analysis of the combined ITS1-2, *tef1* and *tub2* sequence alignment. Maximum likelihood bootstrap values and Bayesian posterior probabilities are shown at the nodes. Ex-type strains are indicated in bold font. The tree was rooted to *Botryosphaeria dothidea* (CBS 115476).

In the case of *Lasiodiplodia* species ML and BY trees (Figure 3) showed exactly the same topology, showing six well supported clades corresponding to the species of *L. mediterranea*, *L. pseudotheobromae*, *L. laeliocattleyae*, *L. theobromae*, *L. viticola*, and *L. brasiliensis*. The unique isolate from the present work assigned to Lasiodiplodia genus (B161), was clearly assigned to the clade of *L. brasiliensis*.

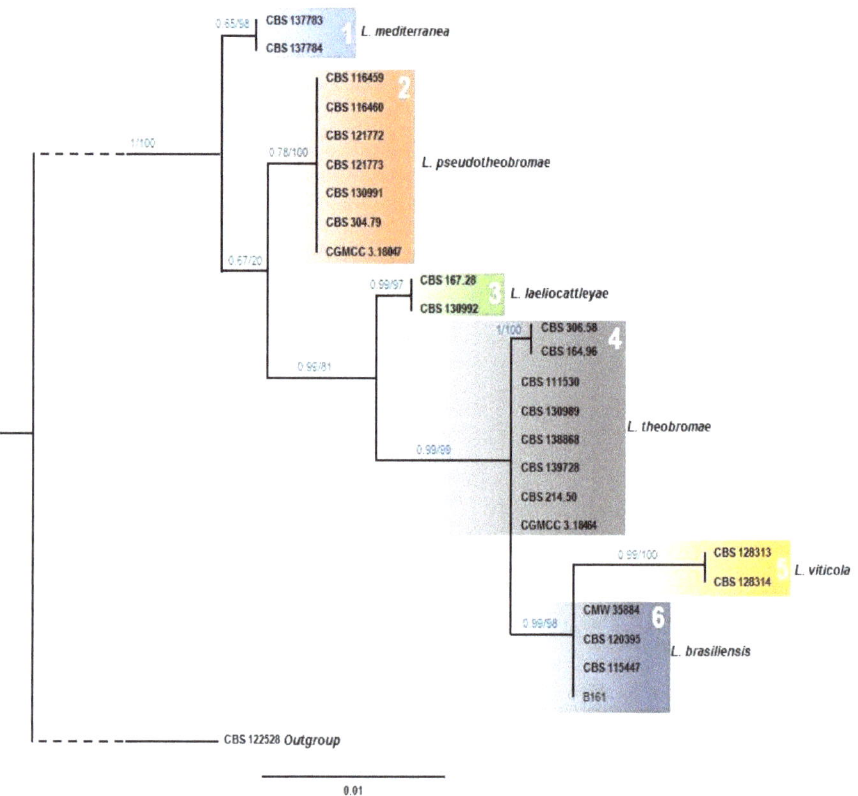

Figure 3. Phylogenetic tree of *Lasiodiplodia* species resulting from a Bayesian analysis of the combined ITS1-2, *tef1* and *tub2* sequence alignment. Maximum likelihood bootstrap values and Bayesian posterior probabilities are shown at the nodes. Ex-type strains are indicated in bold font. The tree was rooted to *Neodeightonia phoenicum* (CBS 122528).

3.3. Symptoms and Distribution of Botryosphaeriaceae Species Associated to Avocado in Tenerife

Some species of the Botryosphaeriaceae family were detected in 26 out of 40 orchards surveyed (Figure 4): *N. luteum*, 16 orchards; *N. cryptoaustrale/stellenboschiana*, 11 orchards; *N. parvum*, 6 orchards; *N. australe*, 2 orchards; *L. brasiliensis*, 1 orchard; and 5 orchards with unidentified species of Botryosphaeriaceae. There were also isolated species of *Alternaria* from 13 orchards (in four of them as the sole fungus, in the other nine along with Botryosphaeriaceae fungi). Other fungal species such as *Aureobasidium* sp., *Cladosporium* sp., *Colletotrichum* sp., *Nigrospora* sp. and *Pestalotiopsis* sp. were also isolated in 10 orchards.

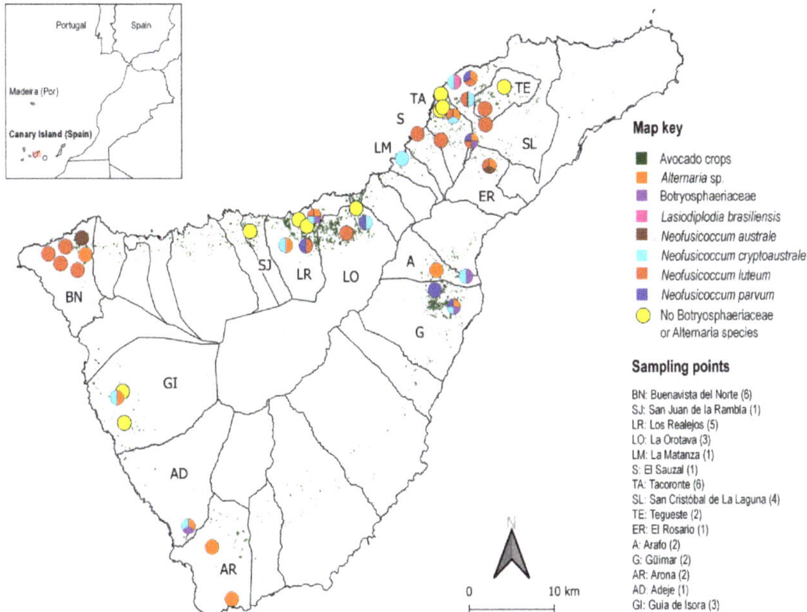

Figure 4. Map showing the sampling points and the species of fungi that were detected in Tenerife Island. Avocado plantations in the Canary Islands are highlighted in green. The map indicates the townships and the abbreviated names of places where samples were collected.

Botryosphaeriaceae fungal species were isolated from samples of necrotic branches (mainly diebacks) and fruits collected in the surveyed orchards, with 47.6% and 30.3% of positive results, respectively. However, no Botryosphaeriaceae fungi were isolated from cankers and the isolates obtained from samples showing panicle blight were not species of Botryosphaeriaceae. In addition, *N. luteum* and *N. australe* were isolated from samples of necrotic branches that could not be geographically located, with the first species in two samples and the second in one. In samples of postharvest fruits collected in 6 sampling points (2 avocado packaging facilities and 4 retail stores), *N. luteum* was detected in 3 of them, *N. cryptoaustrale/stellenboschiana* in 2, and *N. parvum* and *N. australe* in 1 sampling point each one.

3.4. Pathogenicity Tests

Plants inoculated by inserting a mycelial plug in a wound made in the stem at 5 cm from the ground showed symptoms after 3–6 months being inoculated with *L. brasiliensis*, *N. luteum* and *N. parvum*. Figure 5 shows external stem necrosis produced by *L. brasiliensis*, with exudates of whitish sugars characteristic of avocado at the inoculation point or in the parts where Parafilm M was applied to fix the inoculum to the stem. All isolates caused internal necrosis that progressed from the inoculation point inside the plants, although irregularly and with variations in its length depending on the inoculated plant (Figure 6). No substantial differences were observed in the internal symptoms between bagged and non-bagged plants. When *N. australe*, *N. cryptoaustrale/stellenboschiana*, *N. luteum* and *N. parvum* were inoculated by pruning the plants at the top and placing the fungal inoculum onto the pruning-cut wound, external symptoms were clearly visible at 21 days after inoculation. Inoculated plants showed stem discolouration and necrosis that progressed along the outside of the plant stem downwards from the point of inoculation, along with the typical whitish-dry exudate (Figure 7, Table 5), showing the same symptoms of dieback that can be observed in the field. All the inoculated fungi in the different pathogenicity tests were re-isolated from the plant lesions, thus complying with Koch's postulates.

Figure 5. External symptoms in Topa Topa avocado seedlings 3 months after inoculation. Plants were inoculated by inserting a mycelial plug in a wound made in the stem at 5 cm from the ground. (**A**), negative control (plants inoculated with a fungus-free agar fragment); (**B,C**), damage caused by isolate B161 of *Lasiodiplodia brasiliensis*.

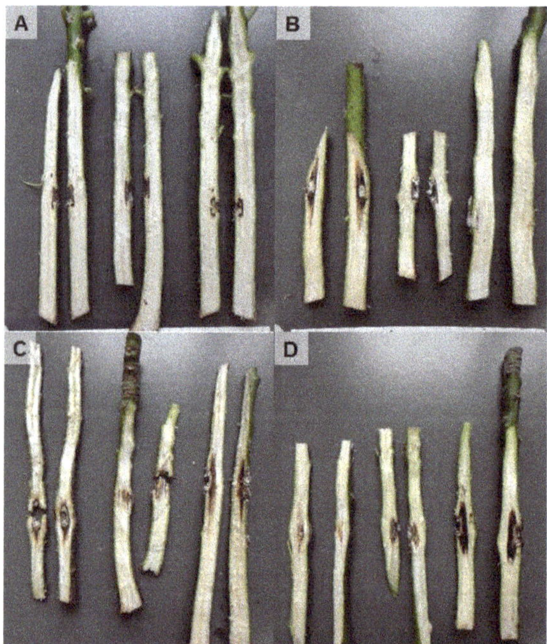

Figure 6. Internal symptoms in Topa Topa avocado seedlings 6 months after inoculation. Plants were inoculated by inserting a mycelial plug in a wound made in the stem at 5 cm from the ground. All plants shown in the figure were bagged for 10 d after inoculation. (**A**), negative control (plants inoculated with a fungus-free agar fragment); plants inoculated with isolates: (**B**), *Neofusicoccum luteum* B153; (**C**), *N. parvum* B156; and (**D**), *Lasiodiplodia brasiliensis* B161.

Figure 7. External symptoms in West Indian avocado seedlings 21 days after inoculation. Plants were pruned at the top, and they were inoculated by placing a mycelial plug onto the pruning-cut wound. Isolates B043, B050: *N. cryptoaustrale/stellenboschiana*. Negative control consists of plants inoculated with a fungus-free agar fragment.

Table 5. External symptoms (length of necrotic lesion) caused by *Neofusicoccum* species in West Indian avocado seedlings 21 days after inoculation in pruning-cut wound.

Species	Isolate	Mean Lesion Length (cm)	Standard Deviation (cm)
Control	-	0.0	0.0
N. australe	B018	4.8	0.2
N. cryptoaustrale/stellenboschiana	B043 [1]	3.7/2.5	3.5/1.3
	B050	2.4	0.3
N. luteum	B004	4.3	0.3
	B047	5.1	1.2
N. parvum	B034	5.6	1.2
	B045	1.4	0.7
	B113 [1]	3.4/3.9	0.8/2.7

[1] Pathogenicity tests with isolates B043 and B113 were repeated twice.

4. Discussion

This work studied the occurrence of damages to the aerial part of avocado plants on the island of Tenerife and the diversity of the associated fungi. The symptom most frequently found was the necrosis in branches (dieback-type symptoms), which produces losses on plantlets in recent plantations and young trees, but also in adult trees. Panicle blight and fruit drop during its early development stages also caused significant losses in production. Cankers on trunks and branches appeared less frequently on field, although, very occasionally, some orchards may have many affected trees. We determined the presence and diversity of fungi associated with these symptoms using sequence analysis of ITS1-2, with particular interest in species of the Botryosphaeriaceae family, for which additional markers, *tef1* and *tub2*, were also used.

Five species of Botryosphaeriaceae were isolated from symptomatic avocados: *Neofusiccocum australe*, *N. cryptoaustrale/stellenboschiana*, *N. luteum*, *N. parvum* and *Lasiodiplodia brasiliensis*, and it is important to highlight that it was frequent to find more than one species of this family in the same orchard. *Neofusicoccum australe, N. cryptoaustrale/ stellenboschiana, N. luteum* and *N. parvum* have been reported in different countries as causal agents of diseases in avocado plants [5–7,10,33,52]. The high frequencies of *N. luteum* or *N. cryptoaustrale/stellenboschiana* found in this study does not correspond to what has been described in avocado-growing areas in the south of the Spanish mainland where *N. parvum* predominates [5]. No relationship was found in our study between the identified fungal species and the geographic location of avocado orchards. In other countries, the various types of symptoms seem to be caused in a general way by different species of the Botryosphaeriaceae family, along with other species of fungi, with the prevalence of one or the other according to countries or regions. Some publications even mention that the incidence and distribution of species varies between areas within the same region [5,10,53].

Moreover, fungi belonging to the Botryosphaeriaceae family have been isolated in the Canary Islands in other crops such as vineyards, almond trees, or mangoes [54,55], and in ornamental plants such as Indian laurels in which they cause significant damage in urban gardens [56]. The presence of Botryosphaeriaceae species infecting other hosts rather than avocado in the Canary Islands could play a role in the epidemiology of the diseases in avocado, as has been recently reported for other woody crops [57,58].

To the best of our knowledge, this study constitutes the first report of *N. cryptoaustrale/ stellenboschiana* infecting avocado plants in Spain and the first time that *N. cryptoaustrale/ stellenboschiana* has been cited in avocado causing symptoms of dieback and stem-end rot, although it had been previously cited causing branch cankers on avocado in Crete (Greece) [52]. However, it was not possible to assign our isolates unequivocally to one of these two species, as reference strains of *N. cryptoaustrale* and *N. stellenboschiana* are grouped in a unique clade. The additional analysis of the *rpb2* marker for *N. cryptoaustrale/ stellenboschiana* showed that all analysed isolates except one presented a single sequence cited by various authors who have worked with this marker in the two species (the entire sequence fully matches for the two species). In addition, a sequence was obtained for isolate B043 that corresponded to the reference CBS:122813 registered in GenBank as KX464014.1. This sequence shows a difference of 24 snp with the others cited for *N. cryptoaustrale* and *N. stellenboschiana*. Therefore, in view of the results obtained it is necessary to clarify the taxonomic situation of both species and determine the origins of the differences, which is beyond the scope of this work. Moreover, results of conidial dimensions obtained for these isolates are similar to the references for both species, and they could not be used to morphologically discriminate them. Similar results were described by other authors who could not differentiate between *N. cryptoaustrale* and *N. stellenboschiana* using ITS1-2, *tef1* or *tub2* markers [59,60].

Neofusicoccum cryptoaustrale/stellenboschiana seems to be quite widespread in avocado in Tenerife, considering the frequency with which it has been found in different locations (11 out of 40 orchards). *Neofusicoccum cryptoaustrale* has been isolated from branches and leaves of living *Eucalyptus* trees in South Africa [45], but also from other species of arboreal plants or other geographical locations such as *Pistacia* spp. [46,61], *Olea europea* [46] or different mangrove trees [62]. *Neofusicoccum stellenboschiana* was also recently described in South Africa isolated from *Vitis vinifera* [46], but also has been cited associated to *Arum italicum* leaf spot [46], *Olea europea* [46], *Persea americana* cankers [52], *Prunus* spp. [46] and *Quercus suber* [63]. Our pathogenicity tests carried out with isolates of this species show its ability to develop disease symptoms in avocado seedlings equivalent to those of *N. parvum* or *N. luteum*, which makes it a species to consider in the epidemiology of this disease in the Canary Islands.

The unique isolate of the genus *Lasiodiplodia* obtained here was initially thought to belong to the most frequent species *L. theobromae*, already cited in avocado in several countries as well as in other plants in the Canary Islands [54,56]. However, its phylogenetic analysis indicates that it belongs to the species *L. brasiliensis*. This species was described in Brazil in 2014 associated with stem-end rot of papaya [64]. It has also been detected in that country associated to dieback in grapevine [65], postharvest fruit rot of custard apple [66] and cankers and dieback of apple trees [67]. Moreover, *L. brasiliensis* causes stem-end rot of mango in China [68]. It is closely related to *L. viticola*, but it differentiates on conidial sizes, which are longer and larger than those described for this last species [69]. Six nucleotides in the ITS1-2 region differentiate *L. brasiliensis* from *L. viticola*, with no difference in the nucleotide sequence of *tef1*. In our work, a single isolate of *L. brasiliensis* was obtained from an avocado plant with symptoms of dieback in an orchard in which *N. cryptoaustrale/stellenboschiana* was also detected. The results of the pathogenicity tests pointed out that *L. brasiliensis* was capable of reproducing symptoms of necrosis in inoculated avocado plants. This is the first time that this species has been reported in avocado and the first time that it has been reported in Spain.

In this study, the number of samples of the main types of symptoms attributed to Botryosphaeriaceae in field conditions was highly unbalanced in favour of branch necrosis (mainly diebacks), with some representation of fruit rot and panicle blight and low representation of cankers on trunks or branches. According to the fungi isolated from these samples, the symptoms of branch necrosis (dieback-type) seem to be caused mainly by species of Botryosphaeriaceae, whereas fungi of other genus were prevalently isolated from panicle blight. No pathogens were consistently isolated from the few samples of cankers that were analysed. No fungal isolates from the Botryosphaeriaceae family were detected, nor were other species of harmful organisms to which cankers could be attributed such as *Phytophthora* species or bacteria (data not shown). Regarding the necrosis in the peduncles of unripe fruits that come off the trees without completing their development, it was possible to isolate species of Botryosphaeriaceae, *Alternaria* sp., but also other species. Although the loss of developing fruit is a natural process in the summer, we have to consider that stress, pests and diseases, such as those produced by Botryosphaeriaceae or other fungi, can cause excessive fruit drop.

The same *Neofusicoccum* species found in avocado orchards (*N. australe*, *N. cryptoaustrale/ stellenboschiana*, *N. luteum* and *N. parvum*) were isolated from post-harvest fruits showing symptoms of stem-end rot or dark spots in other locations on the fruit. In addition, *Colletotrichum* sp. was isolated only from one sample. It has been described that damages due to the Botryosphaeriaceae family are located in the area of insertion of the peduncle in the fruit (stem-end rot) [12,27,70], while those caused by *Colletotrichum* species are characterized by the appearance of darkening of the skin of the fruit (anthracnose) [12,27]. In this work, most of *Neofusicoccum* and *Colletotrichum* isolates were obtained from branches, suggesting that the fungi that infest the branches of avocado trees in Tenerife may also be the ones that cause damage to the fruit in post-harvest.

Regarding the pathogenicity tests, Koch's postulates conducted with isolates of Botryosphaeriaceae obtained in this work confirmed their pathogenicity on avocado plants. Most of the inoculation procedures carried out with Botryosphaeriaceae fungi on avocado plants, but also in other woody plants, are mainly based on the use of a fungal mycelial plug, which is placed into a stem wound performed at a fixed distance from the ground and far from the top of the plant [3,6,9]. Using this procedure, inoculated plants show symptoms of necrosis several months after inoculation, and similar results were obtained in our study. However, when plants were inoculated by placing the inoculum onto pruning-cut wounds at the apex, inoculated plants showed symptoms of stem necrosis in less than a month. This procedure, based on wound inoculation at the apical tip region of the plants, has been described by some authors for pathogenicity tests of Botryosphaeriaceae fungi on avocado [4,11] and other woody plants [71], resulting in brown stem lesions 2–4 weeks after inoculation. Therefore, wound inoculation at the apex of the plants seems to reproduce disease symptoms in a short time, suggesting it could facilitate the study of different aspects of Botryosphaeriaceae disease on avocado such as varietal differences, variation in virulence between species and isolates, as well as to evaluate methods for its control.

In Tenerife, similarly to other avocado-growing regions, fungi of the Botryosphaeriaceae family seem to be involved in avocado dieback, in the premature fall of fruits during their development in the field, and in post-harvest damage. However, further research is needed in relation to panicle blight and avocado cankers in Tenerife. Panicle blight or inflorescence dieback occurs frequently and yields in some years are greatly reduced in the worst affected orchards, but little information is available on this symptom in avocado [19]. Fungi belonging to the family Botryosphaeriaceae have been described to cause cankers on avocado and other woody plants. However, no Botryosphaeriaceae fungi or other pathogens were consistently isolated from cankers in this work. The samples were taken from the cankers just in borders inside of it in adult trees without killing the tree, where other saprophytic organisms may have probably settled. In these old cankers, it is probably necessary to take the samples from more internal points at the zone of advancing necrosis. Furthermore, cankers are not frequent in avocado orchards in Tenerife and, for this reason, only a very low number of cankers could be analysed. Therefore, it is necessary to carry out a greater number of isolations, particularly from cankers, as well as from symptoms in relation to phenological growth stages to clarify their association with the fungal pathogens of avocado.

Supplementary Materials: The following supporting information can be downloaded at: https:// www.mdpi.com/article/10.3390/microorganisms11030585/s1, Figure S1: Symptoms in aerial parts of young and adult avocado plants; Figure S2: Panicle blight and dieback symptoms; Figure S3: Cankers in blanch and trunks, and rots in developing fruits; Figure S4: Symptoms in post-harvest avocado fruits; Figure S5. *Neofusiccocum australe*; Figure S6. *Neofusiccocum cryptoaustrale/*stellenboschiana; Figure S7. *Neofusiccocum luteum*; Figure S8. *Neofusiccocum parvum*; Figure S9. *Lasiodiplodia brasiliense*. Table S1. GenBank and culture collection accession numbers of *Neofusicoccum* species treated in the phylogenies. Table S2. GenBank and culture collection accession numbers of *Lasiodiplodia* species treated in the phylogenies.

Author Contributions: Conceptualization, S.P., M.A.G.-C., A.R.-P. and F.S.; methodology, O.G.-P., M.A.G.-C., A.R.-P. and F.S.; software, D.H., M.A.G.-C. and F.S.; validation, D.H., M.A.G.-C., A.R.-P. and F.S.; formal analysis, D.H., O.G.-P., M.A.G.-C. and F.S.; investigation, D.H., O.G.-P. and F.S.; resources, D.H., S.P. and F.S.; data curation, D.H. and F.S.; writing—original draft preparation, D.H., A.R.-P. and F.S.; writing—review and editing, D.H., M.A.G.-C., A.R.-P. and F.S.; visualization, D.H., and F.S.; supervision, A.R.-P. and F.S. project administration, F.S.; funding acquisition, F.S. All authors have read and agreed to the published version of the manuscript.

Funding: This work was funded by a CAIA Project 2017-0001 "Optimization of avocado production systems" of the Ministry of Agriculture, Livestock and Fisheries of the Government of the Canary Islands. David Hernández-Hernández was recipient of a 2021–2025 PhD grant PRE2020-095122 from the Agencia Estatal de Investigación (AEI).

Data Availability Statement: Not applicable.

Acknowledgments: The authors are grateful for the participation of the Agricultural Extension Services of the Technical Service for Agriculture and Rural Development of the Cabildo Insular de Tenerife, which collaborated in carrying out the surveys and in the collection of samples. To all the farmers who participated in the survey and who allowed the collection of samples in their crops. We also want to thank the Plant Health Service of the General Direction of Agriculture of the Government of the Canary Islands for allowing the use of its facilities and equipment.

Conflicts of Interest: The authors declare no conflict of interest.

References

1. ISTAC Estadística Agraria de Canarias/Series Anuales de Agricultura. Municipios, Islas y Provincias de Canarias. 1999–2021. Available online: http://www.gobiernodecanarias.org/istac/jaxi-istac/tabla.do?uripx=urn:uuid:803af13a-04f6-4789-8e78-48f5fdff64a7&uripub=urn:uuid:ef5f2e5c-e2c4-4c1d-b5ed-c20fe946ce6f%0A%0A (accessed on 30 December 2022).
2. MAPA Encuesta Sobre Superficies y Rendimientos Cultivos (ESYRCE). Encuesta de Marco de Áreas de España. Available online: https://www.mapa.gob.es/es/estadistica/temas/estadisticas-agrarias/agricultura/esyrce/default.aspx (accessed on 30 December 2022).
3. Auger, J.; Palma, F.; Pérez, I.; Esterio, M. First Report of *Neofusicoccum australe* (*Botryosphaeria australis*), as a Branch Dieback Pathogen of Avocado Trees in Chile. *Plant Dis.* **2013**, *97*, 842. [CrossRef]
4. Zea-Bonilla, T.; González-Sánchez, M.A.; Martín-Sánchez, P.M.; Pérez-Jiménez, R.M. Avocado Dieback Caused by *Neofusicoccum parvum* in the Andalucia Region, Spain. *Plant Dis.* **2007**, *91*, 1052. [CrossRef] [PubMed]
5. Arjona-Girona, I.; Ruano-Rosa, D.; López-Herrera, C.J. Identification, Pathogenicity and Distribution of the Causal Agents of Dieback in Avocado Orchards in Spain. *Span. J. Agric. Res.* **2019**, *17*, e1003. [CrossRef]
6. Valencia, A.L.; Gil, P.M.; Latorre, B.A.; Rosales, I.M. Characterization and Pathogenicity of Botryosphaeriaceae Species Obtained from Avocado Trees with Branch Canker and Dieback and from Avocado Fruit with Stem End Rot in Chile. *Plant Dis.* **2019**, *103*, 996–1005. [CrossRef]
7. Hartill, W.F.T.; Everett, K.R. Inoculum Sources and Infection Pathways of Pathogens Causing Stem-End Rots of 'Hass' Avocado (*Persea americana*). *N. Zeal. J. Crop Hortic. Sci.* **2002**, *30*, 249–260. [CrossRef]
8. Twizeyimana, M.; McDonald, V.; Mayorquin, J.S.; Wang, D.H.; Na, F.; Akgül, D.S.; Eskalen, A. Effect of Fungicide Application on the Management of Avocado Branch Canker (Formerly *Dothiorella* Canker) in California. *Plant Dis.* **2013**, *97*, 897–902. [CrossRef] [PubMed]
9. McDonald, V.; Lynch, S.; Eskalen, A. First Report of *Neofusicoccum australe*, *N. luteum*, and *N. parvum* Associated with Avocado Branch Canker in California. *Plant Dis.* **2009**, *93*, 967. [CrossRef]
10. McDonald, V.; Eskalen, A. Botryosphaeriaceae Species Associated with Avocado Branch Cankers in California. *Plant Dis.* **2011**, *95*, 1465–1473. [CrossRef]
11. Alama, I.; Maldonado, E.; Rodríguez-Gálvez, E. *Lasiodiplodia theobromae* Affect the Cultivation of Palto (*Persea americana*) under the Conditions of Piura, Peru. *Universalia* **2006**, *11*, 4–13.
12. Hartill, W.F.T. Post-Harvest Diseases of Avocado Fruits in New Zealand. *N. Zeal. J. Crop Hortic. Sci.* **1991**, *19*, 297–304. [CrossRef]
13. Darvas, J.; Kotze, J. Fungi Associated with Pre-and Postharvest Diseases of Avocado Fruit at Westfalia Estate, South Africa. *Phytophylactica* **1987**, *19*, 83–85.
14. Twizeyimana, M.; Förster, H.; McDonald, V.; Wang, D.H.; Adaskaveg, J.E.; Eskalen, A. Identification and Pathogenicity of Fungal Pathogens Associated with Stem-End Rot of Avocado in California. *Plant Dis.* **2013**, *97*, 1580–1584. [CrossRef] [PubMed]
15. Denman, S.; Crous, P.W.; Taylor, J.E.; Kang, J.C.; Pascoe, I.; Wingfield, M.J. An Overview of the Taxonomic History of Botryosphaeria, and a Re-Evaluation of Its Anamorphs Based on Morphology and ITS RDNa Phylogeny. *Stud. Mycol.* **2000**, *2000*, 129–140.

16. Phillips, A.J.L.; Alves, A.; Abdollahzadeh, J.; Slippers, B.; Wingfield, M.J.; Groenewald, J.Z.; Crous, P.W. The Botryosphaeriaceae: Genera and Species Known from Culture. *Stud. Mycol.* **2013**, *76*, 51–167. [CrossRef]
17. Dissanayake, A.J.; Phillips, A.J.L.; Li, X.H.; Hyde, K.D. Botryosphaeriaceae: Current Status of Genera and Species. *Mycosphere* **2016**, *7*, 1001–1073. [CrossRef]
18. Slippers, B.; Wingfield, M.J. Botryosphaeriaceae as Endophytes and Latent Pathogens of Woody Plants: Diversity, Ecology and Impact. *Fungal Biol. Rev.* **2007**, *21*, 90–106. [CrossRef]
19. Dann, E.; Prabhakaran, A.; Bransgrove, K. Panicle Blight (Flower Dieback). Available online: https://avocado.org.au/public-articles/ta31v3panicle/ (accessed on 30 December 2022).
20. Burgess, T.I.; Tan, Y.P.; Garnas, J.; Edwards, J.; Scarlett, K.A.; Shuttleworth, L.A.; Daniel, R.; Dann, E.K.; Parkinson, L.E.; Dinh, Q. Current Status of the Botryosphaeriaceae in Australia. *Australas. Plant Pathol.* **2019**, *48*, 35–44. [CrossRef]
21. Peterson, R.A. Susceptibility of Fuerte Avocado Fruit at Various Stages of Growth, to Infection by Anthracnose and Stem End Rot Fungi. *Aust. J. Exp. Agric.* **1978**, *18*, 158–160. [CrossRef]
22. Firmino, A.C.; Fischer, I.H.; Tozze Júnior, H.J.; Dias Rosa, D.; Luiz Furtado, E. Identificação de Espécies de *Fusicoccum* Causadoras de Podridão Em Frutos de Abacate. *Summa Phytopathol.* **2016**, *42*, 100–102. [CrossRef]
23. Montealegre, J.R.; Ramírez, M.; Riquelme, D.; Armengol, J.; León, M.; Pérez, L.M. First Report of *Neofusicoccum australe* in Chile Causing Avocado Stem-End Rot. *Plant Dis.* **2016**, *100*, 2532. [CrossRef]
24. Tapia, L.; Larach, A.; Riquelme, N.; Guajardo, J.; Besoain, X. First Report of *Neofusicoccum luteum* Causing Stem-End Rot Disease on Avocado Fruits in Chile. *Plant Dis.* **2020**, *104*, 2027. [CrossRef]
25. Qiu, F.; Xu, G.; Zhou, J.; Zheng, F.Q.; Zheng, L.; Miao, W.G.; Wang, W.L.; Xie, C.P. First Report of *Botryosphaeria dothidea* Causing Stem-End Rot in Avocado (*Persea americana*) in China. *Plant Dis.* **2019**, *104*, 286. [CrossRef]
26. Ramirez-Gil, J.G. Avocado Wilt Complex Disease, Implications and Management in Colombia. *Rev. Fac. Nac. Agron. Colomb.* **2018**, *71*, 8525–8541. [CrossRef]
27. Ramírez-Gil, J.G.; López, J.H.; Henao-Rojas, J.C. Causes of Hass Avocado Fruit Rejection in Preharvest, Harvest, and Packinghouse: Economic Losses and Associated Variables. *Agronomy* **2019**, *10*, 8. [CrossRef]
28. Kebede, M.; Belay, A. Fungi Associated With Post-Harvest Avocado Fruit Rot at Jimma Town, Southwestern Ethiopia. *J. Plant Pathol. Microbiol.* **2019**, *10*, 2. [CrossRef]
29. Guarnaccia, V.; Aiello, D.; Cirvilleri, G.; Polizzi, G.; Susca, A.; Epifani, F.; Perrone, G. Characterisation of Fungal Pathogens Associated with Stem-End Rot of Avocado Fruit in Italy. *Acta Hortic.* **2016**, *1144*, 133–139. [CrossRef]
30. Molina-Gayosso, E.; Silva-Rojas, H.V.; García-Morales, S.; Ávila-Quezada, G. First Report of Black Spots on Avocado Fruit Caused by *Neofusicoccum parvum* in Mexico. *Plant Dis.* **2011**, *96*, 287. [CrossRef]
31. Trakunyingcharoen, T.; Cheewangkoon, R.; To-anun, C. Phylogenetic Study of the Botryosphaeriaceae Species Associated with Avocado and Para Rubber in Thailand. *Chiang Mai J. Sci.* **2015**, *42*, 104–116.
32. Ni, H.F.; Liou, R.F.; Hung, T.H.; Chen, R.S.; Yang, H.R. First Report of a Fruit Rot Disease of Avocado Caused by *Neofusicoccum mangiferae*. *Plant Dis.* **2009**, *93*, 760. [CrossRef]
33. Ni, H.; Chuang, M.; Hsu, S.; Lai, S.; Yang, H. Survey of *Botryosphaeria* spp., Causal Agents of Postharvest Disease of Avocado, in Taiwan. *J. Taiwan Agric. Res.* **2011**, *60*, 157–166.
34. Truett, G.E.; Heeger, P.; Mynatt, R.L.; Truett, A.A.; Walker, J.A.; Warman, M.L. Preparation of PCR-Quality Mouse Genomic Dna with Hot Sodium Hydroxide and Tris (HotSHOT). *Biotechniques* **2000**, *29*, 52–54. [CrossRef] [PubMed]
35. Collado-Romero, M.; Mercado-Blanco, J.; Olivares-García, C.; Valverde-Corredor, A.; Jiménez-Díaz, R.M. Molecular Variability within and among *Verticillium dahliae* Vegetative Compatibility Groups Determined by Fluorescent Amplified Fragment Length Polymorphism and Polymerase Chain Reaction Markers. *Phytopathology* **2006**, *96*, 485–495. [CrossRef]
36. White, T.J.J.; Bruns, T.D.; Lee, S.B.; Taylor, J.W. Amplification and Direct Sequencing of Fungal Ribosomal RNA Genes for Phylogenetics. *PCR Protoc.* **1990**, 315–322. [CrossRef]
37. Alves, A.; Crous, P.W.; Correia, A.; Phillips, A.J.L. Morphological and Molecular Data Reveal Cryptic Speciation in *Lasiodiplodia theobromae*. *Fungal Divers.* **2008**, *28*, 1–13.
38. Glass, N.L.; Donaldson, G.C. Development of Primer Sets Designed for Use with the PCR to Amplify Conserved Genes from Filamentous Ascomycetes. *Appl. Environ. Microbiol.* **1995**, *61*, 1323–1330. [CrossRef] [PubMed]
39. Sakalidis, M.L.; Hardy, G.E.S.J.; Burgess, T.I. Use of the Genealogical Sorting Index (GSI) to Delineate Species Boundaries in the *Neofusicoccum Parvum-neofusicoccum* Ribis Species Complex. *Mol. Phylogenet. Evol.* **2011**, *60*, 333–344. [CrossRef]
40. Tamura, K.; Stecher, G.; Kumar, S. MEGA11: Molecular Evolutionary Genetics Analysis Version 11. *Mol. Biol. Evol.* **2021**, *38*, 3022–3027. [CrossRef]
41. Altschul, S.F.; Gish, W.; Miller, W.; Myers, E.W.; Lipman, D.J. Basic Local Alignment Search Tool. *J. Mol. Biol.* **1990**, *215*, 403–410. [CrossRef]
42. Sayers, E.W.; Bolton, E.E.; Brister, J.R.; Canese, K.; Chan, J.; Comeau, D.C.; Connor, R.; Funk, K.; Kelly, C.; Kim, S.; et al. Database Resources of the National Center for Biotechnology Information. *Nucleic Acids Res.* **2022**, *50*, D20–D26. [CrossRef]
43. Mussi-Dias, V.; Valmir dos Santos, A.; das Graças Machado Freire, M. Pycnidia and Conidia Quantification of *Lasiodiplodia* Using a Culture Medium Enriched with Sugarcane Bagasse. *J. Agric. Res.* **2016**, *2*, 1–16.
44. Marques, M.W.; Lima, N.B.; De Morais, M.A.; Barbosa, M.A.G.; Souza, B.O.; Michereff, S.J.; Phillips, A.J.L.; Câmara, M.P.S. Species of *Lasiodiplodia* Associated with Mango in Brazil. *Fungal Divers.* **2013**, *61*, 181–193. [CrossRef]

45. Crous, P.W.; Wingfield, M.J.; Guarro, J.; Cheewangkoon, R.; van der Bank, M.; Swart, W.J.; Stchigel, A.M.; Cano-Lira, J.F.; Roux, J.; Madrid, H.; et al. Fungal Planet Description Sheets: 154–213. *Persoonia-Mol. Phylogeny Evol. Fungi* **2013**, *31*, 188–296. [CrossRef] [PubMed]
46. Yang, T.; Groenewald, J.Z.; Cheewangkoon, R.; Jami, F.; Abdollahzadeh, J.; Lombard, L.; Crous, P.W. Families, Genera, and Species of *Botryosphaeriales*. *Fungal Biol.* **2017**, *121*, 322–346. [CrossRef]
47. Nguyen, L.T.; Schmidt, H.A.; Von Haeseler, A.; Minh, B.Q. IQ-TREE: A Fast and Effective Stochastic Algorithm for Estimating Maximum-Likelihood Phylogenies. *Mol. Biol. Evol.* **2015**, *32*, 268–274. [CrossRef] [PubMed]
48. Miller, M.A.; Pfeiffer, W.; Schwartz, T. The CIPRES Science Gateway. In Proceedings of the 1st Conference of the Extreme Science and Engineering Discovery Environment on Bridging from the eXtreme to the Campus and beyond—XSEDE '12, Chicago, IL, USA, 16–20 July 2012; ACM Press: New York, NY, USA, 2012; p. 1.
49. Tavaré, S. Some Probabilistic and Statistical Problems in the Analysis of DNA Sequences. *Am. Math. Soc. Lect. Math. Life Sci.* **1986**, *17*, 57–86.
50. Stamatakis, A. RAxML Version 8: A Tool for Phylogenetic Analysis and Post-Analysis of Large Phylogenies. *Bioinformatics* **2014**, *30*, 1312–1313. [CrossRef] [PubMed]
51. Ronquist, F.; Teslenko, M.; Van Der Mark, P.; Ayres, D.L.; Darling, A.; Höhna, S.; Larget, B.; Liu, L.; Suchard, M.A.; Huelsenbeck, J.P. Mrbayes 3.2: Efficient Bayesian Phylogenetic Inference and Model Choice across a Large Model Space. *Syst. Biol.* **2012**, *61*, 539–542. [CrossRef]
52. Guarnaccia, V.; Polizzi, G.; Papadantonakis, N.; Gullino, M.L. *Neofusicoccum* Species Causing Branch Cankers on Avocado in Crete (Greece). *J. Plant Pathol.* **2020**, *102*, 1251–1255. [CrossRef]
53. Eskalen, A.; McDonald, V. Geographical Distribution of Botryosphaeriaceae and *Phomopsis/Diaporthe* Canker Pathogens of Avocado in California. *Calif. Avocado Soc.* **2010**, *93*, 87–98.
54. Gallo Llobet, L.; Hernández Hernández, J.M.; del Jaizme Vega, M.C.; Sala Mayato, L. Especies Fúngicas Enontradas En Los Cultivos de Canarias En El Período 1974–1984. In *Proceedings of the III Congreso Nacional de Fitopatología*; Sociedad Española de Fitopatología, Ed.; The Canary Islands: Tenerife, Spain, 1984; pp. 43–50.
55. Márquez, M.P.; Santana, M.L.; Albalat, A.R.; Armengol, J. Muestreo e Identificación de Hongos Asociados a Las Enfermedades de La Madera de La Vid En Lanzarote (Canarias). *Enoviticultura* **2020**, *65*, 16–26.
56. Benito Hernández, P. Botryosphaeriaceae Asociadas al Decaimiento de *Ficus microcarpa* en Gran Canaria. Master Thesis, Universitat Politècnica de València, Valencia, Spain, 2018.
57. Mojeremane, K.; Lebenya, P.; Plessis, I.L.D.; Van Der Rijst, M.; Mostert, L.; Armengol, J.; Halleen, F. Cross Pathogenicity of *Neofusicoccum* Australe and *Neofusicoccum stellenboschiana* on Grapevine and Selected Fruit and Ornamental Trees. *Phytopathol. Mediterr.* **2020**, *59*, 581–593. [CrossRef]
58. Sessa, L.; Abreo, E.; Bettucci, L.; Lupo, S. Botryosphaeriaceae Species Associated with Wood Diseases of Stone and Pome Fruits Trees: Symptoms and Virulence across Different Hosts in Uruguay. *Eur. J. Plant Pathol.* **2016**, *146*, 519–530. [CrossRef]
59. Van Dyk, M.; Spies, C.F.J.; Mostert, L.; Halleen, F. Survey of Trunk Pathogens in South African Olive Nurseries. *Plant Dis.* **2021**, *105*, 1630–1639. [CrossRef] [PubMed]
60. Spies, C.F.J.; Mostert, L.; Carlucci, A.; Moyo, P.; van Jaarsveld, W.J.; du Plessis, I.L.; van Dyk, M.; Halleen, F. Dieback and Decline Pathogens of Olive Trees in South Africa. *Persoonia Mol. Phylogeny Evol. Fungi* **2020**, *45*, 196–220. [CrossRef] [PubMed]
61. Linaldeddu, B.T.; Maddau, L.; Franceschini, A.; Alves, A.; Phillips, A.J.L. Botryosphaeriaceae Species Associated with Lentisk Dieback in Italy and Description of *Diplodia insularis* sp. Nov. *Mycosphere* **2016**, *7*, 962–977. [CrossRef]
62. Osorio, J.A.; Crous, C.J.; de Beer, Z.W.; Wingfield, M.J.; Roux, J. Endophytic Botryosphaeriaceae, Including Five New Species, Associated with Mangrove Trees in South Africa. *Fungal Biol.* **2017**, *121*, 361–393. [CrossRef]
63. Mahamedi, A.E.; Phillips, A.J.L.; Lopes, A.; Djellid, Y.; Arkam, M.; Eichmeier, A.; Zitouni, A.; Alves, A.; Berraf-Tebbal, A. Diversity, Distribution and Host Association of Botryosphaeriaceae Species Causing Oak Decline across Different Forest Ecosystems in Algeria. *Eur. J. Plant Pathol.* **2020**, *158*, 745–765. [CrossRef]
64. Netto, M.S.B.; Assunção, I.P.; Lima, G.S.A.; Marques, M.W.; Lima, W.G.; Monteiro, J.H.A.; de Queiroz Balbino, V.; Michereff, S.J.; Phillips, A.J.L.; Câmara, M.P.S. Species of *Lasiodiplodia* Associated with Papaya Stem End Rot in Brazil. *Fungal Divers.* **2014**, *67*, 127–141. [CrossRef]
65. Correia, K.C.; Silva, M.A.; de Morais, M.A.; Armengol, J.; Phillips, A.J.L.; Câmara, M.P.S.; Michereff, S.J. Phylogeny, Distribution and Pathogenicity of *Lasiodiplodia* Species Associated with Dieback of Table Grape in the Main Brazilian Exporting Region. *Plant Pathol.* **2016**, *65*, 92–103. [CrossRef]
66. Cardoso, J.E.; Lima, J.S.; Viana, F.M.P.; Ootani, M.A.; Araújo, F.S.A.; Fonseca, W.L.; Lima, C.S.; Martins, M.V.V. First Report of *Lasiodiplodia brasiliense* Causing Postharvest Fruit Rot of Custard Apple (*Annona squamosa*) in Brazil. *Plant Dis.* **2017**, *101*, 1542. [CrossRef]
67. Martins, M.V.V.; Lima, J.S.; Hawerroth, F.J.; Ootani, M.A.; Araujo, F.S.A.; Cardoso, J.E.; Serrano, L.A.L.; Viana, F.M.P. First Report of *Lasiodiplodia brasiliense* Causing Disease in Apple Trees in Brazil. *Plant Dis.* **2018**, *102*, 1027. [CrossRef]
68. Zhang, H.; Wei, Y.X.; Qi, Y.X.; Pu, J.J.; Liu, X.M. First Report of *Lasiodiplodia brasiliense* Associated with Stem-End Rot of Mango in China. *Plant Dis.* **2018**, *102*, 679. [CrossRef]

69. Urbez-Torres, J.R.; Peduto, F.; Striegler, R.K.; Urrea-Romero, K.E.; Rupe, J.C.; Cartwright, R.D.; Gubler, W.D. Characterization of Fungal Pathogens Associated with Grapevine Trunk Diseases in Arkansas and Missouri. *Fungal Divers.* **2012**, *52*, 169–189. [CrossRef]
70. Menge, J.A.; Ploetz, R.C. Diseases of Avocado. In *Diseases of Tropical Fruit Crops*; CABI: Wallingford, UK, 2003; pp. 35–71.
71. Rodríguez-Gálvez, E.; Hilário, S.; Lopes, A.; Alves, A. Diversity and Pathogenicity of *Lasiodiplodia* and *Neopestalotiopsis* Species Associated with Stem Blight and Dieback of Blueberry Plants in Peru. *Eur. J. Plant Pathol.* **2020**, *157*, 89–102. [CrossRef]

 *microorganisms*

Article

Ralstonia solanacearum Facing Spread-Determining Climatic Temperatures, Sustained Starvation, and Naturally Induced Resuscitation of Viable but Non-Culturable Cells in Environmental Water

Belén Álvarez [1,2,†], María M. López [1] and Elena G. Biosca [2,*]

[1] Departamento de Bacteriología, Instituto Valenciano de Investigaciones Agrarias (IVIA), 46113 Valencia, Spain
[2] Departamento de Microbiología y Ecología, Universitat de València (UV), 46100 Valencia, Spain
* Correspondence: elena.biosca@uv.es
† Present address: Departamento de Investigación Aplicada y Extensión Agraria, Instituto Madrileño de Investigación y Desarrollo Rural, Agrario y Alimentario (IMIDRA), 28800 Madrid, Spain.

Abstract: *Ralstonia solanacearum* is a bacterial phytopathogen affecting staple crops, originally from tropical and subtropical areas, whose ability to survive in temperate environments is of concern under global warming. In this study, two *R. solanacearum* strains from either cold or warm habitats were stressed by simultaneous exposure to natural oligotrophy at low (4 °C), temperate (14 °C), or warm (24 °C) temperatures in environmental water. At 4 °C, the effect of temperature was higher than that of oligotrophy, since *R. solanacearum* went into a viable but non-culturable (VBNC) state, which proved to be dependent on water nutrient contents. Resuscitation was demonstrated *in vitro* and *in planta*. At 14 °C and 24 °C, the effect of oligotrophy was higher than that of temperature on *R. solanacearum* populations, displaying starvation-survival responses and morphological changes which were stronger at 24 °C. In tomato plants, starved, cold-induced VBNC, and/or resuscitated cells maintained virulence. The strains behaved similarly regardless of their cold or warm areas of origin. This work firstly describes the natural nutrient availability of environmental water favoring *R. solanacearum* survival, adaptations, and resuscitation in conditions that can be found in natural settings. These findings will contribute to anticipate the ability of *R. solanacearum* to spread, establish, and induce disease in new geographical and climatic areas.

Keywords: bacterial wilt; global warming; environmental stress; VBNC; pathogenicity

Citation: Álvarez, B.; López, M.M.; Biosca, E.G. *Ralstonia solanacearum* Facing Spread-Determining Climatic Temperatures, Sustained Starvation, and Naturally Induced Resuscitation of Viable but Non-Culturable Cells in Environmental Water. *Microorganisms* **2022**, *10*, 2503. https://doi.org/10.3390/microorganisms10122503

Academic Editors: Michael J. Bidochka and James F. White

Received: 31 October 2022
Accepted: 12 December 2022
Published: 16 December 2022

Publisher's Note: MDPI stays neutral with regard to jurisdictional claims in published maps and institutional affiliations.

1. Introduction

The plant pathogen *Ralstonia solanacearum* is a relevant species and former constituent of the *R. solanacearum* species complex [1–3]. It causes severe wilt disease and economic losses in solanaceous and other basic crops for human consumption worldwide as well as in important ornamentals [4–8]. The pathogen frequently cannot be effectively controlled due to its high pathogenic potential and persistence in natural settings. Bacterial wilt control in the field has frequently been addressed by conventional methods, mainly agrochemicals and/or cultural practices, with variable results, and often with environmental impact [9,10]. Alternatively, biological control methods are being explored, such as bacteriophage-based treatments. The use of lytic bacteriophages may be an eco-sustainable strategy because of their specificity and bactericidal activity, although until now no bacteriophage-based product is commercially available against *R. solanacearum* [10]. Therefore, this pathogen poses a threat to the maintenance of global food security. In fact, the species has a quarantine status in the European Union (EU), the USA and Canada [5,11,12], and is considered a priority pathogen in agriculture for control and containment [13]. A major concern is that *R. solanacearum* seems to hold great potential for geographical expansion even under environmentally unsuitable conditions, as it appears to infect plants and persist during variable periods in soil or surface water as a free-living form or associated to plant material or non-host roots [4,14–17]. This is despite the exposure to abiotic stresses compromising the endurance of the bacterium, such as sustained oligotrophy and sub-optimal temperatures. In water systems, it can be consistently detected for years after its introduction, maintaining pathogenicity [16–22]. This creates a problem for farmers, as water is a scarce resource, particularly under the current conditions of global warming. According to EU and other countries' regulations, there is a ban on the irrigation of host plants with *R. solanacearum*-contaminated water as long as the bacterium is detected. The procedure for detection is mainly based on the molecular identification of *R. solanacearum* colonies isolated from the water. Global climatic changes are thought to increase extreme climate events and the prevalence of abiotic stresses. It is therefore decisive to understand the impact of prevailing environmental factors on *R. solanacearum* persistence in watercourses to be able to foresee changes regarding water-borne dissemination of the pathogen [6,8,12] and colonization of new geographical and climatic areas.

In that sense, knowledge about the effects of environmental temperatures on microorganisms is crucial to understand bacterial growth and adaptations facing global warming, as well as pathogen virulence and expression of symptoms in the plant [23]. Further, increased temperature is frequently associated with increased severity of the bacterial wilt disease [14,24,25]. Likewise, in watercourses, *R. solanacearum* population levels seasonally varied according to a range of temperatures [15–17], and persistence was also variable in agricultural water microcosms [26]. *R. solanacearum*, considered a pathogen of tropical and subtropical regions, is capable of causing bacterial wilt in temperate latitudes [27]. However, results from epidemiological studies are contradictory regarding the ability of *R. solanacearum* to survive in cold conditions away from plants [28]. Although the pathogen becomes viable but non-culturable (VBNC) in pure water by prolonged exposure to 4 °C [26,28,29], the fact that up to date the dynamics of this process has not been monitored in more realistic environmental water microcosms limits the extent to which the results can be extrapolated to the field [28].

Not only temperature, but also the limited nutrient availability characteristic of environmental water has long been claimed to affect bacterial survival in natural settings [30,31]. Weather events are likely to increase the impact of nutrient availability on bacterial communities in the environment. Without climate adaptation strategies, bacteria will probably have to either disseminate or stay and face starvation. *R. solanacearum* can persist in water for different periods [26,29,32–35], remaining pathogenic up to four years in environmental water at a favorable temperature through strategies, such as starvation-survival responses, the VBNC state, transition to coccoid cells, and aggregation [6], mechanisms evolved by non-sporulating bacteria facing adverse environmental conditions [30,31,36,37]. However, the effects of the simultaneous exposure to environmental abiotic stresses, such as different climatic temperatures and starvation on this pathogen survival and adaptation capability in natural environmental water have not yet been clarified.

Furthermore, after induction to the VBNC state by exposure to low temperature, transition to a fully culturable and pathogenic state or resuscitation may occur in favorable environmental conditions [37,38], although evidence for this requires a clear-cut distinction between true resuscitation (reversion from non-culturability to culturability) and regrowth (multiplication of a few culturable cells that had remained undetected). In fact, attempts to resuscitate cold-induced VBNC *R. solanacearum* in water have been carried out by the addition of hydrogen peroxide-degrading compounds, such as catalase to standard solid media to release the pathogen from oxidative stress [28,29]. However, *R. solanacearum* resuscitation by a simple reversal of low temperature has not been documented so far, despite being the main environmental VBNC inducing factor in natural settings, and a key factor in bacterial wilt outbreaks. This finding could have relevant epidemiological consequences under global warming, since increases in water temperature could lead to increased geographical expansion and/or incidence of water-borne infections resulting from resuscitated cells of this phytopathogenic bacterial species.

This work addressed, for the first time, the simultaneous effect of different temperatures and starvation on stress induction of two *R. solanacearum* strains (former *R. solanacearum* phylotype II) from either cold or warm habitats, and their responses in environmental water microcosms. Further, the capability of these *R. solanacearum* strains to resuscitate and keep pathogenic under conditions that can be found in natural settings was also firstly demonstrated. Knowledge derived from this work will help to foresee tendencies in *R. solanacearum* persistence and dissemination in aquatic systems within the frame of global warming, as well as their capability for establishment and disease induction in new geographical and climatic areas.

2. Materials and Methods

2.1. Bacterial Strains and Culture Conditions

Two bacterial strains of the present species *R. solanacearum* [2,3] isolated from either warm or cold habitats were used: strain IVIA-1602.1, from a diseased potato tuber from Canary Islands (Spain), and strain IPO-1609, from a diseased potato plant from The Netherlands [29], both are race 3, biovar 2 of the former *R. solanacearum* phylotype II. They were kept at $-80\ ^\circ C$ in a 30% (*v/v*) glycerol medium and routinely grown on the non-selective Yeast Peptone Glucose Agar (YPGA) [39] for 72 h at 29 $^\circ C$. In stress induction assays, bacterial culturability was tested on YPGA and the Semiselective Medium South Africa (SMSA) agar developed for *R. solanacearum* isolation [40] after incubation at 29 $^\circ C$ for 72 h. SMSA medium was also used to re-isolate the pathogen from the host (tomato plants). Both media are frequently used for *R. solanacearum* isolation, since the colonies of the pathogen can be easily recognized as typically smooth. YPGA contains filtered-sterilized glucose, from which *R. solanacearum* produces a large amount of extracellular polysaccharide. Colonies are fluidal with pearly cream-white whorls. With respect to SMSA, semiselectivity is mainly based on the action of four antibiotics (penicillin, polymyxin, chloramphenicol and bacitracin), triphenyl-tetrazolium chloride, and crystal violet. Colonies are fluidal with reddish whorls.

2.2. Characteristics of Environmental and Distilled Water Samples

River water samples were collected according to [16] from four different locations in Spain, and nutrient contents were separately determined for each of them. In the different water samples, organic matter levels were from 2 to 3.73% (*w/v*), and the main ion concentrations ranged as follows (values per liter): Na^+, 9.7–9.9 mg; K^+, 2.1–2.9 mg; Ca^{2+}, 10.1–13.0 mg; Mg^{2+}, 3.9–5.0 mg; dissolved Fe, 0.24–0.27 mg; Mn, 0.06–0.11 mg; Cu, <0.024 mg; dissolved Zn^{2+}, <0.018 mg; CO_3^{2-}, <1.8 mg; NO^{3-}, 4.37–5.93 mg; P_2O_5, 0.374–0.583 mg, and Cl^-, 9.4–11.4 mg. Salt contents in the samples were correspondent with conductivity values from 151 to 168 μSiemens/cm at 20 $^\circ C$, and pH values were from 7.48 to 7.83. Distilled water, used for comparative purposes in some assays, had no organic matter and only trace mineral ions: at 20 $^\circ C$, conductivity was ≤ 20 μSiemens/cm, and pH value was 7. All water samples were stored in the refrigerator.

2.3. Preparation and Monitoring of Stressed R. solanacearum in Water Microcosms

All water samples were autoclaved and filtered through 0.22-μm-pore-size membranes and used for microcosm preparation and inoculation with either of the strains IVIA-1602.1 or IPO-1609 at a range of 5×10^6–1×10^7 CFU (colony-forming units)/mL similarly to [6]. To induce stressed *R. solanacearum* populations, cells in microcosms were incubated at 4 $^\circ C$, 14 $^\circ C$, and 24 $^\circ C$ without shaking for 40 days or until loss of culturability. The temperature of 4 $^\circ C$ was selected because it had induced the VBNC state in *R. solanacearum* (former *R. solanacearum* phylotype II) in non-environmental pure water [26,28,29]. The temperatures of 14 $^\circ C$ and 24 $^\circ C$ were within the range in which *R. solanacearum* had been detected in environmental water [16,17]. Initially, microcosms were prepared with each of the four river water samples from different locations, and the survival of *R. solanacearum* monitored at 4 $^\circ C$ and 24 $^\circ C$. Based on the results obtained, one river water sample was selected for a comparative study on the survival of *R. solanacearum* in microcosms of river water *versus* distilled water at 4 °C, 14 °C, and 24 °C. Microcosms from river water samples and distilled water were prepared in triplicate.

Sampling from each microcosm was performed at inoculation time (day 0) and at 1, 2, 4, 8, 14, 28, and 40 days post-inoculation (dpi) to monitor:

2.3.1. Total, Viable, and Culturable Bacterial Populations

Microscopic counts of total and viable *R. solanacearum* cells were done by a direct viable count (DVC) method [41], extended to 16 h [26] and subsequent staining with either the polyclonal antiserum 1546-H IVIA against *R. solanacearum* or acridine orange [16,42]. Plate counts of culturable cells were done on two media, the general YPGA and the semiselective SMSA, both recommended by EU legislation to isolate the pathogen from environmental samples [43]. *R. solanacearum* colonies for culturable cell counts were confirmed by PCR as described [43] with primers Ps-1 and Ps-2 based on the sequence of the *16S rRNA* gene.

2.3.2. Cell Morphology

Bacterial cell shape was observed by specific immunofluorescence staining with the polyclonal antiserum 1546-H IVIA against *R. solanacearum* [16]. Cell morphology was observed with a Nikon Eclipse E800 microscope at a magnification of $\times 1000$. Pictures were taken with an adapted digital camera DXM1200 using ACT-1 version 2.62 software, and no processing of the images was performed. At each sampling time and for each temperature and environmental water microcosm, the number of bacilli and/or cocci from at least 20 random fields was counted (approximately 300 cells).

2.3.3. Pathogenicity

The ability of starved *R. solanacearum* cells incubated in the environmental water microcosms at 4 °C, 14 °C and 24 °C to induce disease was tested from each triplicate microcosm at each sampling time on groups of 72 tomato plants cv. 'Roma' aged three weeks (two plants per microcosm, six plants per water sample at each temperature). Inoculations were performed by injecting into the stem volumes of 10 μL directly taken from the microcosms. Plant inoculations were carried out according to EU Legislation [43]. Positive and negative controls were performed on groups of 12 tomato plants cv. 'Roma' (six plants per positive/negative control) at each sampling time. In the case of positive controls, inoculations were performed by injecting 10 μL of a freshly growing cell suspension from either of the two *R. solanacearum* strains. Each suspension was previously washed and adjusted in sterile 10 mM phosphate buffered saline solution (PBS), pH 7.2 (NaCl, 8 g/L; $PO_4H_2Na \cdot 2H_2O$, 0.4 g/L; $PO_4HNa_2 \cdot 12H_2O$, 2.7 g/L) to $OD_{600nm} = 0.1$ (approximately 10^8 CFU/mL), and diluted to a final concentration of about 10^7 CFU/mL. In the case of negative controls, inoculations were done by injecting 10 μL of sterile 10 mM PBS. Incubation of the plants and monitoring of disease symptoms were performed in a growth chamber (16 h light, 8 h dark; 26 °C) under quarantine conditions. The pathogen was re-isolated from the wilting plants by cutting 2–3 cm of the stems above the inoculation point, and plating the obtained extracts onto SMSA. The colonies were PCR-identified as described [43] with primers Ps-1 and Ps-2 based on the sequence of the *16S rRNA* gene. Stems from inoculated non-wilted plants were processed in the same way.

2.4. Resuscitation of R. solanacearum Populations from the VBNC State Induced in Environmental Water Microcosms

Assays for resuscitation were performed in three different conditions with VBNC *R. solanacearum* populations from the environmental water microcosms at 4 °C in triplicate. To determine if the appearance of culturable cells was due to true resuscitation instead of regrowth of a few remaining culturable but undetected cells, serial ten-fold dilutions were carried out with the VBNC cells, as described [44,45]. The first series of ten-fold dilutions was performed when the microcosms were containing initially approximately 10^6 viable cells/mL and <10 culturable cells (CFU)/mL, until reaching concentrations of 10^{-2} viable cells/mL and $<10^{-7}$ culturable cells (CFU)/mL. From then and prior to the resuscitation assays, non-culturability was tested at each sampling time by plating volumes of 1 mL directly taken from the microcosms. Resuscitation assays were performed with aliquots (1–10 mL) taken from each environmental water microcosm and their ten-fold dilutions, both *in vitro* and *in planta*, according to three different procedures, as follows:

2.4.1. By Enrichment in a Modified Wilbrink (WB) Broth [16]

Direct aliquots and their ten-fold dilutions in WB broth were incubated at 29 °C with shaking (200 r.p.m.) until appearance of turbidity or for at least one week. Moreover, additional aliquots were taken, transferred to WB broth, and maintained at 4 °C without shaking. Sampling was at time 0 and once a week during a month. To check culturability from turbid tubes, streaks were plated onto YPGA. To test non-culturability from non-turbid tubes, 100-μL volumes were plated onto YPGA. Colonies appeared on plates were PCR-identified. To prove pathogenicity of the colonies on plates, bacterial suspensions were prepared from them and 10-μL volumes were stem-inoculated onto tomato plants (two plants per suspension). Moreover, 10-μL volumes were directly taken from the turbid tubes and inoculated onto stems (two plants per tube) for pathogenicity tests of the cells. Plants were processed for re-isolation and identification of the bacterial pathogen as abovementioned.

2.4.2. By Temperature Upshift in Environmental Water

Briefly, 10-mL aliquots and their ten-fold dilutions in sterile environmental water were incubated at 24 °C without shaking. Sampling was at time zero and each two weeks during a month. To check culturability, 100-μL volumes from the dilutions were daily plated onto YPGA until appearance of colonies or for at least one week. The colonies were PCR-identified and their pathogenicity tested as abovementioned. Moreover, pathogenicity of the cells in the ten-fold dilutions was proved by inoculating 10-μL volumes directly from the dilutions onto the stems (two plants per dilution), which were processed as described.

2.4.3. In Planta

From direct aliquots and their ten-fold dilutions in sterile environmental water, 10-μL volumes were stem-inoculated onto tomato plants aged three weeks (two plants per dilution). Sampling was at time 0 and over one month. Plants were processed for re-isolation and identification of the pathogen as abovementioned.

2.5. Statistical Analysis

Survival assays were performed at three incubation temperatures with environmental water sample and distilled water in triplicate *R. solanacearum*-inoculated microcosms. Total, viable, and culturable data of *R. solanacearum* cell counts were normalized by log-transformation, and mean values analyzed by a linear regression model considering the following factors: incubation temperature, type of water (environmental or distilled), period of incubation, media, and bacterial strain. Differences among means of coccoid percentages at the three temperatures were estimated by variance analysis (ANOVA). A p value < 0.05 was defined as significant.

3. Results

3.1. R. solanacearum Goes into a Nutrient-Dependent Cold-Induced VBNC State in Environmental Water

At low (4 °C) temperature, in environmental or distilled water (Figure 1), total populations of the strain IVIA-1602.1 of *R. solanacearum* remained above their initial inoculation numbers throughout the 40-day experiments, while viability was slightly lower, with declines approximately from 25–30 dpi in both types of water. In contrast, culturable bacterial populations significantly decreased ($p < 0.05$) about one log unit up until eight and four dpi for river and distilled water respectively, pointing out a proportion of cells sensitive to low-temperature conditions. Thereafter, progressively and significant stronger losses in culturability occurred, with values below detection level (10^1 CFU/mL) by 40 ± 7 and 20 ± 3 days, depending on the water sample, in river and distilled water, respectively (Figure 1). These drops in culturable counts with high numbers of cells still viable indicated a majority of the populations becoming VBNC. The strain IVIA-1602.1 displayed similar trends in the microcosms of the other environmental water samples at 4 °C, only with differences in non-culturability between environmental and distilled water ($p < 0.05$). The non-selective medium YPGA and the semiselective SMSA medium yielded similar results for each of the water samples ($p > 0.05$).

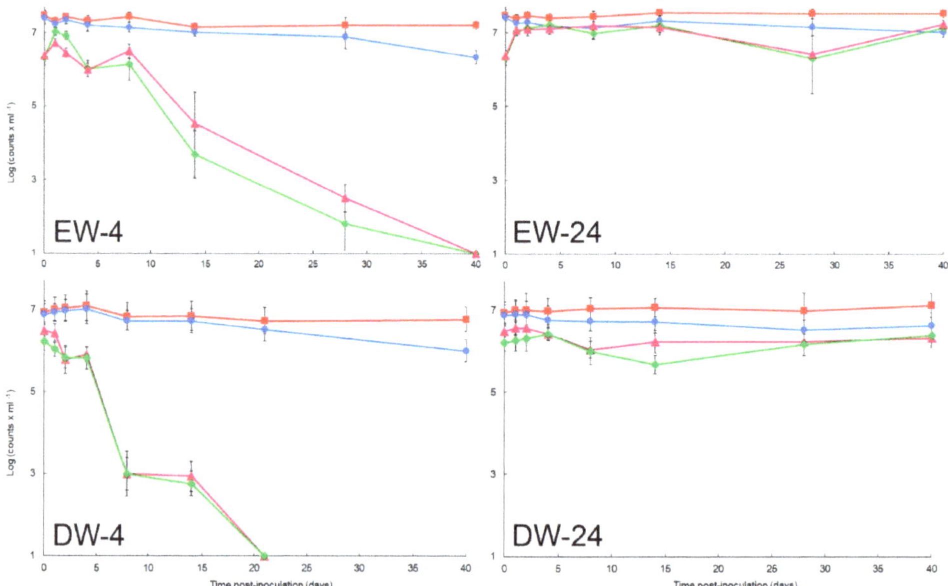

Figure 1. Effect of low temperature under nutrient-limiting conditions on survival of *Ralstonia solanacearum* strain IVIA-1602.1 during 40-day periods in water. Microcosms of: EW−4, environmental water at 4 °C; DW−4, distilled water at 4 °C; EW−24, environmental water at 24 °C, and DW−24, distilled water at 24 °C. Total (■), viable (●), and culturable cells on SMSA (▲) and YPGA (♦) media. Data from one representative environmental water sample have been plotted. Points are mean ± standard deviation of triplicate microcosms.

At temperate (14 °C) and warm (24 °C) temperatures, trends in total, viable and culturable populations were similar ($p > 0.05$) (Figure 1 and Figure S1), and so only those at 24 °C have been plotted in Figure 1. At both temperatures, total populations of the strain IVIA-1602.1 remained above 10^7 cells/mL in environmental water and around this value in distilled water, and viability was slightly lower in both types of water for the 40-day experiments. During the period, culturability remained roughly at 10^7 CFU/mL in environmental water while in distilled water culturable cells stabilized below this value (Figure 1). Assays with the other water samples yielded analogous results, also on both media ($p > 0.05$).

Similarity in trends of culturable data from the microcosms of the four water samples inoculated with the *R. solanacearum* strains could be observed by the statistical analyses. For comparative purposes, increments of culturable data at 4 °C and 24 °C were jointly calculated with respect to the initial value and plotted with time to assess the effect of water sample (Figure 2, left) and the effect of media (Figure 2, right).

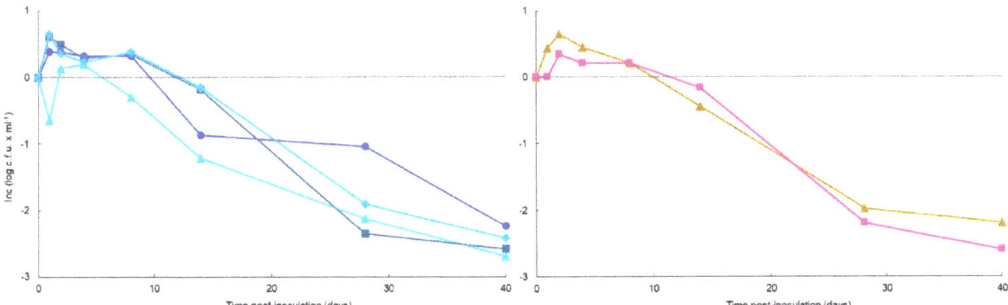

Figure 2. Similarity in trends of culturable cell counts from *Ralstonia solanacearum* strain IVIA-1602.1-inoculated environmental water microcosms throughout 40-day periods at 4 °C and 24 °C. Inc stands for Increments, which were calculated with the differences between mean values of culturable data at both temperatures with respect to the values at time zero (Inc zero). (**Left**): comparison of culturable data on SMSA medium among environmental water (EW)—1 (♦), EW—2 (●), EW—3 (▲), and EW—4 (■). (**Right**): comparison between culturable data on SMSA (■), and YPGA (▲) media for the four environmental water samples.

Population dynamics of total, viable, and culturable cells of the strain IPO-1609 of *R. solanacearum* were similar to those of the strain IVIA-1602.1 ($p > 0.05$) in triplicate microcosms from environmental water samples (Figure S2).

3.2. R. solanacearum Changes Their Shape in Environmental Water with Increased Temperatures

Cells of the strain IVIA-1602.1 were examined in triplicate environmental water sample microcosms at 4 °C, 14 °C, and 24 °C. Data are plotted in Figure 3.

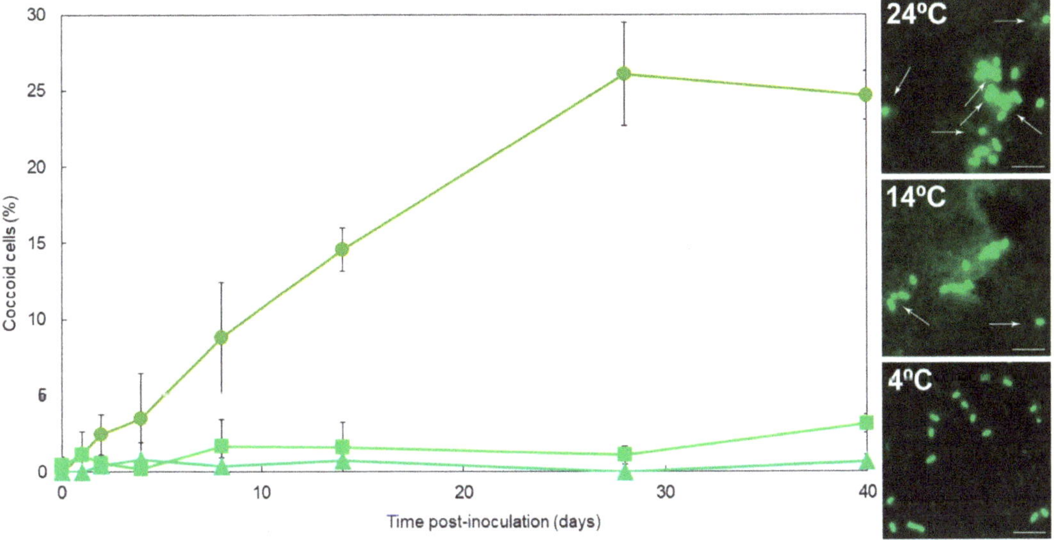

Figure 3. Proportions of coccoid cells appearing in *Ralstonia solanacearum* strain IVIA-1602.1 populations starved in the environmental water microcosms during 40-day periods at 24 °C, 14 °C, and 4 °C. (**Left**) Symbols for each temperature: 24 °C (●), 14 °C (■), and 4 °C (▲). Data from one representative environmental water sample have been plotted. Points are mean ± standard deviation of triplicate microcosms. (**Right**) Representative fluorescence microscopy images of proportions of coccoid cells (white arrows) of the bacterium at each temperature. Scale bars: 5 µm.

At 4 °C and throughout the 40-day experiments, bacterial cells showed the typical *R. solanacearum* bacillar morphology and coccoid cells were seldom detected, with a frequency <1% depending on the water sample (Figure 3).

At 14 °C, a great majority of *R. solanacearum* cells kept bacillar shape. Coccoids were observed in a low proportion, with constant percentages around a value between 1–3% throughout the first 28 days, and then a slight increase up to values ranging 2–6% by 40 dpi, depending on the water sample (Figure 3).

At 24 °C *R. solanacearum* bacilli remained a majority but, coccoid cells were more frequent: depending on the water sample, percentages by the first week were around 8–10%, progressively increasing to 13–16% by the second week, and then up to 22–32% by 28 dpi which stabilized to the end of the 40 days (Figure 3).

Among the low, temperate, and warm temperatures, the average percentage of coccoids significantly increased with temperature ($p < 0.05$). Cell shape of the strain IPO-1609 showed the same trends in one-off trials at each of the three temperatures.

3.3. Starved and/or Cold-Induced VBNC R. solanacearum in Environmental Water Keeps Virulent in Planta

During the 40-day periods of incubation at 4 °C, 14 °C, and 24 °C in the water microcosms under starvation conditions, aliquots were taken at different times to be inoculated in planta. *R. solanacearum* cells of strain IVIA-1602.1 incubated at 4 °C and inoculated in tomato stems induced disease in 98–100% of the plants (Figure 4). Similar wilting percentages were obtained with cells from microcosms at 14 °C and at 24 °C (Figure 4) and were comparable to those of the strain IPO-1609 in one-off trials. At the three temperatures and depending on the water sample, viable *R. solanacearum* cells inoculated per plant were about 10^5 throughout the 40-day sampling periods, and only from approximately 28 dpi at 4 °C there was a slight decline to values around 10^4 viable cells per plant (Figure 4). At 4 °C, culturable cells inoculated per plant ranged from 10^5 to 10^4 in the initial dpi depending on the water sample. Then, they were decreasing until 10^4–10^3 CFU per plant by the first week, and progressively to <10 CFU per plant by 28 dpi and to undetectable levels by 40 dpi (Figure 4). At 14 °C and 24 °C, culturable cells were 10^5–10^4 per plant throughout the sampling periods (Figure 4). Plants started to show symptoms within 8–11 dpi and completely wilted within four weeks. The pathogens were re-isolated on SMSA agar from the diseased plants and PCR-identified. Positive control plants yielded 100% wilting. Negative control plants did not show any symptoms.

3.4. R. solanacearum Resuscitates from the Cold-Induced VBNC State in Environmental Water and Is Fully Pathogenic in the Host

Assays carried out to assess the resuscitation capability of the VBNC *R. solanacearum* cells of strain IVIA-1602.1 yielded similar results in triplicate microcosms from environmental water samples. Data are summarized in Table 1, in the three different conditions. The viability of the VBNC cells when the microcosms were containing approximately 10^6 viable cells/mL and <10 culturable cells (CFU)/mL is illustrated in Figure 5.

3.4.1. By Enrichment in WB Broth

From cold-induced VBNC cells of *R. solanacearum* strain IVIA-1602.1, and after the temperature upshift with shaking and nutrients, monitoring was of: (i) turbidity by *R. solanacearum* growth in the direct aliquots and their serial ten-fold dilutions, (ii) culturability on YPGA, and (iii) pathogenicity in the host. These were observed in all the direct aliquots and their serial ten-fold dilutions up to 10^{-6}, corresponding to 1 VBNC cell/mL, at time 0 of the VBNC induction of the *R. solanacearum* populations (Table 1). Thereafter, the resuscitation capability of these VBNC cells in the microcosms was decreasing with time until reaching about two orders of magnitude by one month from the VBNC induction. Time for resuscitation (estimated as time for observation of turbidity by *R. solanacearum* growth) was 24 h for direct aliquots, 36 h for dilutions 10^{-1}, 10^{-2}, and 10^{-3}, 48 h for dilutions 10^{-4} and 10^{-5}, and four days for dilutions 10^{-6}. These rates of growth were maintained throughout the experimental period. Culturability of the cells in the turbid dilutions was positive in all cases, and colonies were PCR-identified as *R. solanacearum*. However, cells from aliquots in WB broth maintained at 4 °C without shaking remained non-culturable. Pathogenicity assays were positive in all plants, either when inoculated directly from the turbid dilutions or from the colonies on the plates. The pathogen was re-isolated from the wilted plants and PCR-identified. Sensitivity of the detection of resuscitated *R. solanacearum* cells from the VBNC state was 1 VBNC cell/mL.

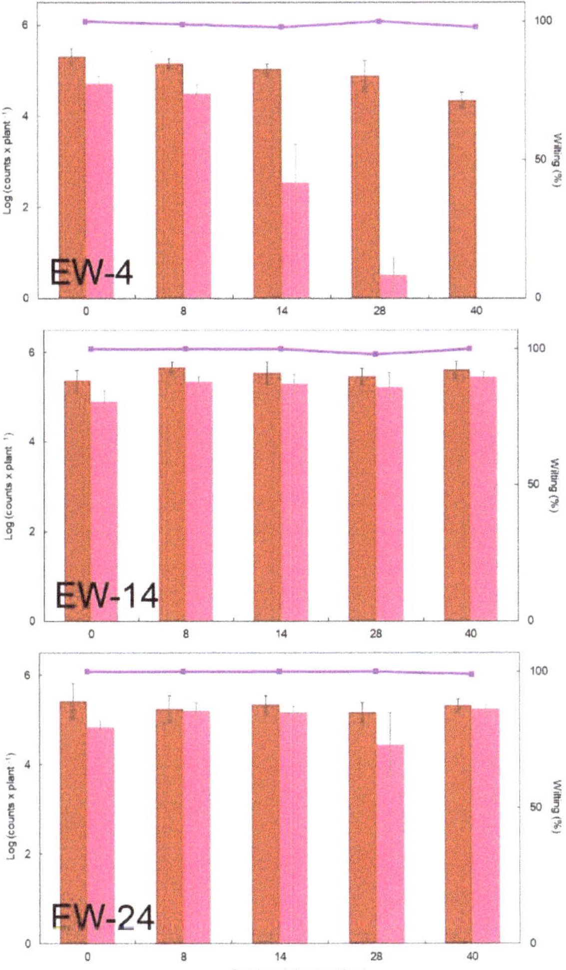

Figure 4. Pathogenicity of *Ralstonia solanacearum* strain IVIA-1602.1 previously starved in environmental water microcosms during 40 days, in tomato plants cv. 'Roma'. Viable (red bars) and culturable (pink bars) cells per plant, and percentage of wilted plants (■). Only assays performed in at least weekly intervals from one representative environmental water (EW) at 4 °C (EW−4), 14 °C (EW−14) and 24 °C (EW−24) have been plotted. Points are mean ± standard deviation (SD) of triplicate microcosms. Absolute value for 100% wilting refers to 24 plants (6 × 4 sets). At the three temperatures, SD of wilting values for most of the points was zero, and ± 2.0% in some cases. Control plants inoculated with freshly grown *R. solanacearum* strain IVIA-1602.1 developed 100% wilting, while those inoculated with PBS were negative.

Table 1. Resuscitation of *Ralstonia solanacearum* strain IVIA-1602.1 previously induced to the VBNC state in environmental water microcosms at 4 °C.

Time in the VBNC State (Weeks)	Culturable Cells (CFU/mL)		Resuscitation Assays *In Vitro*–In Enrichment Conditions								
			Direct	−1	−2	−3	−4	−5	−6	−7	−8
0	3	Turbidity	3/3	3/3	3/3	3/3	3/3	3/3	3/3	0/3	0/3
		Culturability	3/3	3/3	3/3	3/3	3/3	3/3	3/3	0/3	0/3
		Pathogenicity	3/3	3/3	3/3	3/3	3/3	3/3	3/3	0/3	0/3
1	0	Turbidity	3/3	3/3	3/3	3/3	3/3	3/3	0/3	0/3	
		Culturability	3/3	3/3	3/3	3/3	3/3	3/3	0/3	0/3	
		Pathogenicity	3/3	3/3	3/3	3/3	3/3	3/3	0/3	0/3	
2	0	Turbidity	3/3	3/3	3/3	3/3	3/3	3/3	0/3		
		Culturability	3/3	3/3	3/3	3/3	3/3	3/3	0/3		
		Pathogenicity	3/3	3/3	3/3	3/3	3/3	3/3	0/3		
3	0	Turbidity	3/3	3/3	3/3	3/3	3/3	2/3	0/3		
		Culturability	3/3	3/3	3/3	3/3	3/3	2/3	0/3		
		Pathogenicity	3/3	3/3	3/3	3/3	3/3	2/3	0/3		
4	0	Turbidity	3/3	3/3	3/3	3/3	3/3	0/3	0/3		
		Culturability	3/3	3/3	3/3	3/3	3/3	0/3	0/3		
		Pathogenicity	3/3	3/3	3/3	3/3	3/3	0/3	0/3		
			In Vitro–In Environmental Water								
			Direct	−1	−2	−3	−4	−5	−6	−7	−8
0	3	Culturability	3/3	3/3	3/3	3/3	3/3	0/3	0/3	0/3	0/3
		Pathogenicity	3/3	3/3	3/3	3/3	3/3	0/3	0/3	0/3	0/3
2	0	Culturability	3/3	3/3	3/3	1/3	0/3	0/3			
		Pathogenicity	3/3	3/3	3/3	1/3	0/3	0/3			
4	0	Culturability	3/3	2/3	0/3	0/3	0/3				
		Pathogenicity	3/3	2/3	0/3	0/3	0/3				
			In Planta								
			Direct	−1	−2	−3	−4	−5			
0	3	Pathogenicity	3/3	3/3	3/3	3/3	0/3	0/3			
4	0	Pathogenicity	3/3	3/3	2/3	0/3	0/3	0/3			

Data from one representative environmental water at 4 °C (EW−4) are summarized. Direct stands for direct aliquots, and the negative numbers stand for the serial ten-fold dilutions. Pathogenicity assays were considered positive when at least one of the two tomato plants cv. 'Roma' inoculated per microcosm showed bacterial wilt symptoms. *R. solanacearum* was re-isolated from wilted plants and PCR-identified. Control plants inoculated with *R. solanacearum* strain IVIA-1602.1 at each sampling time yielded 100% wilting in stem inoculation. Plants inoculated with PBS were negative. Data represent results from triplicate microcosms.

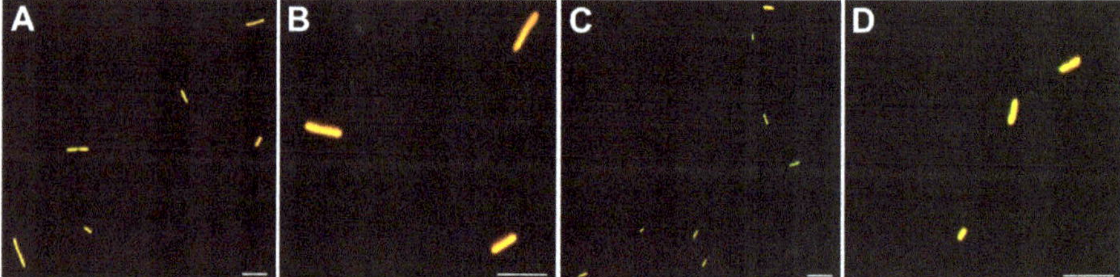

Figure 5. Viability of *Ralstonia solanacearum* strain IVIA-1602.1 populations at 40 days post-inoculation in the environmental water microcosms. (**A,B**) starved cells at 24 °C, and (**C,D**) starved and cold-induced VBNC cells at 4 °C. Viability was similar between starved (**A,B**) and VBNC (**C,D**) *R. solanacearum* cells, although in (**C,D**) cells were no culturable on solid media. For VBNC cells, this time was considered time zero of VBNC induction of the *R. solanacearum* populations in the resuscitation assays. Viability was measured by the DVC method and subsequent staining with acridine orange [41,42], and estimated in 10^6 VBNC cells/mL. Scale bars: 5 μm.

3.4.2. By Temperature Upshift in Environmental Water

From cold-induced VBNC cells of *R. solanacearum* strain IVIA-1602.1 and after the temperature upshift, monitoring was performed regarding: (i) culturability on YPGA and (ii) pathogenicity in the host. These were observed in cells from the direct aliquots and their serial ten-fold dilutions up to 10^{-4}, corresponding to 10^2 VBNC cells/mL, and up to 10^{-3}, corresponding to 10^3 VBNC cells/mL, respectively, at time 0 of the VBNC induction (Table 1). Then, the resuscitation capability of the cells in the microcosms was decreasing with time until about three orders of magnitude by one month from the induction (Table 1). Time for resuscitation (estimated as time for observation of culturability after plating) was 48 h from the temperature upshift. This was observed after plating 100-μL volumes from the same direct aliquots incubated during 24 h, 48 h and 72 h from the temperature upshift, and then on plates for 3 days at 29 °C. Sampling at 24 h yielded no growth on the plates, sampling at 48 h turned out in countable colonies, and continuous bacterial growth was observed after sampling at 72 h from the temperature upshift, reaching in all cases 10^6 CFU/mL. Colonies were PCR-identified as *R. solanacearum*. Pathogenicity assays to test cells from the colonies were positive in all plants. Pathogenicity assays to test cells from the direct aliquots and their serial ten-fold dilutions were positive in at least one of the two inoculated plants per microcosm. The pathogen was re-isolated and PCR-identified from the wilted plants. Sensitivity of the detection of resuscitated *R. solanacearum* cells from the VBNC state was 10^2 VBNC cells/mL.

3.4.3. *In Planta*

From cold-induced VBNC cells of *R. solanacearum* strain IVIA-1602.1 and after the temperature upshift in the host, pathogenicity was monitored. This was observed in cells from the direct aliquots and their serial ten-fold dilutions up to 10^{-3}, corresponding to 10^3 VBNC cells/mL (10 VBNC cells/plant), at time 0 of the VBNC induction, and then the resuscitation capability was decreasing until about one order of magnitude by one month from the induction (Table 1). The pathogen was re-isolated and PCR-identified from the wilted plants. Sensitivity of the detection of resuscitated *R. solanacearum* cells from the VBNC state was 10^3 VBNC cells/mL.

4. Discussion

To anticipate the spread of the disease, current prevention and control strategies against bacterial wilt should take into account knowledge on the potential behavior of *R. solanacearum* in response to global warming. In this work, adaptations by strains of the pathogen from different climatic regions were observed under exposure to environmental temperatures in oligotrophic freshwater, which allowed survival without losing wilting capacity.

With respect to low temperatures, this is the first report of viable *R. solanacearum* populations induced to the VBNC state in environmental water, in conditions more approaching those of natural settings. Previous work described either: (i) a loss in *R. solanacearum* culturability under low temperature in natural or distilled, ultrapure water, but without determining the viability of the bacterial populations, thus without confirming the presence of VBNC cells [26,27,32,46]; or (ii) the VBNC induction in distilled, ultrapure, non-environmental water [28,29]. The fact that other cold-adapted water bacteria are not likely to be cold-induced VBNC [47], contrarily to what was observed in this work with *R. solanacearum* from different climates, suggests that this pathogen is not naturally cold-adapted, even when introduced to cold habitats. Thus, this work demonstrated that, in *R. solanacearum*, low temperature plays a major role than starvation in inducing the VBNC state (Figure 6), contrarily to what has been reported for other bacteria [47]. However, starvation-induced stress proteins could have protected *R. solanacearum* from temperature damage, since it was less vulnerable to cyclic cold stress in pure water than in host tissue [27]. Moreover, the VBNC *R. solanacearum* cold-induction period occurred more slowly in environmental water, pointing to an effect of water nutrient contents, namely trace organic matter and some dissolved salts available for the cells but absent in distilled water, and so nutrient concentrations not supporting *R. solanacearum* growth would act as an additional stress contributing to the cold-induced VBNC state. In the field, latent VBNC cells maintain structure, biology, and significant gene expression, and global climate change might be resuscitating them when low temperature is the inducing factor, leading to increased outbreaks [48]. Similar to *R. solanacearum*, a lower mineral salt concentration markedly shortened the VBNC *Vibrio parahaemolyticus* induction period [49]. Bacterial species, such as *V. vulnificus* and *Aeromonas hydrophila*, also behaved similarly to *R. solanacearum* under low-temperature and nutrient-limiting conditions [50,51], whilst others, such as *Campylobacter jejuni* and *Erwinia amylovora*, displayed different responses [52–55].

At temperate and warm temperatures in environmental water, *R. solanacearum* populations displayed starvation-survival responses as described at 24 °C [6] and similar in terms of population levels. Lack of unculturability was in agreement with previous work reporting the isolation and persistence of the pathogen in environmental water at temperatures allowing *R. solanacearum* multiplication [15–17,26]. The presence of organic matter and salts in environmental water contributed to stimulate *R. solanacearum* survival, similarly to *Aerobacter aerogenes* [56], *E. amylovora* [53,54], and *Leuconostoc mesenteroides* [57], where trace minerals facilitated culturability, since mineral salts can affect not only cell growth, but also cell survival during nutrient limitation conditions [31].

Morphological changes are a visible indicator of adaptation to the environment [31,54,58]. Starved *R. solanacearum* cells transformed from the typical bacilli into coccoids, since shape rounding off and size reduction allow nutrients to be sequestered more efficiently [31]. This was observed in different proportions according to temperature. Although cells entering the VBNC state often exhibit dwarfing [37,59], *R. solanacearum* coccoids were seldom observed during this process, probably because low temperature rapidly causes decrease in *R. solanacearum* metabolism and uptake of water nutrients, with constitutive expression of genes associated with survival and stress response for a stable maintenance of their transcript level [60]. Likewise, copper-induced VBNC *R. solanacearum* cells were unchanged in size [61]. Therefore, at both starvation- and survival-inducing temperatures, the transition to coccoids would be mostly influenced by nutrient limitation and to a lesser extent by low temperature, as reported elsewhere [62]. At these two temperatures, the proportions of coccoids differed, with significantly higher numbers at warm temperature, probably to improve the speed for exchange of material with the surrounding environment to hold a faster energy-consuming metabolism, which becomes a requirement at elevated temperatures [58,63]. Thus, in natural nutrient-deprived environments, the stress of oligotrophy would be less intense for the pathogen at temperatures around 14 °C than at values nearer to the optimum as 24 °C, and so temperature would be modulating this adaptation to oligotrophy (Figure 6), acting on cell metabolism rate and nutrient requirement frequency. Moreover, in the presence of indigenous microbiota, *R. solanacearum* survived longer at 14 °C than at 24 °C in oligotrophic environmental water [35], and the culturability of *R. solanacearum* strain IPO-1609 was favored at 12 °C and 20 °C rather than at 28 °C in agricultural water in both the presence and absence of other aquatic microorganisms [26]. Similar to *R. solanacearum*, a number of bacterial species decreased their sizes with increasing environmental temperatures [51,52,63]. Notwithstanding, this cannot be considered a general bacterial behavior [47,50,58,63].

Although *R. solanacearum* has frequently been described as cold tolerant [14,28], the strains introduced to either cold or warm areas were apparently better temperate-adapted than cold-adapted as considered [12], and similarly to [28], where data indicated that *R. solanacearum* had no special adaptation to survive cold temperatures in water under controlled conditions. Likewise, the cold-water-adapted *Vibrio tasmaniensis* did not enter the VBNC state at 4 °C while the warm-water-adapted *V. shiloi* did [47].

R. solanacearum resuscitation from the cold-induced VBNC state was observed after stress removal by placement of the VBNC cells in three different favorable conditions, including the host plant, and all of them implying, at least, an upshift in temperature. In enrichment conditions, culturability on solid medium and pathogenicity of the resuscitated *R. solanacearum* cells from the dilutions were both confirmed. Restoration of culturability was more dependent on the temperature upshift and shaking than the presence of nutrients, since VBNC cells in enrichment liquid medium at 4 °C were not able to form colonies, and so they maintained their VBNC status. In environmental water and after the temperature upshift, culturability and pathogenicity of the resuscitated cells were similarly confirmed, the only resuscitation-inducing factor here being the temperature upshift, which has not been reported for *R. solanacearum* up until now. That would explain the seasonal variation of *R. solanacearum* populations in environmental water [16,17]. If temperature is so critical, increases in water temperature corresponding to rising global surface temperatures will likely lead to a wider geographic distribution of *R. solanacearum* and a higher incidence of infections in planta resulting from resuscitated cells of the pathogen, as it is being observed in *Vibrio* species [48]. Likewise, a simple reversal of temperature was sufficient to allow the resuscitation of other bacterial species [37,38,59]. In contrast, it was not effective to resuscitate the close *R. pseudosolanacearum* (former *R. solanacearum* phylotype I) in soil and water [60,64], since the addition of hydrogen peroxide-degrading compounds, such as catalase or sodium pyruvate, was necessary. The number of bacterial cells resuscitated by temperature upshift in environmental water was equal to the initial inoculum, similarly to [37,44,60]. Resuscitation in planta of the VBNC *R. solanacearum* cells was evidenced by the occurrence of wilting symptoms, and progressively declined over time, accordingly to [29]. Virulence in tomato plants was also observed in revived cells of *R. pseudosolanacearum* after exiting a cold-induced VBNC state in pure water [64]. In all the three different resuscitation conditions, the resuscitated *R. solanacearum* cells displayed similar phenotypes to the original culturable cells, including virulence in tomato plants, as described [60]. Moreover, also in the three different conditions, a decrease in the proportion of VBNC cells capable of resuscitation occurred over time, this process being dependent on the age of the VBNC cells, as stated [38]. In that respect, several authors agree to consider the existence of gradual stages within the VBNC state, namely a reversible non-culturable stage where cells can be resuscitated and an irreversible non-culturable stage, where cells cannot be resuscitated, although they keep respiratory activity [28,60,64] (Figure 6). Among these conditions, in this work, the enrichment was the most effective for *R. solanacearum* resuscitation and the most sensitive for their detection. This is probably because it combines a temperature upshift with nutrients and shaking, which supplies with oxygen and disperses oxidative compounds (peroxides, other free radicals) accumulated extra- or intracellularly either produced in the cells in response to low temperature stress or commonly present in rich culture media [37,65]. Reversal of adverse VBNC-state-inducing factors can be efficiently applied to the detection of *R. solanacearum* resuscitated cells.

On the basis of all these results, starved and/or cold-stressed and/or cold-induced VBNC *R. solanacearum* cells could be present in environmental water, being a threat to secure crop production as they are not easily detected [61], can survive cold temperature fluctuations [27], and can revert to a fully pathogenic state just by a temperature upshift, which can be favored within the frame of global climate change conditions. All of these *R. solanacearum* survival forms maintained their capacity for in planta multiplication and colonization, causing disease symptoms in the host, as observed elsewhere for similar time periods [6,16,29]. Not only can global warming contribute to the pathogen spread and virulence, but crops resistant to *R. solanacearum* at moderate temperatures can also become more susceptible at high ambient temperatures [14,24], increasing the probability of infections. The bacterial wilt disease is most severe on plants at temperature values ranging from 25 °C to 35 °C [14,28,66].

On the other hand, taking into account the temperature interval of 4–10 °C applied by EU legislation [43] to transport suspected water samples, temperatures above 4 °C up to around 10 °C would be more advisable than 4 °C, since cultivation-based methods are required to confirm pathogen detection. For the inspection of environmental samples, it should be determined whether to also test for these VBNC cells [64] to improve the sensitivity of the detection. This is a relevant point since the early detection of *R. solanacearum* in irrigation water and its eradication would contribute to improve any integrated management program of the bacterial wilt disease [61].

Overall, *R. solanacearum* strains from either cold or warm origin were able to adapt to a combined effect of temperature and oligotrophy. At low temperature, the delay in the induction of the VBNC state in environmental water suggested a protective effect of water nutrient contents on bacterial cells and pointed out the relevance of performing survival studies in conditions better approaching those in the environment. At temperate and warm temperatures, adaptations to oligotrophy were starvation–survival responses and morphological changes influenced by temperature. It appeared that, when temperature was the main stress (cold conditions), nutrient deprivation acted as an additional stress, contributing to accelerate the effect of temperature, and conversely, when oligotrophy was the main stress (temperate and warm conditions), temperature increased the effect of oligotrophy (Figure 6). In all conditions, *R. solanacearum* cells remained pathogenic and capable of resuscitation by a simple reversal of temperature.

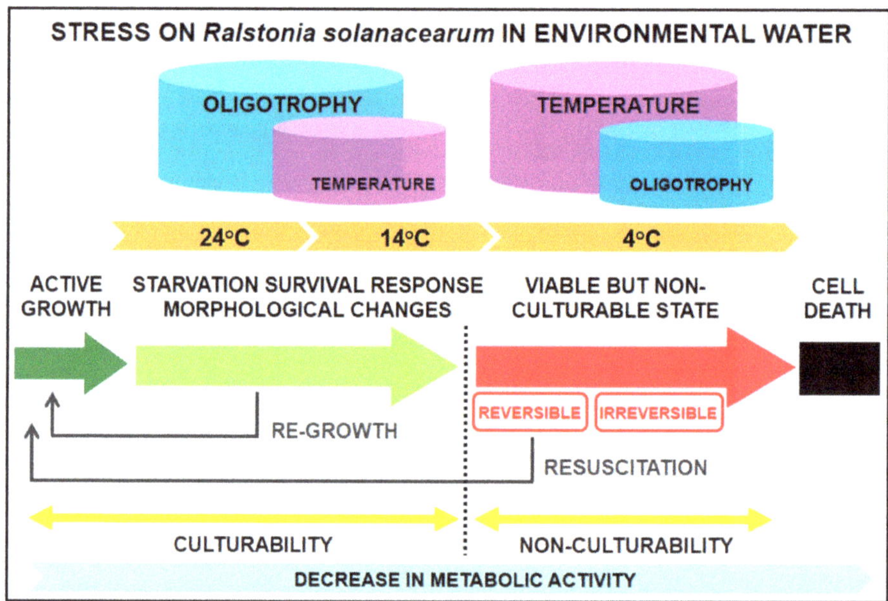

Figure 6. Influence of temperature and oligotrophy on *Ralstonia solanacearum* stress induction in environmental water and stress responses by the bacterium. Adapted to *R. solanacearum* from a proposed model by [67]. See text for details.

Under circumstances of global warming, understanding *R. solanacearum* adaptations to environmental abiotic stresses can help to design strategies to prevent and control their spread and dissemination in waterways and other natural settings. This is particularly important in the case of *R. solanacearum*-contaminated water, since it cannot be used for irrigation, contributing to the global problem of the increased water scarcity in the environment due to climate change, which has serious implications, among others, for food production and health.

Supplementary Materials: The following supporting information can be downloaded at: https://www.mdpi.com/article/10.3390/microorganisms10122503/s1, Figure S1: Effect of temperature under nutrient-limiting conditions on survival of *Ralstonia solanacearum* strain IVIA-1602.1 during 40-day periods in environmental water at 14 °C. Total (■), viable (●), and culturable cells on SMSA

(▲) and YPGA (♦) media. Points are mean ± standard deviation of triplicate microcosms; Figure S2: Effect of low temperature under nutrient-limiting conditions on survival of *Ralstonia solanacearum* strain IPO-1609 during 40-day periods in water. Microcosms of environmental water at: 4 °C (top), 14 °C (middle), and 24 °C (bottom). Total (■), viable (●), and culturable cells on SMSA (▲) and YPGA (♦) media. Points are mean ± standard deviation of triplicate microcosms.

Author Contributions: Conceptualization, B.Á., M.M.L. and E.G.B.; methodology, B.Á., M.M.L. and E.G.B.; validation, B.Á., M.M.L. and E.G.B.; formal analysis, B.Á. and E.G.B.; investigation, B.Á., M.M.L. and E.G.B.; resources, M.M.L. and E.G.B.; writing—original draft preparation, B.Á.; writing—review and editing, B.Á., M.M.L. and E.G.B.; supervision, M.M.L. and E.G.B.; project administration, M.M.L. and E.G.B.; funding acquisition, M.M.L. and E.G.B. All authors have read and agreed to the published version of the manuscript.

Funding: This research was funded by the EU project QLK 3-CT-2000-01598, the Spanish project RTA2015-00087-C02 financed by MCIN/AEI/10.13039/501100011033, the Spanish Instituto Nacional de Investigación y Tecnología Agraria y Alimentaria (INIA), and "ERDF A way of making Europe", as well as the Laboratory of Reference for Plant Pathogenic Bacteria of the Spanish Ministry of Agriculture (MAPA) at Instituto Valenciano de Investigaciones Agrarias (IVIA), the University of Valencia (UV) project UV-INV-AE 112-66196, and the University of Valencia-Research Support to BACPLANT group GIUV2015-219.

Data Availability Statement: Not applicable.

Acknowledgments: The authors thank JL Palomo and the Consejería de Agricultura de Castilla-León for sending water samples, JD van Elsas for strain IPO-1609, and E Carbonell, J Pérez (Instituto Valenciano de Investigaciones Agrarias, Moncada, Valencia, Spain) and JL Díez (Hospital Universitario y Politécnico La Fe, Valencia, Spain) for the statistical analyses.

Conflicts of Interest: The authors declare no conflict of interest. The funders had no role in the design of the study; in the collection, analyses, or interpretation of data; in the writing of the manuscript; or in the decision to publish the results.

References

1. Fegan, M.; Prior, P. How Complex Is the "*Ralstonia solanacearum* Species Complex"? In *Bacterial wilt Disease and the Ralstonia solanacearum Species Complex*; Allen, C., Prior, P., Hayward, A.C., Eds.; APS Press: St. Paul, MN, USA, 2005; pp. 449–461.
2. Safni, I.; Cleenwerck, I.; De Vos, P.; Fegan, M.; Sly, L.; Kappler, U. Polyphasic taxonomic revision of the *Ralstonia solanacearum* species complex: Proposal to emend the descriptions of *Ralstonia solanacearum* and *Ralstonia syzygii* and reclassify current *R. syzygii* strains as *Ralstonia syzygii* subsp. *syzygii* subsp. nov., *R. solanacearum* phylotype IV strains as *Ralstonia syzygii* subsp. *indonesiensis* subsp. nov., banana blood disease bacterium strains as *Ralstonia syzygii* subsp. *celebesensis* subsp. nov. and *R. solanacearum* phylotype I and III strains as *Ralstonia pseudosolanacearum* sp. nov. *Int. J. Syst. Evol. Microbiol.* **2014**, *64*, 3087–3103. [PubMed]
3. Prior, P.; Ailloud, F.; Dalsing, B.L.; Remenant, B.; Sanchez, B.; Allen, C. Genomic and proteomic evidence supporting the division of the plant pathogen *Ralstonia solanacearum* into three species. *BMC Genom.* **2016**, *17*, 90. [CrossRef] [PubMed]
4. Elphinstone, J.G. The Current Bacterial Wilt Situation: A Global Overview. In *Bacterial wilt Disease and the Ralstonia solanacearum Species Complex*; Allen, C., Prior, P., Hayward, A.C., Eds.; APS Press: St. Paul, MN, USA, 2005; pp. 9–28.
5. Swanson, J.K.; Yao, J.; Tans-Kersten, J.; Allen, C. Behaviour of *Ralstonia solanacearum* race 3 biovar 2 during latent and active infection of geranium. *Phytopathology* **2005**, *95*, 136–143. [CrossRef] [PubMed]
6. Álvarez, B.; López, M.M.; Biosca, E.G. Survival strategies and pathogenicity of *Ralstonia solanacearum* phylotype II subjected to prolonged starvation in environmental water microcosms. *Microbiology* **2008**, *154*, 3590–3598. [CrossRef] [PubMed]
7. Álvarez, B.; Vasse, J.; Le-Courtois, V.; Trigalet-Démery, D.; López, M.M.; Trigalet, A. Comparative behavior of *Ralstonia solanacearum* biovar 2 in diverse plant species. *Phytopathology* **2008**, *98*, 59–68. [CrossRef]
8. Álvarez, B.; Biosca, E.G.; López, M.M. On the Life of *Ralstonia solanacearum*, a Destructive Bacterial Plant Pathogen. In *Current Research, Technology and Education Topics in Applied Microbiology and Microbial Biotechnology*; Méndez-Vilas, A., Ed.; Formatex: Badajoz, Spain, 2010; pp. 267–279.
9. Yuliar, N.Y.A.; Toyota, K. Recent trends in control methods for bacterial wilt diseases caused by *Ralstonia solanacearum*. *Microbes Environ.* **2015**, *30*, 1–11. [CrossRef]
10. Álvarez, B.; Biosca, E.G. Bacteriophage-based bacterial wilt biocontrol for an environmentally sustainable agriculture. *Front. Plant Sci.* **2017**, *8*, 1218. [CrossRef]
11. Anonymous. Commission Implementing Regulation (EU) 2019/2072 of 28 November 2019 establishing uniform conditions for the implementation of Regulation (EU) 2016/2031 of the European Parliament and the Council, as regards protective measures against pests of plants, and repealing Commission Regulation (EC) N° 690/2008 and amending Commission Implementing Regulation (EU) 2018/2019. *OJEU* **2019**, *L319*, 1–279.

12. Brown, D. Ralstonia Solanacearum and Bacterial Wilt in the Post-Genomics Era. In *Plant Pathogenic Bacteria. Genomics and Molecular Biology*; Jackson, R.W., Ed.; Caister Academic Press: London, UK, 2009; pp. 175–202.

13. United States Department of Agricultue. Plant Pests and Diseases Programs. Available online: http://www.aphis.usda.gov (accessed on 11 December 2022).

14. Hayward, A.C. Biology and epidemiology of bacterial wilt caused by *Pseudomonas solanacearum*. *Annu. Rev. Phytopathol.* **1991**, *29*, 65–87. [CrossRef]

15. Wenneker, M.; Verdel, M.S.W.; Groeneveld, R.M.W.; Kempenaar, C.; van Beuningen, A.R.; Janse, J.D. *Ralstonia* (*Pseudomonas*) *solanacearum* race 3 (biovar 2) in surface water and natural weed hosts: First report on stinging nettle (*Urtica dioica*). *Eur. J. Plant Pathol.* **1999**, *105*, 307–315. [CrossRef]

16. Caruso, P.; Palomo, J.L.; Bertolini, E.; Álvarez, B.; López, M.M.; Biosca, E.G. Seasonal variation of *Ralstonia solanacearum* biovar 2 populations in a Spanish river: Recovery of stressed cells at low temperatures. *Appl. Environ. Microbiol.* **2005**, *71*, 140–148. [CrossRef]

17. Hong, J.; Ji, P.; Momol, M.T.; Jones, J.B.; Olson, S.M.; Pradhanang, P.; Guven, K. Ralstonia Solanacearum Detection in Tomato Irrigation Ponds and Weeds. In *Acta Horticulturae*; ISHS: Miami, FL, USA, 2005; pp. 309–311.

18. Elphinstone, J.G.; Stanford, H.; Stead, D.E. Survival and transmission of *Ralstonia solanacearum* in aquatic plants of *Solanum dulcamara* and associated surface water in England. *EPPO Bull.* **1998**, *28*, 93–94. [CrossRef]

19. Stevens, P.; van Elsas, J.D. Genetic and phenotypic diversity of *Ralstonia solanacearum* biovar 2 strains obtained from Dutch waterways. *Antonie Van Leeuwenhoek* **2010**, *97*, 171–188. [CrossRef]

20. Cruz, L.; Eloy, M.; Quirino, F.; Oliveira, H.; Tenreiro, R. Molecular epidemiology of *Ralstonia solanacearum* strains from plants and environmental sources in Portugal. *Eur. J. Plant Pathol.* **2012**, *133*, 687–706. [CrossRef]

21. Parkinson, N.; Bryant, R.; Bew, J.; Conyers, C.; Stones, R.; Alcock, M.; Elphinstone, J. Application of variable-number tandem-repeat typing to discriminate *Ralstonia solanacearum* strains associated with English watercourses and disease outbreaks. *Appl. Environ. Microbiol.* **2013**, *79*, 6016–6022. [CrossRef]

22. Caruso, P.; Biosca, E.G.; Bertolini, E.; Marco-Noales, E.; Gorris, M.T.; Licciardello, C.; López, M.M. Genetic diversity reflects geographical origin of *Ralstonia solanacearum* strains isolated from plant and water sources in Spain. *Int. Microbiol.* **2017**, *20*, 155–164. [CrossRef]

23. Singh, D.; Yadav, D.K.; Sinha, S.; Choudhary, G. Effect of temperature, cultivars, injury of root and inoculums load of *Ralstonia solanacearum* to cause bacterial wilt of tomato. *Arch. Phytopathol. Plant Prot.* **2014**, *47*, 1574–1583. [CrossRef]

24. Prior, P.; Bart, S.; Leclercq, S.; Darrasse, A.; Anais, G. Resistance to bacterial wilt in tomato as discerned by spread of *Pseudomonas* (*Burholderia*) *solanacearum* in the stem tissues. *Plant Pathol.* **1996**, *45*, 720–726. [CrossRef]

25. Bittner, R.J.; Arellano, C.; Mila, A.L. Effect of temperature and resistance of tobacco cultivars to the progression of bacterial wilt, caused by *Ralstonia solanacearum*. *Plant Soil* **2016**, *408*, 299–310. Available online: http://www.jstor.org/stable/44136958 (accessed on 11 December 2022). [CrossRef]

26. van Elsas, J.D.; Kastelein, P.; de Vries, P.M.; van Overbeek, L.S. Effects of ecological factors on the survival and physiology of *Ralstonia solanacearum* bv. 2 in irrigation water. *Can. J. Microbiol.* **2001**, *47*, 842–854. [CrossRef]

27. Scherf, J.M.; Milling, A.; Allen, C. Moderate temperature fluctuations rapidly reduce viability of *Ralstonia solanacearum* race 3 biovar 2 in infected geranium, tomato, and potato. *Appl. Environ. Microbiol.* **2010**, *76*, 7061–7067. [CrossRef] [PubMed]

28. Milling, A.; Meng, F.; Denny, T.P.; Allen, C. Interactions with hosts at cool temperatures, not cold tolerance, explain the unique epidemiology of *Ralstonia solanacearum* race 3 biovar 2. *Phytopathology* **2009**, *99*, 1127–1134. [CrossRef] [PubMed]

29. van Overbeek, L.S.; Bergervoet, J.H.W.; Jacobs, F.H.H.; van Elsas, J.D. The low-temperature-induced viable-but-nonculturable state affects the virulence of *Ralstonia solanacearum* biovar 2. *Phytopathology* **2004**, *94*, 463–469. [CrossRef] [PubMed]

30. Roszak, D.B.; Colwell, R.R. Survival strategies of bacteria in the natural environment. *Microbiol. Rev.* **1987**, *51*, 365–379. [CrossRef]

31. Morita, R.Y. Bacteria in Oligotrophic Environments. In *Starvation-Survival Lifestyle*; Reddy, C.A., Chakrabarty, A.M., Demain, A.L., Tiedje, J.M., Eds.; Chapman and Hall: New York, NY, USA, 1997; pp. 368–385.

32. Kelman, A. Factors influencing viability and variation in cultures of *Pseudomonas solanacearum*. *Phytopathology* **1956**, *46*, 16–17.

33. Tanaka, Y.; Noda, N. Studies on the factors affecting survival of *Pseudomonas solanacearum* E.F. Smith, the causal agent of tobacco wilt disease. *Bull. Okayama Tob. Exp. Stn.* **1973**, *32*, 81–91.

34. Wakimoto, S.; Utatsu, I.; Matsuo, N.; Hayashi, N. Multiplication of *Pseudomonas solanacearum* in pure water. *Ann. Phytopathol. Soc. Jpn.* **1982**, *48*, 620–627. [CrossRef]

35. Álvarez, B.; López, M.M.; Biosca, E.G. Influence of native microbiota on survival of *Ralstonia solanacearum* phylotype II in river water microcosms. *Appl. Environ. Microbiol.* **2007**, *73*, 7210–7217. [CrossRef]

36. Chaiyanan, S.; Chaiyanan, S.; Grim, C.; Maugel, T.; Huq, A.; Colwell, R.R. Ultrastructure of coccoid viable but non-culturable *Vibrio cholerae*. *Environ. Microbiol.* **2007**, *9*, 393–402. [CrossRef]

37. Oliver, J.D. Recent findings on the viable but nonculturable state in pathogenic bacteria. *FEMS Microbiol. Rev.* **2010**, *34*, 415–425. [CrossRef]

38. Li, L.; Mendis, N.; Trigui, H.; Oliver, J.D.; Faucher, S.P. The importance of the viable but non-culturable state in human bacterial pathogens. *Front. Microbiol.* **2014**, *5*, 258. [CrossRef] [PubMed]

39. Ridé, M. Bactéries Phytopathogènes et Maladies Bactériennes des Végétaux. In *Les Bactérioses ET Les Viroses Des Arbres Fruitiers*; Ponsot, M., Ed.; Viennot-Bourgin: Paris, France, 1969; pp. 4–59.

40. Elphinstone, J.G.; Hennessy, J.; Wilson, J.K.; Stead, D.E. Sensitivity of different methods for the detection of *Ralstonia solanacearum* in potato tuber extracts. *EPPO Bull.* **1996**, *26*, 663–678. [CrossRef]

41. Kogure, K.; Simidu, U.; Taga, N. A tentative direct microscopic method for counting living marine bacteria. *Can. J. Microbiol.* **1979**, *25*, 415–420. [CrossRef]

42. Oliver, J.D. Heterotrophic bacterial populations of the Black sea. *Biol. Oceanogr.* **1987**, *4*, 83–97.

43. Anonymous. Commission Directive 2006/63/EC of 14 July 2006: Amending Annexes II to VII to Council Directive 98/57/EC on the control of *Ralstonia solanacearum* (Smith) Yabuuchi et al. *Off. J. Eur. Communities* **2006**, *L206*, 36–106.

44. Whitesides, M.D.; Oliver, J.D. Resuscitation of *Vibrio vulnificus* from the viable but nonculturable state. *Appl. Environ. Microbiol.* **1997**, *63*, 1002–1005. [CrossRef]

45. Ordax, M.; Marco-Noales, E.; López, M.M.; Biosca, E.G. Survival strategy of *Erwinia amylovora* against copper: Induction of the viable-but-nonculturable state. *Appl. Environ. Microbiol.* **2006**, *72*, 3482–3488. [CrossRef]

46. Stevens, P.; van Overbeek, L.S.; van Elsas, J.D. *Ralstonia solanacearum* DPGI-1 strain KZR-5 is affected in growth, response to cold stress and invasion of tomato. *Microb. Ecol.* **2011**, *61*, 101–112. [CrossRef]

47. Vattakaven, T.; Bond, P.; Bradley, G.; Munn, C.B. Differential effects of temperature and starvation on induction of the viable-but-nonculturable state in the coral pathogens *Vibrio shiloi* and *Vibrio tasmaniensis*. *Appl. Environ. Microbiol.* **2006**, *72*, 6508–6513. [CrossRef]

48. Oliver, J.D. The viable but nonculturable state for bacteria: Status update. *Microbe* **2016**, *11*, 159–164. [CrossRef]

49. Wong, H.C.; Wang, P. Induction of viable but nonculturable state in *Vibrio parahaemolyticus* and its susceptibility to environmental stresses. *J. Appl. Microbiol.* **2004**, *96*, 359–366. [CrossRef] [PubMed]

50. Biosca, E.G.; Amaro, C.; Marco-Noales, E.; Oliver, J.D. Effect of low temperature on starvation-survival of the eel pathogen *Vibrio vulnificus* biotype 2. *Appl. Environ. Microbiol.* **1996**, *62*, 450–455. [CrossRef] [PubMed]

51. Mary, P.; Chihib, N.E.; Charafeddine, O.; Defives, C.; Hornez, J.P. Starvation survival and viable but nonculturable states in *Aeromonas hydrophila*. *Microb. Ecol.* **2002**, *43*, 250–258. [CrossRef] [PubMed]

52. Rollins, D.M.; Colwell, R.R. Viable but nonculturable stage of *Campylobacter jejuni* and its role in survival in the natural aquatic environment. *Appl. Environ. Microbiol.* **1986**, *52*, 531–538. [CrossRef] [PubMed]

53. Biosca, E.G.; Marco-Noales, E.; Ordax, M.; López, M.M. Long-Term Starvation-Survival of *Erwinia amylovora* in Sterile Irrigation Water. In *Acta Horticulturae*; Bazzi, C., Mazzucchi, U., Eds.; ISHS: Bologna, Italy, 2006; pp. 107–112.

54. Santander, R.D.; Oliver, J.D.; Biosca, E.G. Cellular, physiological and molecular adaptive responses of *Erwinia amylovora* to starvation. *FEMS Microbiol. Ecol.* **2014**, *88*, 258–271. [CrossRef]

55. Santander, R.D.; Biosca, E.G. *Erwinia amylovora* psychrotrophic adaptations: Evidence of pathogenic potential and survival at temperate and low environmental temperatures. *PeerJ* **2017**, *5*, e3931. [CrossRef]

56. Trulear, M.G.; Characklis, W.G. Dynamics of biofilm processes. *J. Water Pollut. Control Fed.* **1982**, *54*, 1288–1301.

57. Kim, D.; Thomas, S.; Fogler, H.S. Effects of pH and trace minerals on long-term starvation of *Leuconostoc mesenteroides*. *Appl. Environ. Microbiol.* **2000**, *66*, 976–981. [CrossRef]

58. Shi, B.; Xia, X. Morphological changes of *Pseudomonas pseudoalcaligenes* in response to temperature selection. *Curr. Microbiol.* **2003**, *46*, 120–123. [CrossRef]

59. Su, X.; Sun, F.; Wang, Y.; Hashmi, M.Z.; Guo, L.; Ding, L.; Shen, C. Identification, characterization and molecular analysis of the viable but nonculturable *Rhodococcus biphenylivorans*. *Sci. Rep.* **2015**, *5*, 18590. [CrossRef]

60. Kong, H.G.; Bae, J.Y.; Lee, H.J.; Joo, H.J.; Jung, E.J.; Chung, E.; Lee, S. Induction of the viable but nonculturable state of *Ralstonia solanacearum* by low temperature in the soil microcosm and its resuscitation by catalase. *PLoS ONE* **2014**, *9*, e109792. [CrossRef]

61. Um, H.Y.; Kong, H.G.; Lee, H.J.; Choi, H.K.; Park, E.J.; Kim, S.T.; Murugiyan, S.; Chung, E.; Kang, K.Y.; Lee, S. Altered gene expression and intracellular changes of the viable but nonculturable state in *Ralstonia solanacearum* by copper treatment. *Plant Pathol. J.* **2013**, *29*, 374–385. [CrossRef]

62. Klancnik, A.; Zorman, T.; Možina, S.S. Effects of low temperature, starvation and oxidative stress on the physiology of *Campylobacter jejuni* cells. *Croat. Chem. Acta* **2008**, *81*, 41–46.

63. Christophersen, J. Basic Aspects of Temperature Action on Microorganims. In *Temperature and Life*; Precht, H., Christophersen, J., Hensel, H., Larcher, W., Eds.; Springer: Berlin/Heidelberg, Germany, 1973; pp. 3–59.

64. Imazaki, I.; Nakaho, K. Temperature-upshift-mediated revival from the sodium-pyruvate-recoverable viable but nonculturable state induced by low temperature in *Ralstonia solanacearum*: Linear regression analysis. *J. Gen. Plant Pathol.* **2009**, *75*, 213–226. [CrossRef]

65. Kong, I.S.; Bates, T.C.; Hülsmann, A.; Hassan, H.; Smith, B.E.; Oliver, J.D. Role of catalase and oxyR in the viable but nonculturable state of *Vibrio vulnificus*. *FEMS Microbiol. Ecol.* **2004**, *50*, 133–142. [CrossRef]

66. Fajinmi, A.A.; Fajinmi, O.B. An overview of bacterial wilt disease of tomato in Nigeria. *Agric. J.* **2010**, *5*, 242–247. [CrossRef]

67. Shleeva, M.O.; Bagramyan, K.; Telkov, M.V.; Mukamolova, G.V.; Young, M.; Kell, D.B.; Kaprelyants, A.S. Formation and resuscitation of "non-culturable" cells of *Rhodococcus rhodochrous* and *Mycobacterium tuberculosis* in prolonged stationary phase. *Microbiology* **2002**, *148*, 1581–1591. [CrossRef]

Article

Chemotactic Responses of *Xanthomonas* with Different Host Ranges

Marta Sena-Vélez [1,2] , Elisa Ferragud [2], Cristina Redondo [2], James H. Graham [3] and Jaime Cubero [2,*]

1. Laboratoire de Biologie des Ligneux et des Grandes Cultures (LBLGC) EA 1207, L'institut National de Recherche pour L'agriculture, L'alimentation et L'environneme (INRAE) USC1328, Orléans University, BP 6759, CEDEX 2, 45067 Orléans, France
2. Instituto Nacional de Investigación y Tecnología Agraria y Alimentaria (INIA/CSIC), 28040 Madrid, Spain
3. Citrus Research and Education Center (CREC), University of Florida, 700 Experiment Station Road, Lake Alfred, FL 33850-2299, USA
* Correspondence: cubero@inia.csic.es; Tel.: +34-913474162

Abstract: *Xanthomonas citri* pv. *citri* (*Xcc*) (*X. citri* subsp. *citri*) type A is the causal agent of citrus bacterial canker (CBC) on most *Citrus* spp. and close relatives. Two narrow-host-range strains of *Xcc*, A^W and A*, from Florida and Southwest Asia, respectively, infect only Mexican lime (*Citrus aurantifolia*) and alemow (*C. macrophylla*). In the initial stage of infection, these xanthomonads enter via stomata to reach the apoplast. Herein, we investigated the differences in chemotactic responses for wide and narrow-host-range strains of *Xcc* A, *X. euvesicatoria* pv. *citrumelonis* (*X. alfalfae* subsp. *citrumelonis*), the causal agent of citrus bacterial spot, and *X. campestris* pv. *campestris*, the crucifer black rot pathogen. These strains of *Xanthomonas* were compared for carbon source use, the chemotactic responses toward carbon compounds, chemotaxis sensor content, and responses to apoplastic fluids from *Citrus* spp. and Chinese cabbage (*Brassica pekinensis*). Different chemotactic responses occurred for carbon sources and apoplastic fluids, depending on the *Xanthomonas* strain and the host plant from which the apoplastic fluid was derived. Differential chemotactic responses to carbon sources and citrus apoplasts suggest that these *Xanthomonas* strains sense host-specific signals that facilitate their location and entry of stomatal openings or wounds.

Keywords: *Citrus*; xanthomonads; bacterial motility; MCPs; methyl-accepting chemotaxis proteins

Citation: Sena-Vélez, M.; Ferragud, E.; Redondo, C.; Graham, J.H.; Cubero, J. Chemotactic Responses of *Xanthomonas* with Different Host Ranges. *Microorganisms* **2023**, *11*, 43. https://doi.org/10.3390/microorganisms11010043

Academic Editors: Soledad Sacristán and James F. White

Received: 28 October 2022
Revised: 5 December 2022
Accepted: 19 December 2022
Published: 22 December 2022

1. Introduction

Citrus bacterial canker (CBC) is one of the most important bacterial diseases of citrus in the tropical and subtropical areas of the world. CBC is characterized by the appearance of necrotic, erumpent lesions on leaves, fruits, and stems and may cause premature defoliation and fruit drop in most *Citrus* species and close citrus relatives in the family *Rutaceae* [1–3]. Distinct types of CBC have been described caused by different bacteria within the genus *Xanthomonas*. The symptoms of these canker diseases are similar, and initially, all the causal bacterial strains were classified within the same species of the genus [1,2,4–7]. The most studied and widespread CBC type is the Asiatic citrus canker or type A, which comprises two pathotypes, A* and A^W, that have been characterized as genetically slightly distinct from the *Xanthomonas citri* pv. *citri* (*Xcc*) type A [8–11]. A* and A^W occur in Southwest Asia and Florida, respectively, and have narrow host ranges that include Mexican lime (*C. aurantifolia*) and alemow (*C. macrophylla*). Although in the field, these strains only cause disease on lime, when they are infiltrated into the leaves of other citrus species, they produce atypical lesions, slightly raised with no rupture of the epidermis [8,9]. Furthermore, the *Xcc* A^W strain is able to cause a hypersensitive response on Duncan grapefruit (*C. paradisi*) when infiltrated [8].

Chemotaxis is the mechanism enabling bacteria to sense stimuli, such as nutrients, light, or temperature, that attract them to the site that is optimally suited for host colonization and infection [12]. Several studies on plant pathogens have demonstrated the importance of this mechanism; for example, jasmonate is a plant signal that attracts *Dickeya dadantii* to wounds, facilitating entry of the host and enhancing the infection process [13,14]. Chemotaxis- and motility-related genes were overexpressed during the epiphytic stage of the interaction on bean leaves with *Pseudomonas syringae* but not after reaching the apoplast [15]. In *X. campestris* pv. *campestris* (*Xc*), *cheY* and XAC0324 genes have been associated with chemotaxis in host leaf colonization, although once the bacteria reach the apoplastic space, the participation of chemotaxis is unimportant for symptom development in cabbage [16].

Methyl-accepting chemotaxis proteins (MCPs) are protein receptors present on both, the bacterial membrane and the cytoplasm, able to sense environmental clues and trigger a motile response to favor bacterial fitness and survival in the environment [17,18]. Diverse MCPs were identified in *Xanthomonas* spp., including *Xcc* [19,20].

The role of chemotaxis in the *Xcc* infection progress in Duncan grapefruit has been suggested [21], and the requirement for active bacterial motility and chemotaxis on the plant surface to locate and specifically colonize the host apoplastic site is supported also by indirect evidence. *Xcc* is dispersed by wind and rain; on the leaf surface, *Xcc* is able to swim short distances reaching the plant interior through stomata or wounds. This process is facilitated by the action of wind but also happens in its complete absence [22–24]. If there is water on the leaf surface, bacterial movement and entry into stomata or wounds may be mediated by chemotaxis. This hypothesis was reinforced by confocal laser scanning microscopy visualization of *Xcc* A on citrus, which showed bacterial accumulation at the edge of the stomata immediately after the spray-inoculation of leaves [25]. Furthermore, wide-host-range *Xcc* and *X. euvesicatoria* pv. *citrumelonis* (*Xec*) were detected in the apoplast of Swingle citrumelo leaves, while the non-host strain *Xcc* A^W type was not present. In contrast, all these citrus strains were found to extensively colonize the apoplast of Mexican lime leaves [26]. These events suggest the requirement for chemotaxis and active bacterial motility on the plant surface to locate and colonize the apoplastic site.

Studies have elucidated some of the pathogenesis mechanisms that contribute to host range differences in CBC strains [27–30], but they did not address, most of the time, early events in the infection process, including motility mediated by chemotaxis [31].

In this study, we characterized the chemotactic responses of types A and A* or A^W of *Xcc* and compared their behavior with *Xec*, the causal agent of citrus bacterial spot (CBS), a disease of citrus nursery plants, and *Xc*, the cause of crucifer black rot (CBR) and whose chemotactic role in leaf colonization has been demonstrated [16]. Our aim was to identify the profiles of compounds that act as attractants or repellents for *Xanthomonas* strains and to relate these profiles to carbon source use, MCP content and host range. Furthermore, the chemotactic response to apoplastic fluids from citrus and non-citrus hosts was evaluated in order to determine whether the chemotaxis signals may somehow explain the host specificity of *Xanthomonas* strains at an early stage of infection.

2. Materials and Methods

2.1. Bacterial Strains, Culture Media, and Growth Conditions

Representative bacterial strains from each xanthomonad group used in this study and their natural hosts are listed in Table 1. Two wide-host-range strains (A type) of *Xanthomonas citri* pv. *citri* (*Xcc*) and three narrow-host-range strains (A* and A^W) were evaluated along with *X. euvesicatoria* pv. *citrumelonis* (*Xec*) and *X. campestris* pv. *campestris* (*Xc*), a non-citrus pathogen.

Bacterial strains were routinely grown on Luria Bertani broth (LB; 10 g of tryptone, 5 g L^{-1} of yeast extract, and 5 g of sodium chloride) or on LB plates (1.5% bacteriological agar) at 27 °C for 48 h.

Table 1. Strains and hosts of *Xanthomonas* spp. used in the study.

Strain	Taxon, Disease and Disease Type	Natural Host
Xcc 306	*Xanthomonas citri* pv. *citri*, CBC [a] A	*Citrus sinensis*
Xcc 62	*Xanthomonas citri* pv. *citri*, CBC A	*Citrus paradisi*
Xcc Iran2	*Xanthomonas citri* pv. *citri*, CBC A*	*Citrus aurantifolia*
Xcc Iran10	*Xanthomonas citri* pv. *citri*, CBC A*	*Citrus aurantifolia*
Xcc 12879	*Xanthomonas citri* pv. *citri*, CBC A^w	*Citrus aurantifolia*
Xec F1	*Xanthomonas euvesicatoria* pv. *citrumelonis*, CBS [b]	*Citrus* spp.
Xc 1609	*Xanthomonas campestris* pv. *campestris*, CBR [c]	*Brassica* spp.

[a] CBC: citrus bacterial canker; [b] CBS: citrus bacterial spot; [c] CBR: crucifer black rot.

2.2. Carbon Source Use by Xanthomonas Strains

Biolog GN2 MicroPlateTM was used for analysis of carbon source use following the manufacturer's instructions (Biolog Inc. Hayward, CA, USA). Bacterial strains were grown on LB agar plates and incubated for 48 h at 27 °C. Bacterial colonies were then harvested and suspended in 0.85% NaCl and adjusted to 0.3 absorbance at 600 nm. Each Biolog microplate well was seeded with 150 µL of the bacterial suspension and incubated for 24 h at 27 °C without shaking. Tetrazolium oxidation activity was measured at 0 and 24 h in a microplate reader set at 570 nm absorption.

The assay was repeated at least two times with two replicates per assay. Carbon source use was calculated by subtracting the time 0 absorbance from each well reading. Substrate well readings were further adjusted against the substrate blank well, and each activity value was the average of the assays, with two replicates per assay. Wells with $\geq$160% of activity compared to the blank were considered positive and $\leq$130% of activity considered negative. Values from 129% to 159% were considered non-informative and dropped from further analysis. Data from informative and discriminatory tests were converted to binary form, and similarity coefficients for pairs of strains were calculated with PAST v.4.03 software (University of Oslo, Oslo, Sweden) [32] using the Jaccard coefficient and subjected to the unweighted pair group method (UPGMA). Bootstrap values (based on 1000 replicates) were indicated at the nodes.

2.3. Chemotactic Response of Xanthomonas Strains to Carbon Compounds

A new microtiter plate assay was developed based on a capillary protocol previously described [33]. Pipette tips containing 5 µL of the carbon source were inserted into 48 wells of a microtiter plate, each filled with 200 µL of a 10^8 CFU mL^{-1} bacterial suspension. To measure chemotaxis, the number of bacteria able to enter the tip for 1 hour was estimated by means of serial dilutions of the tip's content. Bacteria used in chemotaxis studies were in the logarithmic phase to ensure active motility; briefly, a colony was harvested from the LB plate, suspended in 5 mL of LB broth, and incubated at 27 °C and shaking o/n at 150 rpm, and then this preculture was diluted in 30 mL of LB broth to a final concentration of 0.01 OD at 600 nm and cultured up to the logarithmic phase in the conditions described before. Bacteria were washed twice with 10 mM MgCl$_2$. The carbon source was considered a chemoattractant or chemorepellent when the average number of bacteria that entered the tip in six replicates from at least two assays was significantly higher or lower ($p < 0.05$) than the control with 10 mM MgCl$_2$. The assay was validated using *D. dadantii* strain 3937, the causal agent of potato soft rot, whose chemotactic profile has been previously described [14,34]. Data from the microtiter plate assay were converted to binary form, and similarity coefficients for pairs of strains were calculated, as described before.

2.4. Chemotactic Response of Xanthomonas Strains to Apoplastic Fluids

Apoplastic fluids were extracted, as previously described for *Solanum lycopersicum* [35]. Briefly, weighed leaves were vacuum-infiltrated with sterile distilled water, introduced into a 5 mL tip, and then centrifuged at 4000× *g* for 20 min. After centrifugation, the suspension containing the apoplastic fluid was recovered in 1.5 mL tubes and centrifuged at 3000× *g* to remove leaf debris. Apoplastic fractions were sterilized by passing them through a 0.2 μm filter. To evaluate the effect of leaf apoplastic fluids, microtiter plate assays were performed, as described before, with fractions of 200, 100, 50, 12.5, 6.3, and 3.1 mg of leaf per mL of sterile distilled water.

The experiment was performed using the microtiter plate assay described in the previous section. To establish an apoplastic fluid threshold concentration for every strain separately, data were analyzed using the Dunnet test on JMP software (SAS Institute Inc., Cary, NC, USA); this test compares using a t-test every apoplastic concentration with the control, the homogeneous environment (water in this experiment.) The apoplastic fluid concentration was considered a chemoattractant or chemorepellent when the average number of bacteria that entered the tip in six replicates from at least two assays was significantly ($p < 0.05$) higher or lower than the water control. When $p > 0.05$, no response was considered.

2.5. Detection of Methyl-Accepting Chemotaxis Proteins

Profiles of MCPs for *Xanthomonas* species and pathovars used in this study were determined *in silico* based on the analysis of a selection of complete representative genomes from the database of each xanthomonad group studied (Table 2) and the search of homologous sequences for 28 MCPs available, as previously described [19]. Sequence homology searches were conducted using Geneious Prime v.2022.1.1 (Biomatters, Auckland, New Zealand).

Table 2. Genomes used for in silico MCP analysis.

Strain	Species/Pathovar	Type	Accession/Assembly
Xcc C40	*X. citri* pv. *citri*	A	CCWX01
Xcc 5208	*X. citri* pv. *citri*	A	NZ_CP009028.1
Xcc 306	*X. citri* pv. *citri*	A	NC_003919.1
Xcc gd2	*X. citri* pv. *citri*	A	NZ_CP009019.1
Xcc jx5	*X. citri* pv. *citri*	A	NZ_CP009010.1
Xcc UI6	*X. citri* pv. *citri*	A	NZ_CP008990.1
Xcc NT17	*X. citri* pv. *citri*	A	NZ_CP008993.1
Xcc BL18	*X. citri* pv. *citri*	A	NZ_CP009023.1
Xcc MN10	*X. citri* pv. *citri*	A	NZ_CP009002.1
Xcc MN11	*X. citri* pv. *citri*	A	NZ_CP008999.1
Xcc DAR73886	*X. citri* pv. *citri*	A*	GCA_016801635.1
Xcc DAR84832	*X. citri* pv. *citri*	A*	GCA_016801615.1
Xcc 12879	*X. citri* pv. *citri*	A^w	NC_020815.1
Xcc AW13	*X. citri* pv. *citri*	A^w	NZ_CP009031.1
Xcc AW14	*X. citri* pv. *citri*	A^w	NZ_CP009034.1
Xcc AW16	*X. citri* pv. *citri*	A^w	NZ_CP009040.1
Xec F1	*X. euvesicatoria* pv. *citrumelonis*	NA[b]	GCA_000225915.1
Xec FDC1637 [a]	*X. euvesicatoria* pv. *citrumelonis*	NA	GCA_005059795.1
Xc CN15	*X. campestris* pv. *campestris*	NA	GCA_000403575.2
Xc MAFF302021	*X. campestris* pv. *campestris*	NA	GCA_009177345.1
Xc ATCC33193	*X. campestris* pv. *campestris*	NA	GCA_000007145.1
Xc ICMP20180	*X. campestris* pv. *campestris*	NA	GCA_001186415.1
Xc SB80	*X. campestris* pv. *campestris*	NA	GCA_021459985.1

[a] All sequences corresponded to full complete genomes, except *Xec* FDC1637, which enclosed 124 contigs. [b] Not applicable (NA).

To classify xanthomonads studied according to their MCP profile, cluster analysis was performed as before, using PAST v.4.03 software (University of Oslo, Oslo, Sweden) [32].

To confirm the MCP content in xanthomonads used in this study, conventional PCR was conducted according to the genomic analysis and using selected primers previously described for MCPs that were not conserved and showed variability within the *Xanthomonas* genus such as XAC3271, XAC3768, XCV1702, XCV1778, XCV1942, XCV1944, XCV1947, XCV1951, and XCC0324 [19]. Two extra set of primers were designed based on the genes

XCAW2504 (MSV_XCAW2504F: ATGCTGTCGGAAATGCAGGA and MSV_XCAW2504R: AGGTGCTTGATCTCCTTGGC) and XCAW2508 (MSV_XCAW02508F: GCGTCGCTCAAT-AACGTCAC and MSV_XCAW02508R: GATGCTGCTTTCGTACTGCG) that were identified in *Xcc* 12879 and corresponded to XCV1933 and XCV1938, which primers described previously [19] did not give positive results from some *Xcc* A strains in a preliminary work in our group. PCR was carried out in a final volume of 25 µL containing 2 mM MgCl$_2$, 0.2 mM of dNTPs (each), 2 units of DNA polymerase (Biotools, Madrid, Spain), and 0.2 mM of each primer. For fragments longer than 1000 bp, FastStart Taq-DNA polymerase from (Roche, Basel, Switzerland) was used to a final volume of 25 µL containing 2 mM of MgCl$_2$, 0.1 mM of dNTPs (each), 2 units of FastStart Taq-DNA polymerase, and 0.2 mM of each primer. The amplification conditions consisted of 94 °C for 1 min, annealing temperatures described by Mhedbi-Hajri [19] and 57 °C for XCAW2504 and XCAW2508 for 1 min and 72 °C for 1 min for 40 cycles, plus an initial step of 95 °C for 10 min and a final step of 72 °C for 10 min. PCR products (10 µL) were run in 1.5% (*w/v*) agarose gels stained with ethidium bromide and visualized under a UV transilluminator. Water was used as a negative control. The presence or absence of the PCR product for each MCP was converted to binary form and cluster analysis performed, as described before.

3. Results

3.1. Carbon Source Use by Xanthomonas Strains

Carbon source use was analyzed for bacterial strains listed in Table 1 with Biolog GN2 Microplate TM (Biolog Inc. Hayward, CA, USA) following the manufacturer instructions. Readings were made at 0 and 24 h post-inoculation (hpi) to detect the earliest metabolic response. The use of carbon sources that differentiate *Xanthomonas* strains studied is presented in Table 3.

Table 3. Biolog activity of carbon sources that differentiates species and strains of *Xanthomonas* pathogenic for citrus and crucifers.

Strains/Additive [a]	*Xcc* 306	*Xcc* 62	*Xcc* 12879	*Xcc* Iran2	*Xcc* Iran10	*Xec* F1	*Xc* 1609
Dextrin	−	+	+	+	+	+	+
Glycogen	+[a]	+	NI	+	+	+	NI
Tween 80	−[b]	−	−	−	−	−	+
L-Arabinose	−	−	−	−	NI	NI	−
D-Arabitol	−	−	−	−	−	NI	−
L-Fucose	NI[c]	+	+	+	+	+	+
α-D-Lactose	−	−	−	+	+	NI	−
Lactulose	NI	+	+	+	+	+	+
D-Melobiose	−	NI	−	+	+	+	+
D-Raffinose	−	−	−	NI	NI	NI	NI
Sucrose	+	+	+	+	+	NI	+
Turanose	−	−	−	+	+	NI	NI
Succinic Acid Mono-Methyl-Ester	NI	+	+	+	+	+	+
Cis-Aconitic Acid	+	+	+	+	+	+	−
D-Gluconic Acid	−	−	−	−	NI	−	−
α-Hidroxybutyric Acid	−	−	NI	+	+	+	+
β-Hidroxybutyric Acid	−	−	−	−	NI	NI	NI
α-Keto Butyric Acid	−	+	+	+	+	+	+
D,L-Lactic Acid	−	−	NI	+	+	+	+
Malonic Acid	NI	NI	+	+	+	+	+
Propionic Acid	−	−	+	+	+	−	+
D-Saccharic Acid	−	−	−	−	−	−	+
Succinamic Acid	NI	+	+	+	+	+	NI
L-Alaninamide	+	+	+	+	+	+	NI
D-Alanine	−	−	+	+	+	+	−
L-Alanine	NI	−	+	+	+	+	NI

Table 3. *Cont.*

Strains/Additive [a]	*Xcc* 306	*Xcc* 62	*Xcc* 12879	*Xcc* Iran2	*Xcc* Iran10	*Xec* F1	*Xc* 1609
L-Alanyl-Glicine	+	+	+	+	+	+	−
L-Asparagine	−	−	NI	−	NI	−	−
L-Aspartic Acid	−	−	NI	+	+	−	NI
Glycyl-L-Aspartic Acid	−	−	+	+	+	−	−
Glycyl-L-Glutamic Acid	+	+	+	+	+	+	NI
Hydroxy-L-Proline	−	−	−	−	NI	−	NI
Urocanic Acid	−	−	−	−	NI	−	−
Uridine	−	−	−	−	+	−	+
D,L-α-Glycerol Phosphate	−	−	NI	+	+	+	−
α-D-Glucose-1-Phosphate	−	NI	−	+	+	NI	−
D-Glucose-6-Phosphate	−	−	−	+	+	+	−

Wells with ≥160% of activity at 24 h compared to the blank were considered positive (+) [a] and ≤130% of activity considered negative (−) [b]. Values from 129% to 159% were considered non-informative and dropped from further analysis (NI) [c].

Tween 40, N-acetyl-D-glucosamine, D-cellobiose, D-fructose, D-galactose, gentiobiose, α-D-glucose, maltose, D-mannose, D-psicose, D-trehalose, pyruvic acid methyl-ester, α-keto glutaric acid, succinic acid, bromosuccinic acid, L-glutamic acid, L-proline, L-serine, L-threonine, and glycerol were used by all strains tested. The compounds not metabolized by any of the strains were α-cyclodextrin, N-acetyl-D-galactosamine, adonitol, m-inositol, D-mannitol, β-methyl-D-glucoside, L-rhamnose, D-sorbitol, xylitol, citric acid, formic acid, D-galactonic acid lactone, D-galacturonic acid, D-glucosaminic acid, D-glucuronic acid, γ-hidroxybutyric acid, ρ-hydroxy phenylacetic acid, itaconic acid, quinic acid, sebacic acid, glucuronamide, L-histidine, L-leucine, L-ornithine, L-phenylalanine, L-pyroglutamic acid, D-serine, D,L-carnitine, γ-amino butyr acid, inosine, thymidine, phenyethyl-amine, and 2-aminoethanol. Compared with *Xc*, citrus strains used cis-aconitic acid and L-alanyl-glycine. *Xc* specifically used Tween 80, D-saccharic acid, and uridine. Among the citrus strains, *Xcc* A, A*, and A^W strains used sucrose. *Xcc* 306 was atypical compared with all other *Xcc* A strains in that no activity was detected for dextrin, L-fucose, lactulose, and α-keto butyric acid; meanwhile, *Xcc* A* Iran10 was the only strain that responded to uridine. Glycyl-L-aspartic acid, propionic acid, D-alanine, and L-alanine were used by A* and A^W but not by wide-host-range A strains. *Xcc* A* strains were the only one that responded to α-D-lactose, turanose, and L-aspartic acid. In addition, *Xcc* A*, as did *Xec* and *Xc*, used D-melobiose, α-hydroxybutiric acid, and D,L-lactic acid.

To study the overall relatedness of the metabolic response among the xanthomonads evaluated, cluster analysis was performed by transforming the data from carbon source use to binary form (uninformative carbon sources were dropped from the analysis). The analysis demonstrated that citrus strains were grouped in the same cluster and separated from *Xc*. Moreover, *Xcc* A strains were clustered according to the host range, i.e., separated from strains *Xcc*, A*, A^W, and *Xec* (Figure 1A), and the two *Xcc* strains showed their diversity.

Because a possible relationship between carbon source use and host range was elucidated, the putative role of chemical compounds in chemotaxis was studied later.

3.2. Chemotactic Response of Xanthomonas Strains to Carbon Compounds

To define the chemotactic profile of *Xanthomonas* strains, a new chemotaxis assay, in which several compounds were concurrently tested with a large number of technical replicates, was developed. In this assay, the quantity of bacteria entering a pipette tip containing the carbon source was used to assess the chemotactic response independently of bacterial growth. This experimental approach has the same principle as the protocols described previously [33,36,37]. This assay was validated with *D. dadantii* 3937, and the chemotactic response obtained matched those previously reported: 10 mM cysteine was repellent, and 10 mM sodium citrate, 10 mM glucose, and 1 and 200 mM serine were attractants [14,34].

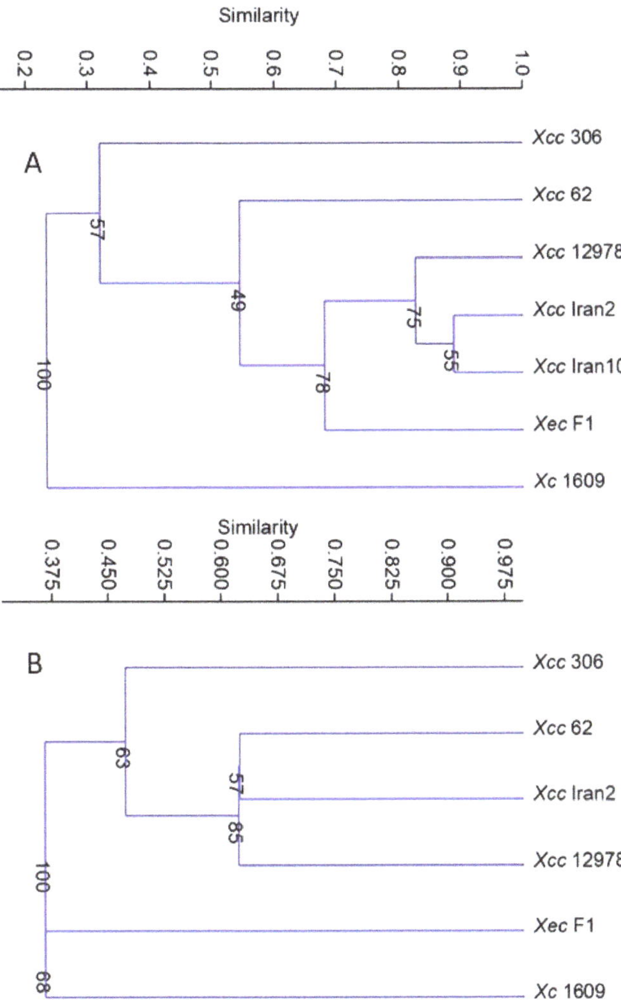

Figure 1. Dendrograms showing relationships among *Xanthomonas* strains based on data from (**A**) Biolog GN2 activity and (**B**) chemotaxis assay.

To determine the chemotactic responses of the *Xanthomonas* studied, 19 compounds were tested (see Table 4); from these chemicals, the metabolic response was determined using Biolog GN2 for 14 of them, and therefore, solely sodium citrate, xylose, arginine, cumaric acid, and cysteine's metabolic responses were not considered.

All *Xanthomonas* strains evaluated responded similarly to 10 mM cysteine as a repellent and 10 mM sucrose, 0.2% glycerol, and 200 mM serine as attractants (Table 4).

Table 4. Chemotactic response of the species and strains of *Xanthomonas* pathogenic on citrus and crucifers.

Additive	*Xcc* 306	*Xcc* 62	*Xcc* Iran2 A*	*Xcc* 12879 A^W	*Xec* F1	*Xc* 1609
Sodium Citrate 10 mM	+ [a]	0 [b]	+	+	0	0
Fructose 10 mM	0	+	0	+	− [b]	0
Galactose 10 mM	0	+	0	+	0	+
Glucose 10 mM	0	0	0	+	−	0
Maltose 10 mM	0	+	+	+	0	+
Sucrose 10 mM	+	+	+	+	+	+
Xylose 10 mM	0	0	0	0	−	−
Arginine 10 mM	0	+	0	0	0	0
Arginine 100 mM	0	+	+	+	+	+
Alanine 10 mM	0	+	+	0	0	−
Alanine 250 mM	+	+	+	+	0	+
Cysteine 10 mM	− [c]	−	−	−	−	−
Leucine 10 mM	+	0	0	0	0	−
Leucine 150 mM	+	+	+	+	−	0
Serine 10 mM	0	0	0	0	−	−
Serine 200 mM	+	+	+	+	+	+
Glycerol 0.2%	+	+	+	+	+	+
Mannitol 0.2%	+	+	+	+	−	0
Galacturonic Acid 10 mM	+	+	0	0	+	0
Glucuronic Acid 10 mM	0	0	0	0	0	+
Citric Acid 10 mM	−	0	0	−	0	0
Succinic Acid 10 mM	0	+	+	0	+	0
Cumaric Acid 10 mM	0	+	0	+	+	0

[a] Chemoattractant (+); [b] no response; [c] chemorepellent (−).

Interestingly, the repellent cysteine has not been detected in the phloem sap of most *Citrus* spp. [38], and sucrose was previously reported as an attractant for other *Xanthomonas* spp. [36,39]. *Xc* differed from citrus strains in that 10 mM alanine and 10 mM leucine acted as repellents and 10 mM glucuronic acid as an attractant. The responses that differentiated *Xcc* strains from *Xec* and *Xc* were 150 mM leucine and 0.2% mannitol as attractants for *Xcc* strains (repellent for *Xec* and no response for *Xc*) and 10 mM xylose and 10 mM serine as repellents for *Xec* and *Xc*, while no response was observed for *Xcc* strains. In addition, among citrus pathogenic strains, *Xec* was the sole strain showing a repellent response toward fructose and glucose, two well-known carbon sources for bacteria; 200 mM alanine did not show any chemotactic effect in *Xec*, while it was an attractant for the *Xcc* strains tested. As previously reported for *Ralstonia solanacearum* strains [40], chemotactic responses varied within *Xcc* strains; *Xcc* 306 and *Xcc* 62 were the only strains attracted to 10 mM galacturonic acid, along with *Xec*. *Xcc* 62 was the only *Xcc* A strain attracted to 10 mM arginine, and *Xcc* 306 was the only showing no response to 100 mM arginine or being attracted by leucine at 10 mM. Cluster analysis based on chemoattraction grouped *Xcc* A strains with the narrow-host-range strains *Xcc* A* Iran2 and *Xcc* A^W 12879 and separated them from *Xec* and *Xc*. Within the *Xcc* subgroup, *Xcc* 62 was more closely related to *Xcc* A* Iran2 and *Xcc* A^W 12879 than to *Xcc* 306 (Figure 1B). Chemotactic responses were more similar for narrow-host-range strains, while the wide-host-range strains responses were variable.

3.3. Identification of MCPs in Xanthomonas Species Used in the Study

The analysis of the complete genomes of different *Xanthomonas* species, pathovars, and pathotypes revealed variants in their MCP profiles. Although 28 different MCPs were found in the genome sequences, the number of MCPs varied from 24 in most of the type A *Xcc* strains to 26 in all A*/A^W *Xcc* and *Xec* FDC1637 strains. An MCP pattern composed of 18 genes was shared by all genomes analyzed; meanwhile, citrus-associated and brassica-associated strains shared 22 and 24 MCPs, respectively. Among those common MCPs, XCV1942, XAC3768, and XAC3271 were only present in citrus-associated xanthomonads and XCC0324 was only found in brassica-associated ones (Figure 2). Results also showed that the MCP content differed among citrus xanthomonads; thereby, XAC3271 was only identified in *Xcc*, but it was not found in *Xec*, and although XCAW2504 and XCAW2508 were detected in all *Xcc* A*/A^w strains, they were found in just one *Xcc* A strain and were not identified in any *Xec*.

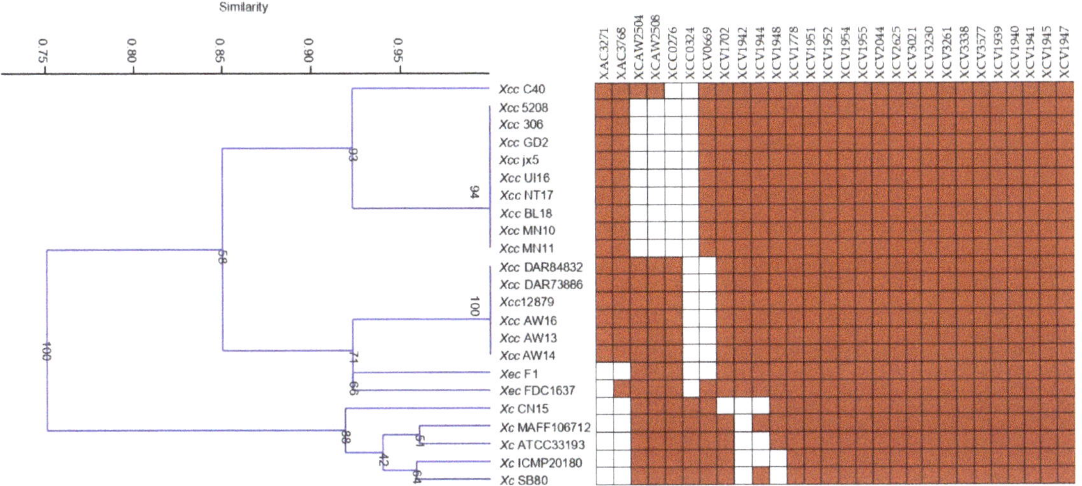

Figure 2. Dendrogram and heat map resulting from MCP identification in 23 completed genome sequences from *Xcc*, *Xce*, and *Xc* strains described in Table 2. Red colour in the heat map means presence of the MCP in the strain.

Cluster analysis of the binary data obtained from MCP analysis revealed major groups according to pathotype and *Xanthomonas* spp. (Figure 2). One cluster included all *Xcc* type A strains separated from A^w/A* that grouped together with *Xec* and more separated from *Xc* (Figure 2).

PCR using primers previously described [19] in addition to those for XCAW2504 and XCAW2508 results confirmed findings from the genomic analysis (Table 5). XCV1942, XAC3768, and XAC3271 were identified in citrus strains but not in *Xc* 1609, and XCC0324 was only found in *Xc* 1609. In addition, some other MCPs were universally distributed in all the strains, in line with genomic results. As well, either A* or A^w *Xcc* strains showed the same MCP/PCR profile; meanwhile, variability among *Xcc* A strains was found in the MCP content (Figure 2, Table 5).

Table 5. PCR amplification of some xanthomonads' MCPs using primers previously described [19] and those designed in this study.

(Red cells in the source indicate presence of the MCP; marked ■ below, blank = absence.)

Strains/Primers		XAC3271	XAC3768	XCCAW2504	XCCAW2508	XCV1702	XCV1778	XCV1942	XCV1944	XCV1947	XCV1951	XCC0324
CBC [a] A type	*Xcc* 306	■	■			■	■	■	■	■	■	
	Xcc 62	■	■	■	■	■	■	■	■	■	■	
CBC A^w type	*Xcc* 12879	■	■	■	■	■	■	■	■	■	■	
CBC A* type	*Xcc* Iran2	■	■	■	■	■	■	■	■	■	■	
	Xcc Iran10	■	■	■	■	■	■	■	■	■	■	
CBS [b]	*Xec* F1			■	■	■	■	■	■	■	■	
CBR [c]	*Xc* 1609					■	■			■	■	■

[a] CBC: citrus bacterial canker, [b] CBS: citrus bacterial spot, [c] CBR: crucifer black rot. Red colour in the heat map means presence of the MCP in the strain

The difference in the presence of specific MCPs was related to the host (citrus vs. crucifer) and the citrus pathogenic species (*Xec* vs. *Xcc* strains); moreover, the minor differences revealed within the *Xcc* A strains were in concordance with their different chemotactic responses to carbon compounds.

3.4. Xanthomonas Strains Are Attracted by Leaf Apoplastic Fluids

To confirm the role of chemotaxis at an early stage of leaf infection, chemotaxis of the different strains was assessed in response to apoplastic fluids from sweet orange (*C. sinensis*) var. 'Valencia Late', Mexican lime (*Citrus aurantifolia*), and Chinese cabbage (*Brassica pekinensis*) var. Kasumi. Our results showed that all apoplastic fluids act as chemoattractants (Table 6). Both cabbage and citrus apoplastic fluids were attractive for all *Xanthomonas* strains.

Table 6. Chemotactic responses toward different concentrations of several apoplastic fluids of *Xanthomonas* pathogenic to citrus and crucifers[a].

Strain/Concentration (mg mL^{-1})		3.12	6.25	12.5	25	50	100	200
Xcc 306 A		+[a]	+	+	+	+	+	+
Xcc 62 A		+	+	+	+	+	+	+
Xcc 12879 A^w	Sweet	+	0	+	+	+	+	+
Xcc Iran2 A*	orange	+	+	+	+	+	+	+
Xec F1		+	+	+	+	+	+	+
Xc 1609		0[b]	+	+	+	+	+	+
Xcc 306 A		0	0	0	0	0	0	+
Xcc 62 A		+	+	+	+	+	+	+
Xcc 12879 A^w	Mexican	0	+	+	+	+	+	+
Xcc Iran2 A*	lime	+	+	+	+	+	+	+
Xec F1		0	0	0	+	+	+	+
Xc 1609		+	+	+	+	+	+	+
Xcc 306 A		0	+	+	+	+	+	+
Xcc 62 A		+	+	+	+	+	+	+
Xcc 12879 A^w	Chinese	+	+	+	+	+	+	+
Xcc Iran2 A*	cabbage	0	+	+	+	+	+	+
Xec F1		0	0	0	+	+	+	+
Xc 1609		0	0	+	+	+	+	+

The apoplastic fluid concentration was considered a chemoattractant (+) [a] when the average number of bacteria that entered the tip in six replicates from at least two assays was significantly ($p < 0.05$) higher compared to the water control. When $p > 0.05$, no response was considered (0) [b].

Nevertheless, the response differed among strains: *Xcc* A 306 was more responsive to sweet orange, *Xec* F1 to Mexican lime, and *Xc* 1609 to Chinese cabbage, indicating a clear difference in the response between citrus and crucifer strains (Figure 3). Moreover, although these strains weakly responded to the lowest concentrations of apoplast fluids from these species (Table 6), their chemoattractive response increased markedly with the apoplastic fluid concentration. The same occurred for the interaction between *Xcc* A 306 and Mexican lime (Figure 3).

To evaluate more precisely the differences among the strains on the different hosts, the variation of the chemotactic response related to the apoplast concentration increase was analyzed. The chemotactic derivative curves in Figure 4 show how the chemotactic response changed as the apoplastic fluid concentration increased.

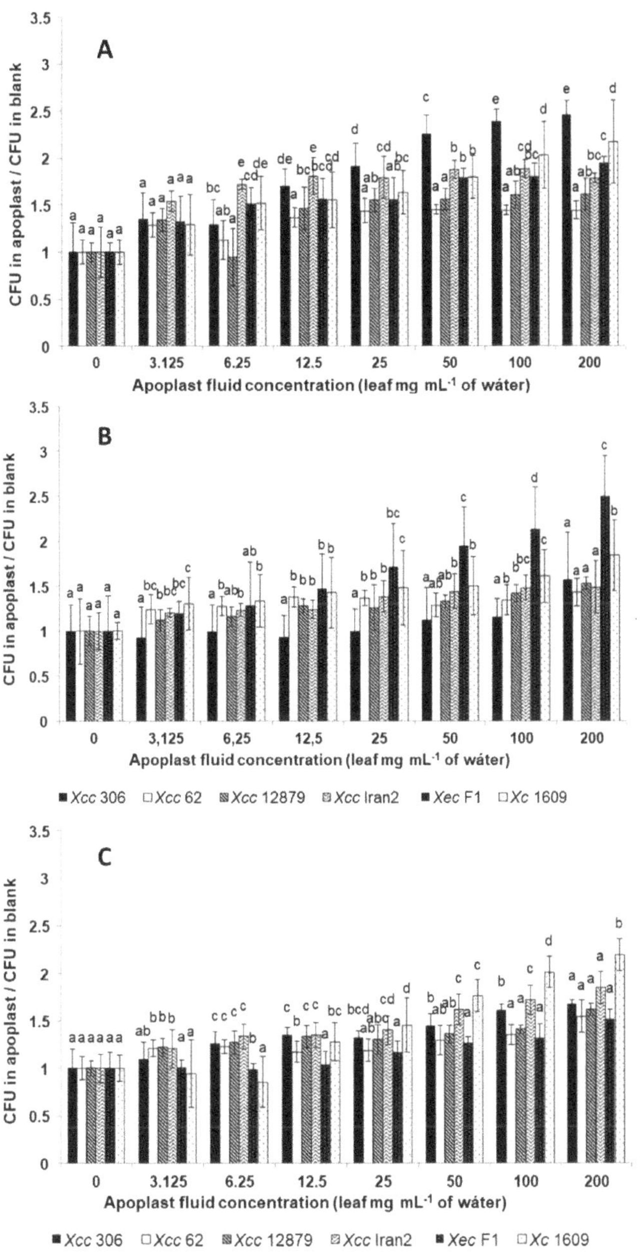

Figure 3. Comparison of the chemotactic response of different *Xanthomonas* strains to apoplastic fluid extracts from leaves of (**A**) sweet orange, (**B**) Mexican lime and (**C**) Chinese cabbage. The graphs show the relative number of bacteria entering the tip in the presence of an apoplastic fluid at different concentrations. The graph shows the mean with the standard deviation (error bars). Means with the same letter within a sample do not differ significantly ($p < 0.05$).

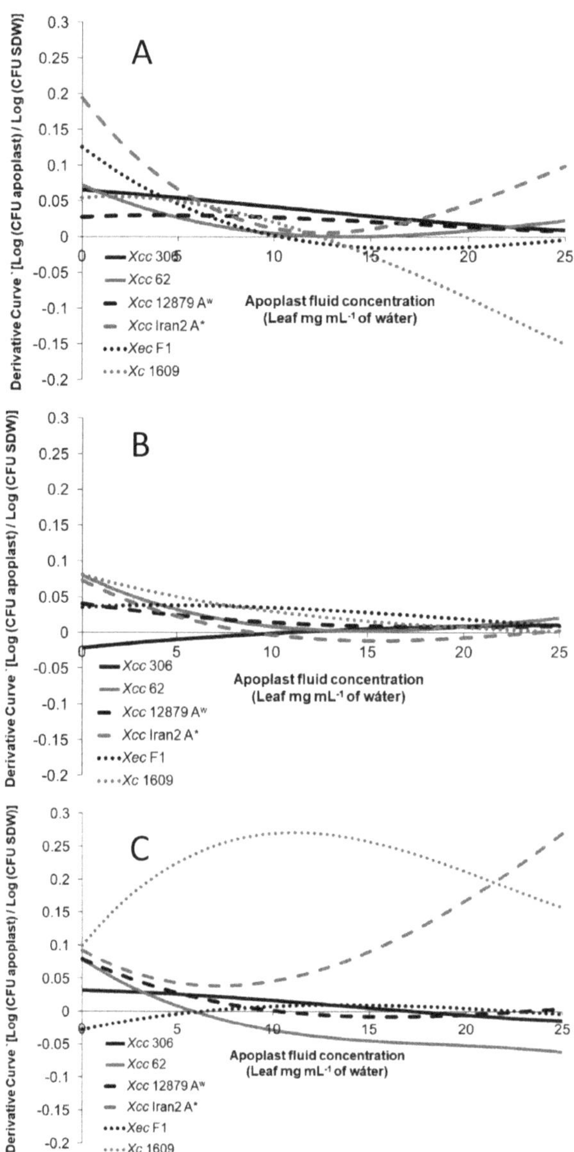

Figure 4. Variation of the chemotaxis response of *Xanthomonas* strains toward leaf apoplastic fluids at different concentrations (from 0 to 25 leaf mg mL^{-1}) from (**A**) sweet orange, (**B**) Mexican lime, and (**C**) Chinese cabbage. The graphs show the derivative curve of the regression curve obtained from data of the chemotaxis assay explained before. The chemotactic responses of the xanthomonads strains fitted onto polynomial regression curves ($r^2 > 0.8$) were derived and the curves plotted. Herein, apoplastic fluid concentrations from 0 to 25 mg mL^{-1} were selected because almost no variation ($p > 0.05$) was observed at higher concentrations, due to a possible saturation of chemoreceptors.

The chemotactic responses of the *Xanthomonas* strains tested toward citrus apoplastic fluids (Figure 4A,B) showed higher response changes at low concentrations for most of the strains. Usually, the chemoattractive response diminished or even declined as the apoplastic fluid concentration increased. However, it is important to note that this reduction in the chemoattractive response does not mean a negative response (chemorepellent) but fewer bacterial cells entering the tip with apoplastic concentration increments.

Citrus pathogenic strains' response toward the Chinese cabbage apoplastic fluid was constant or even negative when the concentration increased, with the exception of *Xcc* Iran2 A* (Figure 4C). The same behavior was observed in *Xc* 1609 toward most citrus apoplastic fluids. This result suggests that on a non-host-plant leaf surface, the xanthomonad chemotactic response would not be as efficient as the pathogen approaches the stomata.

The orange leaf apoplastic fluid produced the most variable response among the *Xanthomonas* strains tested (Figure 4A). The highest variation of the response associated with the concentration was observed for *Xcc* 306, presenting *Xec* F1 and *Xcc* Iran2, A* an intermediate phenotype; meanwhile, the lowest variation was found in *Xcc* 62 and *Xcc* A^W 12879. Moreover, the response of *Xcc* Iran2 A* showed a reduction in the variation at concentrations over 6.25 mg mL^{-1}.

No differences in the chemotactic response toward Mexican lime was observed among *Xcc* 62, *Xcc* A^W 12879, and *Xcc* Iran2 A* (Figure 4B). However, less reaction was observed for *Xcc* 306, although the response increased with the apoplastic fluid concentration.

Xc 1609 was highly responsive toward the Chinese cabbage apoplastic fluid compared with citrus *Xanthomonas* (Figure 4C) and less reactive to citrus apoplastic fluids (Figure 4A,B).

4. Discussion

Chemotaxis plays a key and early role in bacterial attachment, biofilm development, and bacterial regulation in response to the environment [21,41–44]. Moreover, in previous studies, chemotaxis in *Xcc* has been described as a central plant colonization factor at the early stages of the microbe–plant interaction [44,45]. In addition, biofilm formation has been reported as an important step for citrus canker establishment and for *Xcc* to survive on the plant surface [25,46]. In addition, the ability of xanthomonads to form biofilm on citrus has been associated with the host range [26].

To look for the possible link between the chemotactic response and the xanthomonad host range, the metabolic activity on carbon sources was compared to the chemotactic response as well as to the MCP content on *Xcc* pathotypes, *Xec* and *Xc*. The study first showed that, interestingly, CBC wide-host-range strains were able to metabolize fewer metabolites than the narrow-host-range strains. This low ability to metabolize carbon compounds may involve a restriction regarding the environment in which bacteria can multiply and, for instance, a stronger need to colonize the apoplastic space to meet nutritional requirements not available on the leaf surface. This niche restriction might make wide-host-range strains evolve different strategies, such as chemotaxis and virulence factors, to colonize the citrus host interior in order to get access to their nutritional requests. On the contrary, narrow-host-range strains, with higher metabolic capacity, would not require all the same abilities. In previous works by our group, differences in biofilm formation and swimming motility were shown between narrow- and wide-host-range strains of *Xcc*, and this may be related to their different nutritional requests [26].

Our results showed variable chemotaxis responses among the *Xanthomonas* strains tested according to their host range and similar clustering from either their overall metabolic activity or chemotaxis toward chemical compounds, noting that chemotaxis responses in xanthomonads described here might be metabolism dependent in response to effectors addressed to alter energy metabolism or increase intracellular energy. Further analysis is needed to determine the specific role in chemotaxis of particular compounds, their putative synergistic effects, and the impact of their relative concentrations.

Xcc, as many other *Xanthomonas* strains, goes through an epiphytic phase from leaf-deposition until reaching the apoplast [20,31]. During this stage, bacterial sensors, such as MCPs, among others, teach and guide the bacteria where they are and where to go. Herein, the MCP content of *Xcc*, *Xec*, and *Xc* was determined based on the data of available genomes and, besides, partially confirmed by PCR in strains used in the study. Cluster analysis based on the MCP content of *Xcc*, *Xec*, and *Xc* profiles from the complete genome resulted in groups according to the strain host and therefore also according to carbon source use and chemotactic profile toward chemicals. Moreover, differences in the MCP profile among closely related strains, such as *Xcc* pathotypes, were elucidated, resulting in different groups according to the host range in cluster analysis. All the dendrograms from metabolic activity, chemotactic response, or MCP content showed a clear difference between the citrus pathogenic strains and the crucifer black rot strain, as well as changes between wide- and narrow-host-range *Xcc* strains.

The *Xanthomonas* response toward apoplastic fluids of strains with different MCP profiles showed a unique response toward sweet orange, Mexican lime, and Chinese cabbage leaf apoplastic fluids according to the host range of the xanthomonads evaluated. Strains *Xcc* 62, *Xcc* 12879, and *Xcc* Iran2, which responded similarly to apoplastic fluids, showed more similar MCP profiles based on PCR results. It should be noted that strain *Xcc* 62, closely related to *Xcc* 306, presented a chemotactic profile on PCR closer to narrow-host-range CBC strains than to *Xcc* 306. However, strains *Xcc* 62, *Xcc* 12879, and *Xcc* Iran2, even showing the same MCP content on the PCR profile, presented a variable chemotaxis response. This apparent incongruence may be because the limitation of MCP analysis with PCR that was not able to entirely determine the MCP content in these strains due to variability in the PCR primer target sequence or because their chemotaxis may be mediated by several mechanisms besides MCPs, which are variable among CBC strains with different host ranges [47–49]. However, differences in the MCP content among A pathotype strains was supported by the results of genomic analysis performed here, which showed "atypical" profiles in some strains within *Xcc* type A.

Our results suggest that apoplastic fluids, exuding from stomatal openings or leaf wounds, are likely to act as a whole or contain specific chemotactic signals, currently not identified, that would determine the behavior of the pathogen on the leaf surface. These results, along with those from other authors [16,19,21,44], support the role of chemotaxis in the plant–bacteria interaction and the *Xanthomonas* host range. Moreover, our findings are consistent with those in other models, such as *R. solanacearum*, which is more attracted to host root exudates than to non-host exudates [40], or *X. oryzae*, which is attracted toward root exudates based on the susceptibility of the rice cultivar [36]. Our study confirms that apoplastic fluids exuding from stomatal openings or leaf wounds would be detected by the bacteria and that they will trigger a host-dependent chemotactic response leading these xanthomonads toward the host entrances. The apoplastic fluid from the substomatal cavity might be diluted by the natural humidity on the leaf, especially after a raining event that transports the bacteria from one tree to the other, facilitating the bacteria–apoplastic fluid interaction. On the leaf surface, the bacteria would move toward increasing concentrations of the apoplastic fluid as they approach the stomata or wounds. However, at high apoplastic fluid concentrations, as occurring within the apoplast, motility is no longer needed and biofilm formation and effector secretion into plant cells are prompted [26,44,46,50].

To conclude, our work supports the links between the host range of citrus pathogenic *Xanthomonas* strains, the use of carbon sources, and the chemotaxis response to these carbon sources or host leaf apoplastic fluids. Our results indicate a role of leaf apoplastic exudates in chemotaxis and their involvement in the early stages of bacterial infection and host range processes. However, further investigation is needed to determine specific components of the apoplast that underline the chemotaxis mechanism in citrus species or the role of different environmental sensors, including MCPs, in it.

Author Contributions: Conceptualization, M.S.-V., J.C. and J.H.G.; methodology, M.S.-V., J.C., E.F. and C.R.; validation, J.C. and J.H.G.; formal analysis, M.S.-V. and J.C.; draft preparation, M.S.-V.; writing—review and editing, M.S.-V. and J.C.; supervision, J.C.; project administration, J.C.; funding acquisition, J.H.G. and J.C. All authors have read and agreed to the published version of the manuscript.

Funding: This work was kindly supported by the Citrus Advance Technology Program, project CRDF546, as well as through the Instituto Nacional de Investigación y Tecnología Agraria y Alimentaria (INIA), project RTA2008-00048. Marta Sena-Vélez held a PhD fellowship from INIA.

Institutional Review Board Statement: Not applicable.

Informed Consent Statement: Not applicable.

Data Availability Statement: Not applicable.

Acknowledgments: We also thank María Milagros López, Rui Pereira Leite, John Hartung, and Pablo Rodríguez-Palenzuela for supplying some of the bacterial strains and Carmen Martínez and Ana Redondo for supplying some of the plant material used in the study.

Conflicts of Interest: The authors declare no conflict of interest.

References

1. Gottwald, T.R.; Hughes, G.; Graham, J.H.; Sun, X.; Riley, T. The Citrus Canker Epidemic in Florida: The Scientific Basis of Regulatory Eradication Policy for an Invasive Species. *Phytopathology* **2001**, *91*, 30–34. [CrossRef] [PubMed]
2. Graham, J.H.; Gottwald, T.R.; Cubero, J.; Achor, D.S. *Xanthomonas axonopodis* pv. *citri*: Factors affecting successful eradication of citrus canker. *Mol. Plant Pathol.* **2004**, *5*, 1–15. [CrossRef] [PubMed]
3. Licciardello, G.; Caruso, P.; Bella, P.; Boyer, C.; Smith, M.W.; Pruvost, O.; Robene, I.; Cubero, J.; Catara, V. Pathotyping Citrus Ornamental Relatives with *Xanthomonas citri* pv. citri and X. *citri* pv. aurantifolii Refines Our Understanding of Their Susceptibility to These Pathogens. *Microorganims* **2022**, *10*, 986. [CrossRef]
4. Vauterin, L.; Hoste, B.; Kersters, K.; Swings, J. Reclassification of *Xanthomonas*. *Int. J. Syst. Bacteriol.* **1995**, *45*, 472–489. [CrossRef]
5. Vauterin, L.; Rademaker, J.; Swings, J. Synopsis on the taxonomy of the genus *Xanthomonas*. *Phytopathology* **2000**, *90*, 677–682. [CrossRef]
6. Schaad, N.W.; Postnikova, E.; Lacy, G.; Sechler, A.; Agarkova, I.; Stromberg, P.E.; Stromberg, V.K.; Vidaver, A.K. Emended classification of xanthomonad pathogens on citrus. *Syst. Appl. Microbiol.* **2006**, *29*, 690–695. [CrossRef] [PubMed]
7. Schaad, N.W.; Postnikova, E.; Lacy, G.H.; Sechler, A.; Agarkova, I.; Stromberg, P.E.; Stromberg, V.K.; Vidaver, A.K. Reclassification of *Xanthomonas campestris* pv. *citri* (ex Hasse 1915) Dye 1978 forms A, B/C/D, and E as X. *smithii* subsp. *citri* (ex Hasse) sp. nov. nom. rev. comb. nov., X. *fuscans* subsp. *aurantifolii* (ex Gabriel 1989)) sp. nov. nom. r. nom. rev. comb. no. comb. nov., and X. alfalfae subsp. citrumelo (ex Riker and Jones) Gabriel et al., 1989 sp. nov. nom. r. nom. rev. comb. no. comb. nov.; X. *campestris* pv. *malvacearum* (ex Smith 1901) Dye 1978 as X. *smithii* subsp. *smithii* nov. comb. nov. nom. nov.; X. *campestris* pv. *alfalfae* (ex Riker and Jones, 1935) Dye 1978 as X. *alfalfae* subsp. *alfalfae* (ex Riker et al., 1935) sp. nov. nom. r. nom. rev.; and "var. fuscans' of X. *campestris* pv. *phaseoli* (ex Smith, 1987) Dye 1978 as X. *fuscans* subsp. *fuscans* sp. nov. *Syst. Appl. Microbiol.* **2005**, *28*, 494–518. [CrossRef]
8. Sun, X.; Stall, R.E.; Jones, J.B.; Cubero, J.; Gottwald, T.R.; Graham, J.H.; Dixon, W.N.; Schubert, T.S.; Chaloux, P.H.; Stromberg, V.K.; et al. Detection and Characterization of a New Strain of Citrus Canker Bacteria from Key/Mexican Lime and Alemow in South Florida. *Plant Dis.* **2004**, *88*, 1179–1188. [CrossRef]
9. Vernière, C.; Hartung, J.; Pruvost, O.; Civerolo, E.; Alvarez, A.; Maestri, P.; Luisetti, J. Characterization of phenotypically distinct strains of *Xanthomonas axonopodis* pv. *citri* from Southwest Asia. *Eur. J. Plant Pathol.* **1998**, *104*, 477–487. [CrossRef]
10. Pruvost, O.; Goodarzi, T.; Boyer, K.; Soltaninejad, H.; Escalon, A.; Alavi, S.M.; Javegny, S.; Boyer, C.; Cottyn, B.; Gagnevin, L.; et al. Genetic structure analysis of strains causing citrus canker in Iran reveals the presence of two different lineages of *Xanthomonas citri* pv. citri pathotype A*. *Plant Pathol.* **2015**, *64*, 776–784. [CrossRef]
11. Gordon, J.L.; Lefeuvre, P.; Escalon, A.; Barbe, V.; Cruveiller, S.; Gagnevin, L.; Pruvost, O. Comparative genomics of 43 strains of *Xanthomonas citri* pv. *citri* reveals the evolutionary events giving rise to pathotypes with different host ranges. *BMC Genomics* **2015**, *16*, 1098. [CrossRef] [PubMed]
12. Adler, J. Chemotaxis in bacteria. *J. Supramol. Struct.* **1976**, *4*, 305–317. [CrossRef] [PubMed]
13. Río-Álvarez, I.; Muñoz-Gómez, C.; Navas-Vásquez, M.; Martínez-García, P.M.; Antúnez-Lamas, M.; Rodríguez-Palenzuela, P.; López-Solanilla, E. Role of *Dickeya dadantii* 3937 chemoreceptors in the entry to Arabidopsis leaves through wounds. *Mol. Plant Pathol.* **2015**, *16*, 685–698. [CrossRef] [PubMed]
14. Antunez-Lamas, M.; Cabrera, E.; Lopez-Solanilla, E.; Solano, R.; González-Melendi, P.; Chico, J.M.; Toth, I.; Birch, P.; Pritchard, L.; Liu, H.; et al. Bacterial chemoattraction towards jasmonate plays a role in the entry of *Dickeya dadantii* through wounded tissues. *Mol. Microbiol.* **2009**, *74*, 662–671. [CrossRef] [PubMed]
15. Yu, X.; Lund, S.P.; Scott, R.A.; Greenwald, J.W.; Records, A.H.; Nettleton, D.; Lindow, S.E.; Gross, D.C.; Beattie, G.A. Transcriptional responses of *Pseudomonas syringae* to growth in epiphytic versus apoplastic leaf sites. *Proc. Natl. Acad. Sci. USA* **2013**, *110*, E425–E434. [CrossRef]
16. Indiana, A. Rôles du Chimiotactisme et de la Mobilité Flagellaire dans la Fitness des *Xanthomonas*. Ph.D. Thesis, Université d'Angers, Angers, France, 2014.

17. Lacal, J.; García-Fontana, C.; Muñoz-Martínez, F.; Ramos, J.-L.; Krell, T. Sensing of environmental signals: Classification of chemoreceptors according to the size of their ligand binding regions. *Environ. Microbiol.* **2010**, *12*, 2873–2884. [CrossRef]
18. Ud-Din, A.I.M.S.; Roujeinikova, A. Methyl-accepting chemotaxis proteins: A core sensing element in prokaryotes and archaea. *Cell Mol. Life Sci.* **2017**, *74*, 3293–3303. [CrossRef]
19. Mhedbi-Hajri, N.; Darrasse, A.; Pigné, S.; Durand, K.; Fouteau, S.; Barbe, V.; Manceau, C.; Lemaire, C.; Jacques, M.-A. Sensing and adhesion are adaptive functions in the plant pathogenic xanthomonads. *BMC Evol. Biol.* **2011**, *11*, 67. [CrossRef]
20. Garita-Cambronero, J.; Sena-Vélez, M.; Ferragud, E.; Sabuquillo, P.; Redondo, C.; Cubero, J. *Xanthomonas citri* subsp. citri and *Xanthomonas arboricola* pv. *pruni*: Comparative analysis of two pathogens producing similar symptoms in different host plants. Alcaraz LD, editor. *PLoS ONE* **2019**, *14*, e0219797. [CrossRef]
21. Yaryura, P.M.; Conforte, V.P.; Malamud, F.; Roeschlin, R.; de Pino, V.; Castagnaro, A.P.; McCarthy, Y.; Dow, J.M.; Marano, M.R.; Vojnov, A.A. XbmR, a new transcription factor involved in the regulation of chemotaxis, biofilm formation and virulence in *Xanthomonas citri* subsp. *citri*. *Environ. Microbiol.* **2015**, *17*, 4164–4176. [CrossRef]
22. Graham, J.H. Susceptibility of Citrus Fruit to Bacterial Spot and Citrus Canker. *Phytopathology* **1992**, *82*, 452–457. [CrossRef]
23. Bock, C.H.; Graham, J.H.; Gottwald, T.R.; Cook, A.Z.; Parker, P.E. Wind speed and wind-associated leaf injury affect severity of citrus canker on Swingle citrumelo. *Eur. J. Plant Pathol.* **2010**, *128*, 21–38. [CrossRef]
24. Ference, C.M.; Gochez, A.M.; Behlau, F.; Wang, N.; Graham, J.H.; Jones, J.B. Recent advances in the understanding of *Xanthomonas citri* ssp. *citri* pathogenesis and *citrus* canker disease management. *Mol. Plant Pathol.* **2018**, *19*, 1302–1318. [CrossRef] [PubMed]
25. Cubero, J.; Gell, I.; Johnson, E.G.; Redondo, A.; Graham, J.H. Unstable green fluorescent protein for study of *Xanthomonas citri* subsp. *citri* survival on *citrus*. *Plant Pathol.* **2011**, *60*, 977–985. [CrossRef]
26. Sena-Vélez, M.; Redondo, C.; Gell, I.; Ferragud, E.; Johnson, E.; Graham, J.H.; Cubero, J. Biofilm formation and motility of *Xanthomonas* strains with different citrus host range. *Plant Pathol.* **2015**, *64*, 767–775. [CrossRef]
27. Al-Saadi, A.; Reddy, J.D.; Duan, Y.P.; Brunings, A.M.; Yuan, Q.; Gabriel, D.W. All Five Host-Range Variants of *Xanthomonas citri* Carry One pthA Homolog With 17.5 Repeats That Determines Pathogenicity on Citrus, but None Determine Host-Range Variation. *Mol. Plant-Microbe Interact.* **2007**, *20*, 934–943. [CrossRef]
28. Rybak, M.; Minsavage, G.V.; Stall, R.E.; Jones, J.B. Identification of *Xanthomonas citri* ssp. *citri* host specificity genes in a heterologous expression host. *Mol. Plant Pathol.* **2009**, *10*, 249–262. [CrossRef]
29. Escalon, A.; Javegny, S.; Vernière, C.; Noël, L.D.; Vital, K.; Poussier, S.; Hajri, A.; Boureau, T.; Pruvost, O.; Arlat, M.; et al. Variations in type III effector repertoires, pathological phenotypes and host range of *Xanthomonas citri* pv. *citri* pathotypes. *Mol. Plant Pathol.* **2013**, *14*, 483–496. [CrossRef]
30. Teper, D.; Xu, J.; Pandey, S.S.; Wang, N. PthAW1, a Transcription Activator-Like Effector of *Xanthomonas citri* subsp. citri, Promotes Host-Specific Immune Responses. *Mol. Plant-Microbe Interact.* **2021**, *34*, 1033–1047. [CrossRef]
31. An, S.-Q.; Potnis, N.; Dow, M.; Vorhölter, F.-J.; He, Y.-Q.; Becker, A.; Teper, D.; Li, Y.; Wang, N.; Bleris, L.; et al. Mechanistic insights into host adaptation, virulence and epidemiology of the phytopathogen *Xanthomonas*. *FEMS Microbiol. Rev.* **2019**, *44*, 1–32. [CrossRef]
32. Hammer, Ø.; Harper, D.A.T.; Ryan, P.D.; Ryan, D.D.; Ryan, P.D. Past: Paleontological Statistics Software Package for Education and Data Analysis. *Palaeontologia Electron* **2001**, *4*, 9.
33. Han, G.; Cooney, J.J. A modified capillary assay for chemotaxis. *J. Ind. Microbiol.* **1993**, *12*, 396–398. [CrossRef]
34. Antúnez-Lamas, M.; Cabrera-Ordóñez, E.; López-Solanilla, E.; Raposo, R.; Trelles-Salazar, O.; Rodríguez-Moreno, A.; Rodríguez-Palenzuela, P. Role of motility and chemotaxis in the pathogenesis of *Dickeya dadantii* 3937 (ex *Erwinia chrysanthemi* 3937). *Microbiology* **2009**, *155*, 434–442. [CrossRef] [PubMed]
35. Rico, A.; Preston, G.M. *Pseudomonas syringae* pv. *tomato* DC3000 Uses Constitutive and Apoplast-Induced Nutrient Assimilation Pathways to Catabolize Nutrients That Are Abundant in the Tomato Apoplast. *Mol. Plant Microbe Interact.* **2008**, *21*, 269–282. [CrossRef] [PubMed]
36. Feng, T.-Y.; Kuo, T.-T. Bacterial leaf blight of rice plant. VI. Chemotactic responses of *Xanthomonas oryzae* to water droplets exudated from water pores on the leaf of rice plants. *Bot. Bull. Acad. Sin.* **1975**, *16*, 126–136.
37. Palleroni, N.J. Chamber for Bacterial Chemotaxis Experiments. *Appl. Environ. Microbiol.* **1976**, *32*, 729–730. [CrossRef]
38. Killiny, N. Metabolite signature of the phloem sap of fourteen citrus varieties with different degrees of tolerance to *Candidatus* Liberibacter asiaticus. *Physiol. Mol. Plant Pathol.* **2017**, *97*, 20–29. [CrossRef]
39. Kamoun, S.; Kado, C.I. Phenotypic Switching Affecting Chemotaxis, Xanthan Production, and Virulence in *Xanthomonas campestris*. *Appl. Environ. Microbiol.* **1990**, *56*, 3855–3860. [CrossRef]
40. Yao, J.; Allen, C. Chemotaxis Is Required for Virulence and Competitive Fitness of the Bacterial Wilt Pathogen *Ralstonia solanacearum*. *J. Bacteriol.* **2006**, *188*, 3697–3708. [CrossRef]
41. Merritt, P.M.; Danhorn, T.; Fuqua, C. Motility and Chemotaxis in *Agrobacterium tumefaciens* Surface Attachment and Biofilm Formation. *J. Bacteriol.* **2007**, *189*, 8005–8014. [CrossRef]
42. Schmidt, J.; Müsken, M.; Becker, T.; Magnowska, Z.; Bertinetti, D.; Möller, S.; Zimmermann, B.; Herberg, F.W.; Jänsch, L.; Häussler, S. The *Pseudomonas aeruginosa* Chemotaxis Methyltransferase CheR1 Impacts on Bacterial Surface Sampling. *PLoS ONE* **2011**, *6*, e18184. [CrossRef] [PubMed]
43. Kostakioti, M.; Hadjifrangiskou, M.; Hultgren, S.J. Bacterial Biofilms: Development, Dispersal, and Therapeutic Strategies in the Dawn of the Postantibiotic Era. *Cold Spring Harb. Perspect. Med.* **2013**, *3*, a010306. [CrossRef] [PubMed]

44. Moreira, L.M.; Facincani, A.P.; Ferreira, C.B.; Ferreira, R.M.; Ferro, M.I.T.; Gozzo, F.C.; De Oliveira, J.C.F.; Ferro, J.A.; Soares, M.R. Chemotactic signal transduction and phosphate metabolism as adaptive strategies during citrus canker induction by *Xanthomonas citri*. *Funct. Integr. Genom.* **2015**, *15*, 197–210. [CrossRef] [PubMed]
45. Shi, Y.; Yang, X.; Ye, X.; Feng, J.; Cheng, T.; Zhou, X.; Liu, D.X.; Xu, L.; Wang, J. The Methyltransferase HemK Regulates the Virulence and Nutrient Utilization of the Phytopathogenic Bacterium *Xanthomonas citri* subsp. *citri*. *Int. J. Mol. Sci.* **2022**, *23*, 3931. [CrossRef] [PubMed]
46. Rigano, L.A.; Siciliano, F.; Enrique, R.; Sendín, L.; Filippone, P.; Torres, P.S.; Qüesta, J.; Dow, J.M.; Castagnaro, A.P.; Vojnov, A.A.; et al. Biofilm Formation, Epiphytic Fitness, and Canker Development in *Xanthomonas axonopodis* pv. *citri*. *Mol. Plant Microbe Interact.* **2007**, *20*, 1222–1230. [CrossRef]
47. Alexandre, G. Coupling metabolism and chemotaxis-dependent behaviours by energy taxis receptors. *Microbiology* **2010**, *156*, 2283–2293. [CrossRef]
48. Alexandre, G.; Zhulin, I.B. More than one way to sense chemicals. *J. Bacteriol.* **2001**, *183*, 4681–4686. [CrossRef]
49. Egbert, M.D.; Barandiaran, X.E.; Di Paolo, E.A. A minimal model of metabolism-based chemotaxis. *PLoS Comput. Biol.* **2010**, *6*, e1001004. [CrossRef]
50. Yang, T.-C.; Leu, Y.-W.; Chang-Chien, H.-C.; Hu, R.-M. Flagellar Biogenesis of *Xanthomonas campestris* Requires the Alternative Sigma Factors RpoN2 and FliA and Is Temporally Regulated by FlhA, FlhB, and FlgM. *J. Bacteriol.* **2009**, *191*, 2266–2275. [CrossRef]

 *microorganisms*

Article

Gene Overlapping as a Modulator of *Begomovirus* Evolution

Iván Martín-Hernández [1,†] and Israel Pagán [1,2,*]

1   Centro de Biotecnología y Genómica de Plantas UPM-INIA, 28223 Madrid, Spain; imartin@iqfr.csic.es
2   Departamento de Biotecnología—Biología Vegetal, Escuela Técnica Superior de Ingeniería Agronómica, Alimentaria y de Biosistemas, Universidad Politécnica de Madrid, 28045 Madrid, Spain
*   Correspondence: jesusisrael.pagan@upm.es; Tel.: +34-91-067-91-65
†   Current address: Department of Biological Chemical Physics, Rocasolano Institute of Physical Chemistry CSIC, 28006 Madrid, Spain.

Abstract: In RNA viruses, which have high mutation—and fast evolutionary— rates, gene overlapping (i.e., genomic regions that encode more than one protein) is a major factor controlling mutational load and therefore the virus evolvability. Although DNA viruses use host high-fidelity polymerases for their replication, and therefore should have lower mutation rates, it has been shown that some of them have evolutionary rates comparable to those of RNA viruses. Notably, these viruses have large proportions of their genes with at least one overlapping instance. Hence, gene overlapping could be a modulator of virus evolution beyond the RNA world. To test this hypothesis, we use the genus *Begomovirus* of plant viruses as a model. Through comparative genomic approaches, we show that terminal gene overlapping decreases the rate of virus evolution, which is associated with lower frequency of both synonymous and nonsynonymous mutations. In contrast, terminal overlapping has little effect on the pace of virus evolution. Overall, our analyses support a role for gene overlapping in the evolution of begomoviruses and provide novel information on the factors that shape their genetic diversity.

Keywords: overlapping genes; rate of evolution; begomoviruses; ssDNA viruses

Citation: Martín-Hernández, I.; Pagán, I. Gene Overlapping as a Modulator of *Begomovirus* Evolution. *Microorganisms* **2022**, *10*, 366. https://doi.org/10.3390/microorganisms10020366

Academic Editor: Elvira Fiallo-Olivé

Received: 17 December 2021
Accepted: 1 February 2022
Published: 4 February 2022

Publisher's Note: MDPI stays neutral with regard to jurisdictional claims in published maps and institutional affiliations.

1. Introduction

Genomic regions that encode more than one protein, that is, gene overlapping, are commonplace among viruses [1,2]. Such regions have important biological and evolutionary implications. First, they are associated with virus within-host multiplication, between-host transmission, disease severity and strength of host immune response [3–6]. Second, viruses are subjected to strong selection for maintaining smaller genomes because this (i) reduces the chances for deleterious mutations to become fixed in the virus genome, particularly in viruses with high mutation rates; (ii) improves virus fitness due to faster replication; and (iii) optimizes virion formation due to physical limitations imposed by the capsid size [7–9]. Gene overlapping allows increasing the amount of genomic information in viral genomes while controlling for limited capsid space and speeding up the purification of deleterious mutations from the virus population by amplifying their effect, as in overlapping regions these mutations affect more than one gene at the same time [1,9,10].

If gene overlapping is selectively advantageous for viruses, it would be expected to be more frequent: in RNA than in DNA viruses, as the former have (in general) higher mutation rates [11]; in larger than in shorter viral genomes to minimize the chances of deleterious mutations to become fixed [12], and in spherical virions as these generally have smaller inner volumes than other capsid shapes [13]. Although Brades and Linial [9] failed to detect an association between virion shape and frequency of gene overlapping in support of the predictions above, it has been shown that the larger the gene overlapping the greater the reduction in the rate of RNA virus evolution [1], and that gene overlapping appears to be more frequent in DNA viruses, which on average have also larger genome sizes [11], than in RNA viruses [2].

The species of the family *Geminiviridae* of plant viruses are notable exceptions to the virus characteristics associated with gene overlapping. Although geminiviruses are ssDNA viruses, and therefore replicate through high fidelity polymerases [14], and have small genomes, these viruses have a large proportion of their genes with at least one overlapping region [2]. The *Geminiviridae* family is currently divided into nine genera of which the largest one is the genus *Begomovirus*. This is also one of the most numerous genera of plant viruses with more than 400 species [15]. Bipartite begomovirus genomes encode six open reading frames (ORFs): two in the virion strand (AV1 and AV2) and four in the complementary strand (AC1, AC2, AC3 and AC4), with monopartite begomoviruses encoding equivalent proteins with the same names but without the A prefix. The AV1 gene encodes the coat protein (CP), which is essential for genome encapsidation, viral movement and insect transmission. The AV2 gene, which is considered as the pathogenicity gene, is also involved in movement and symptom development, and functions as a suppressor of gene silencing. This ORF is not present in New World bipartite begomoviruses. Viral DNA replication depends on the AC1 gene product (replication initiator protein, Rep). The AC2 gene encodes the transcriptional activator protein (TrAP) that interferes with transcriptional and post-transcriptional gene silencing (TGS and PTGS, respectively), and with the CP expression. The gene encoding for the AC3 protein (replication enhancer protein, REn), enhances viral DNA accumulation, and is involved in interaction with the plant-host retinoblastoma-related (RBR) proteins. Finally, AC4 counteracts PTGS by inhibiting accumulation of siRNA and is considered an important symptom determinant [15]. All of these six ORFs have at least one overlapping region in both mono- and bipartite begomoviruses [15]. Bipartite begomoviruses have two additional non-overlapping ORFs in the B-component: BC1, a movement protein (MP), and BV1, the nuclear shuttle protein (NSP) (Figure 1) [16].

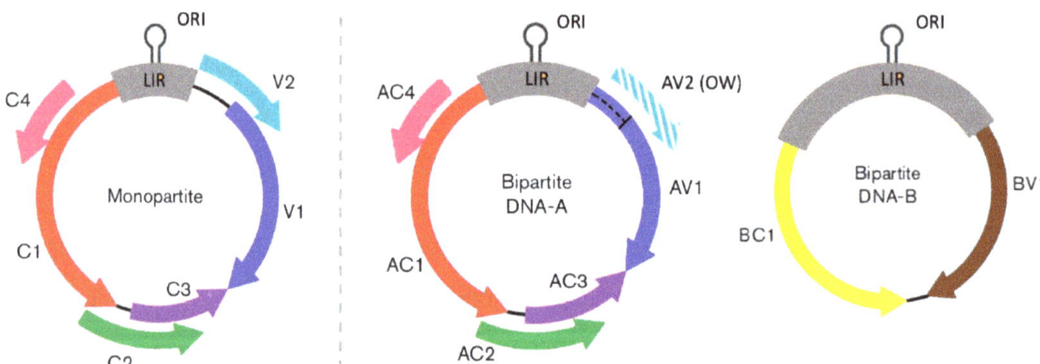

Figure 1. Genome organization of mono- and bipartite begomoviruses. Colored arrows denote the position and orientation of each gene. Monopartite and DNA-A of bipartite begomoviruses encode for: AC1: Replication initiator protein (Rep); AC2: Transcriptional activator protein (TrAP); AC3: Replication enhancer protein (Ren); AC4: Silencing suppressor; AV1: Coat protein (CP); and AV2: Various functions. In bipartite begomoviruses AV2 is only present in Old World (OW) begomoviruses, where AV1 is as long as in monopartite begomoviruses (dashed). DNA-B of bipartite begomoviruses encodes for: BC1: Movement protein (MP) and BC2: Nuclear shuttle protein (NSP). CR, common region. The hairpin which includes the origin of replication (ORI) is indicated in the Long Intergenic Region (LIR) (modified from [17]).

Despite being DNA viruses, begomoviruses have been repeatedly shown to have high evolutionary rates (reviewed by [18]). For instance, *Tomato yellow leaf curl virus* (TYLCV) substitution rate has been estimated to be of 2.88×10^{-4} nucleotide substitutions per site per year, which is in the range of values for RNA viruses [12]. This fast evolutionary rate has been attributed to the effect of oxidative damage in replicated viral genomes, and/or to higher mutation rates than expected for DNA viruses [12]. If so, extensive gene overlapping in begomoviruses may contribute to modulate mutational load, and consequently the rate of virus evolution, as it has been shown for RNA viruses [1]. Experimental evidence supporting this idea is scarce and sometimes contradictory. For instance, a higher variability occurred in the *Tomato yellow leaf curl China virus* (TYLCCNV) AC1-AC4 overlapping (OV) region than in the non-overlapping (NOV) region of AC1 [19], whereas the opposite was observed for *Pepper huasteco yellow vein virus* (PHYVV) [20].

Here, we analyzed the effect of gene overlapping on the rate of begomovirus evolution through comparative genomics and utilizing sequences from 18 species. In particular, we explored whether the following evolutionary parameters vary between OV and NOV regions and among different types of gene overlap: (1) the rate of viral evolution, using overall tree length as a proxy, (2) the frequency of synonymous and nonsynonymous substitutions, (3) selection pressure and (4) magnitude of the effect of gene overlapping in the rate of virus evolution.

2. Materials and Methods

2.1. Sequence Data

Available sequences from begomovirus species were retrieved from GenBank. Sequences from extensively passaged isolates in non-natural hosts were excluded. When possible, we tried to minimize the presence of recombination. Species with more than 10 sequences were retained for analysis, so that we were able to include 18 mono and bipartite begomoviruses, and a total of 8239 sequences. Overall, we analyzed 125 instances of gene overlap ranging between 59 and 423 nt in length: 17 internal overlapping instances, 54 5′-terminal overlapping instances, and 54 3′-terminal overlapping instances. For simplicity, genes are named as for bipartite begomoviruses. Note that we divided sequences from *Bhendi yellow vein mosaic virus* (BYVMV) into two groups: one of sequences originally classified as belonging to this virus, and another originally characterized as *Bhendi yellow vein India virus*. Although both groups are currently considered as belonging to BYVMV [15], we chose to analyze them separately as evolutionary parameters differed between groups (Appendix A). However, analyses merging the two groups did not change our conclusions. We constructed sequence alignments for the 125 overlapping instances, and for the corresponding OV and NOV fragments of each gene. Sequence alignments of the OV regions were adjusted according to the amino acid sequence of each of the two genes involved, thus generating two data sets for each OV region. All alignments were built using MUSCLE 3.7 [21] and adjusted manually according to the amino acid sequences using AliView [22]. Alignments are available as Supplementary Material File S1.

2.2. Estimation of Tree Length

Tree lengths (t) were estimated for the OV and NOV regions of each gene. To do so, we used a maximum likelihood fitting of the General Time Reversible (GTR) nucleotide substitution model as implemented in the HyPhy package [23]. Differences in total tree length between OV and NOV regions were analyzed using a relative ratio test also utilizing HyPhy. Because t is dependent on the number of tree branches (i.e., number of sequences), when values were compared among overlapping instances, t was normalized according to the number of sequences.

2.3. Selection Pressures

Selection pressures for OV and NOV regions were estimated as the difference between the mean number of nonsynonymous (d_N) and synonymous (d_S) nucleotide substitutions per site (d_N/d_S) using the fast unbiased Bayesian approximation (FUBAR), and the fixed effect likelihood (FEL) methods implemented in HyPhy (Appendix A). Because the two methods yielded similar results, only the FUBAR results are shown here. In all cases, d_N/d_S measures were based on neighbor-joining trees inferred using the MG94 nucleotide substitution model. Significant differences between d_N/d_S values in OV and NOV regions, were analyzed using a population level adaptation test [24]. Values of d_N and d_S were also estimated. For each pair of overlapping genes, d_N, d_S, and d_N/d_S estimates were obtained for the two reading frames of the OV region. To do so, we used separated sequence alignments for the two overlapping genes, and we partitioned codons, such that OV and NOV regions could be defined over the full-length sequence of each gene.

2.4. Detection of Recombination

For each pair of overlapping genes, recombination breakpoints were detected using six different methods as implemented in RDP5: RDP, GENECONV, MaxChi, 3Seq, Bootscan, and Chimaera [25]. Only recombination signals detected by at least four methods ($p < 0.05$) were considered as positive. For the purpose of this work, recombinants with breakpoints in the LIR and the V1/C3 limit, which are recombination hotspots [26], were not counted as such as they were not differentially affecting OV and NOV regions of any given gene. Instances with more than 10% of recombinant sequences, regardless of breakpoints were located in OV or NOV regions, were considered to have excessive recombination (Appendix A). Analyses were repeated excluding such instances, but conclusions did not vary. Hence, we present here results obtained using all instances.

2.5. Statistical Analysis

The 125 overlapping instances were used for statistical analysis. Tree lengths (t) were not homoscedastic according to Kolmogorov–Smirnov and Levene's tests. Therefore, this variable was fitted to a gamma distribution; whereas the ratio OV/NOV for t, d_N, d_S, and d_N/d_S, and percentage of overlapping were fitted to a normal distribution, according to Akaike's Information Criteria (R package: RRISKDISTRIBUTIONS; [27]). Consequently, differences in values of these variables between OV and NOV regions and between types of gene overlap were analyzed by generalized linear models (GzLM), using type of region or type of overlapping as factors. Differences in the proportion of genes for which parameters above differed between OV and NOV regions was analyzed by Fisher's exact test [28]. Associations between parameters were tested using Pearson´s correlation tests. All statistical analyses were performed using the statistical software packages SPSS 17.0 (SPSS Inc., Chicago, IL, USA) and R v.3.6.3 [29].

3. Results

3.1. Effect of the Presence and Type of Gene Overlapping on Gene Evolution

We analyzed the effect of gene overlapping on the rate of begomovirus evolution by estimating the total length of the tree (t) inferred for the OV and NOV regions of each gene (Figure 2). In OV regions, t ranged from 0.001 to 5.218, depending on the gene–virus combination, with mean value of 0.797 (median: 0.573). Variation in t for NOV regions ranged between 0.006 and 7.676, with mean value of 1.262 (median: 0.744). A GzLM analysis using type of region (OV and NOV) as a factor indicated that t was significantly smaller in OV than in NOV regions (Wald $\chi^2 = 10.74$; $p = 1 \times 10^{-3}$). In agreement with these results, in most overlapping instances, t was significantly smaller in OV than in NOV regions (93/125, $\chi^2 = 59.54$, $p < 1 \times 10^{-5}$) (Figure 2 and Appendix A). Hence, gene overlapping generally reduces evolutionary rates. However, viruses can generate different types of gene overlap, which arise by different mechanisms and that generally differ in the resulting frameshift [10] and the degree of selective independence of the genes involved [7]. Therefore, it could be hypothesized that evolutionary rates differ by type of gene overlap.

Thus, three types of gene overlap were defined following [1]: (1) internal overlapping, when one of the genes contains the complete sequence of the other; (2) 5′-terminal overlapping, when the OV region is in the 5′-terminal region of the gene; and (3) 3′-terminal overlapping, when the OV region is in the 3′-terminal region of the gene. Genes with terminal overlapping showed significantly lower t values in OV than in NOV regions (Wald $\chi^2 \geq 4.88$; $p \leq 0.027$), with most instances fitting this general observation (42/54, $\chi^2 = 33.33$, $p < 1 \times 10^{-5}$; and 39/54, $\chi^2 = 21.33$, $p < 1 \times 10^{-5}$, for 5′- and 3′-terminal overlapping, respectively). In contrast, in genes with internal overlapping no significant differences between OV and NOV regions were observed (Wald $\chi^2 \leq 1.99$; $p \geq 0.212$), and instances with lower t in OV regions were not significantly more frequent (11/17, $\chi^2 = 2.94$, $p = 0.086$) (Figure 2 and Appendix A). We also analyzed differences in the magnitude of the effect of each type of overlapping in reducing the rate of virus evolution. For that, we calculated the OV/NOV ratio for t values of overlapping instances where this parameter was significantly smaller in OV regions. A GzLM indicated that the magnitude of the effect on t depended on the type of overlapping (Wald $\chi^2 = 3.71$; $p = 0.028$), with terminal ones showing similar effects ($p = 0.314$) and in both cases higher than internal overlapping ($p \leq 0.041$). Same conclusions were obtained when normalized t values were used.

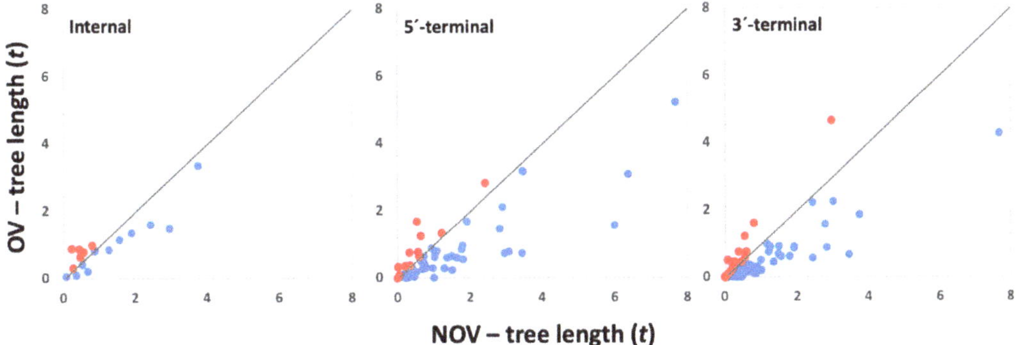

Figure 2. Overlapping and nonoverlapping tree lengths (t) in overlapping genes. Blue dots denote genes in which t is significantly higher in NOV than in OV regions. Red dots denote genes showing the opposite trend. Genes with different types of overlapping (internal, 5′-, and 3′-terminal) are presented in different panels.

Our dataset included mono- and bipartite begomoviruses, which differ in host-virus and virus-virus protein-protein interactions [30]. This may result in differential evolutionary constraints that may modulate how gene overlapping affects virus evolution. Thus, we analyzed whether gene overlapping influenced tree length depending on the begomovirus genome structure. GzLMs using this trait (mono- vs. bipartite) as a factor indicated that it had no effect on t differences between OV and NOV regions (Wald $\chi^2 = 2.24$; $p = 0.137$). In agreement, t was significantly higher in NOV than in OV regions when mono- and bipartite begomoviruses were analyzed separately (Wald $\chi^2 = 4.30$; $p = 0.038$ and Wald $\chi^2 = 8.67$; $p = 3 \times 10^{-3}$, respectively). In both groups of viruses, the same was observed when each type of terminal overlapping was analyzed separately (Wald $\chi^2 \geq 4.90$; $p \leq 0.027$), but not for internal overlapping (Wald $\chi^2 \leq 0.82$; $p \geq 0.366$). The proportion of instances with higher t in NOV that in OV regions was higher than expected by chance in terminal overlapping of both types of genome structures ($\chi^2 \geq 4.92$, $p \leq 0.026$), but not in internal overlapping ($\chi^2 \leq 0.98$, $p \geq 0.173$) (Appendix A).

In sum, these results indicate that the effect of gene overlapping on the rate of begomovirus evolution varies depending on its type; terminal overlapping generally reduces tree length, whereas no clear trend is observed in genes with internal overlapping. On the other hand, the type of genomic structure has little effect on the observed patterns.

3.2. Association between Selection Pressures and Gene Evolution

To further analyze how gene overlapping reduced the rate of evolution, we estimated selection pressures (d_N/d_S) and individual d_N and d_S values for the OV and NOV regions of each gene (Figure 3 and Appendix A). Average d_N/d_S values were 0.35 ± 0.03 and 0.50 ± 0.04 for OV and NOV regions, respectively. A GzLM using type of region as factor indicated that negative selection pressures were significantly stronger in OV than in NOV regions (Wald $\chi^2 = 11.42$; $p < 1 \times 10^{-5}$), and we obtained similar results when each type of overlap was analyzed independently (Wald $\chi^2 \geq 8.18$; $p \leq 2 \times 10^{-4}$, Figure 3 and Appendix A). In agreement, most genes had significantly higher d_N/d_S in NOV than in OV fragments when all types of overlap were considered together (89/125, $\chi^2 = 44.94$, $p < 1 \times 10^{-5}$) and for the three of them independently (14/17, $\chi^2 = 14.24$, $p = 1.6 \times 10^{-4}$; 43/54, $\chi^2 = 37.93$, $p < 1 \times 10^{-5}$; 33/54, $\chi^2 = 5.33$, $p = 0.021$, for internal, 5′-, and 3′-overlapping, respectively) (Figure 3 and Appendix A).

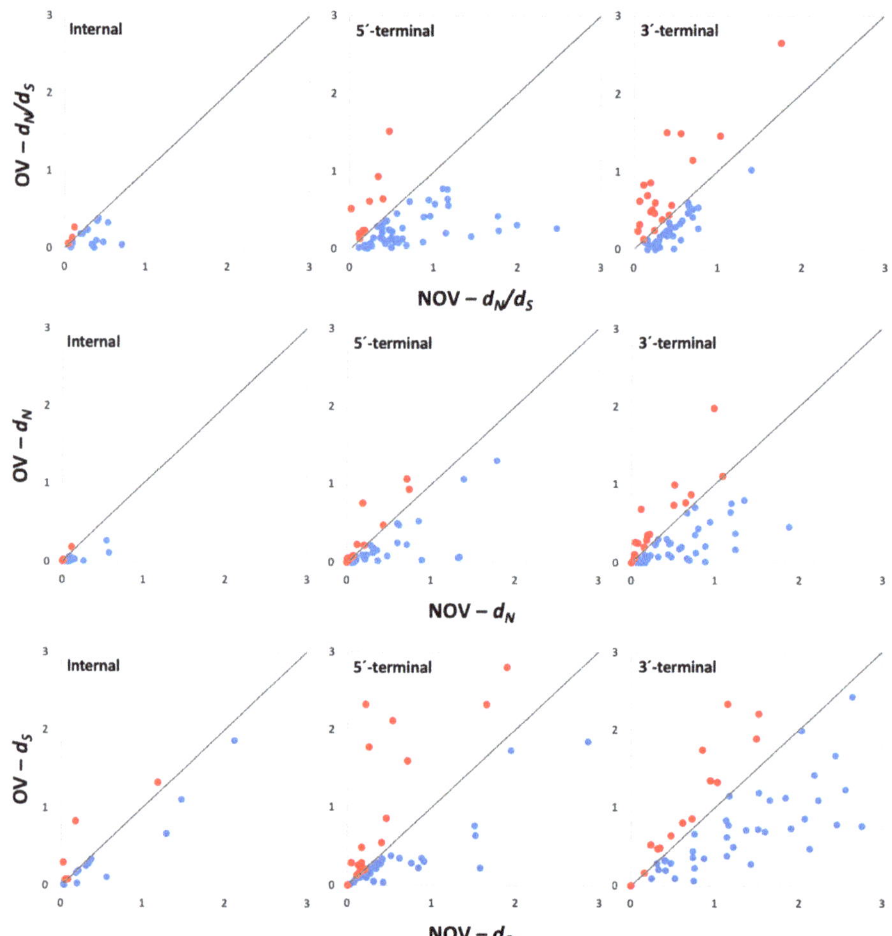

Figure 3. Overlapping and nonoverlapping d_N/d_S (upper line), d_N (middle line) and d_S (lower line) in overlapping genes. Blue dots denote instances in which parameters are significantly higher in NOV than in OV regions. Red dots denote genes showing the opposite trend. Genes with different types of overlapping (internal, 5′-, and 3′-terminal) are presented in different panels.

Similar analysis for d_N indicated significantly lower values in OV than in NOV regions when all genes were considered together (0.24 ± 0.03 and 0.39 ± 0.04, respectively; Wald $\chi^2 = 12.10$; $p < 1 \times 10^{-5}$), and when each type of overlapping was analyzed separately

(Wald $\chi^2 \geq 9.21$; $p \leq 5 \times 10^{-3}$). Also, in most instances, d_N followed this trend (92/125, $\chi^2 = 55.70$, $p < 1 \times 10^{-5}$), with similar results for each type of overlap (13/17, $\chi^2 = 7.53$, $p = 6.1 \times 10^{-3}$; 42/54, $\chi^2 = 33.33$, $p < 1 \times 10^{-5}$; 37/54, $\chi^2 = 14.81$, $p = 1.2 \times 10^{-4}$, for internal, $5'$-, and $3'$-overlapping, respectively) (Figure 3). Finally, d_S was similar in NOV and in OV regions either considering all genes together (0.72 $\pm$ 0.08 and 0.83 $\pm$ 0.07, respectively; Wald $\chi^2 = 2.16$; $p = 0.079$) or analyzing each type of overlap independently (Wald $\chi^2 \leq 3.18$; $p \geq 0.101$) (Figure 3). However, instances with d_S value higher in NOV than in OV regions were more frequent than expected by chance (86/125, $\chi^2 = 35.34$, $p < 1 \times 10^{-5}$), with similar results for each type of overlap (12/17, $\chi^2 = 5.76$, $p = 0.016$; 35/54, $\chi^2 = 9.48$, $p = 2.1 \times 10^{-3}$; 39/54, $\chi^2 = 21.33$, $p < 1 \times 10^{-5}$, for internal, $5'$-, and $3'$-overlapping, respectively) (Figure 3).

When mono- and bipartite begomoviruses were analyzed separately, d_N/d_S and d_N (Wald $\chi^2 \geq 7.70$; $p \leq 6 \times 10^{-3}$, Wald $\chi^2 \geq 4.18$; $p \leq 0.041$, respectively), but not d_S (Wald $\chi^2 \leq 3.00$; $p \geq 0.083$), were always higher in NOV than in OV regions for viruses with both genomic structures. When each type of overlapping was analyzed separately, similar results were obtained for d_N/d_S and d_N (Wald $\chi^2 \geq 4.11$; $p \leq 0.043$, Wald $\chi^2 \geq 6.61$; $p \leq 0.010$ and Wald $\chi^2 \geq 3.60$; $p \leq 0.050$, for internal, $5'$-, and $3'$-overlapping, respectively), and for d_S (Wald $\chi^2 \geq 0.20$; $p \leq 0.652$, Wald $\chi^2 \geq 1.20$; $p \leq 0.274$ and Wald $\chi^2 \geq 2.66$; $p \leq 0.103$, for internal, $5'$-, and $3'$-overlapping, respectively). As above, the proportion of overlapping instances with higher d_N/d_S, d_N and d_S in NOV than in OV regions was generally larger than those showing the opposite trend in viruses with both types of genome structure and in all types of gene overlapping ($\chi^2 \geq 3.63$, $p \leq 0.050$) (Appendix A).

Thus, overlapping genes are generally subjected to stronger purifying selection in OV than in NOV fragments, which seems to be associated with a greater constraint against non-synonymous changes regardless of the type of overlap and, to a lesser extent, with constraints to synonymous changes. Again, the type of genomic structure had no influence in the observed results.

3.3. Association between Proportion of Overlap and Gene Evolution

For RNA viruses, it has been shown that the lengths of the OV region relative to gene length are negatively correlated with these rates in a non-linear manner [1,31]. We analyzed whether this relationship held for begomoviruses by calculating the normalized tree length for the complete sequence of each gene and assessing the strength of association between t and the proportion of gene overlap (Figure 4). As the genome structure had no effect in previous analyses, we did not consider this trait here. On the other hand, we included normalized tree lengths for AC4 (100% overlap), which were not considered previously as in this gene no OV vs. NOV comparison was possible.

The proportion of gene overlap (%) differed among types of overlap (Wald $\chi^2 = 8.74$; $p < 1 \times 10^{-4}$): it was lower in genes with internal (35.67 $\pm$ 0.97) than terminal overlapping (60.59 $\pm$ 3.65 and 59.90 $\pm$ 3.18 for $5'$- and $3'$-terminal overlapping, respectively). Hence, we analyzed the association between per cent of gene overlap and t in the complete sequence of each gene for all genes together and for each type of overlap separately. We performed bivariate analysis considering linear and nonlinear regressions. When a significant association was found, it was best explained by a negative logarithmic relationship between the length of overlap and t (Figure 4). Bivariate analysis revealed a significant negative logarithmic association between these two variables when all instances were considered together ($r = -0.33$; $p < 1 \times 10^{-4}$; Figure 4), with similar results when excluding values for AC4 ($r = -0.32$; $p < 1 \times 10^{-4}$). We also found a significant negative logarithmic association in both types of terminal overlap ($r = -0.37$, $p = 9 \times 10^{-3}$ and $r = -0.31$, $p = 0.027$; for $5'$-, and $3'$-overlapping, respectively), but not for internal ones with ($r = -0.25$, $p = 0.191$; Figure 4) and without ($r = 0.23$, $p = 0.383$) AC4 values. Comparable results were obtained using only those genes for which t values were higher in NOV than in OV regions.

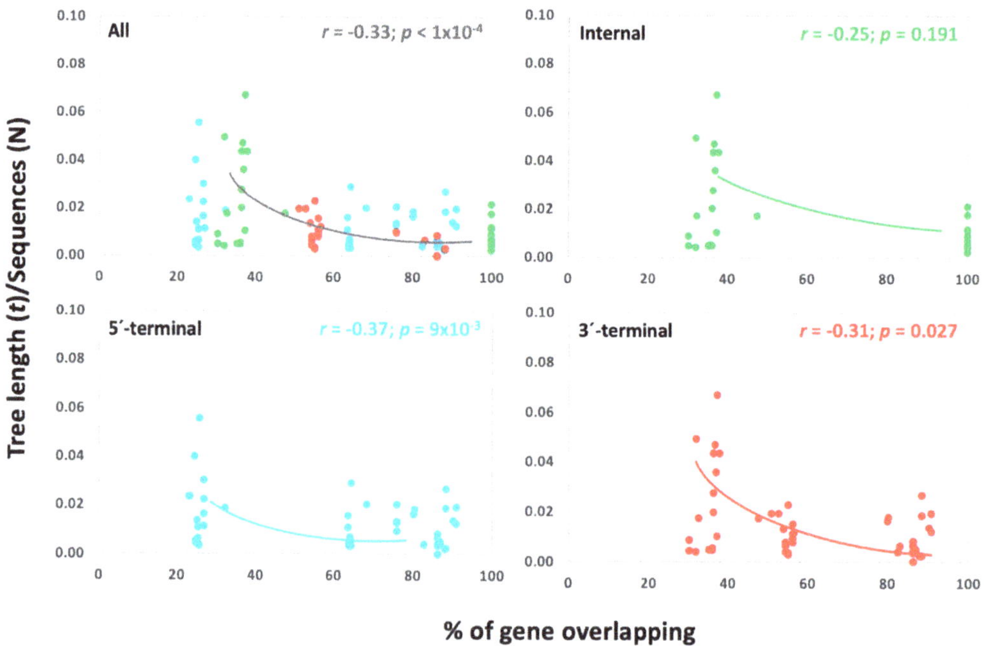

Figure 4. Correlation between tree length (t/number of sequences) and the proportion of gene overlap (length of overlap/total gene length) for all types of overlap considered together (**upper left**), internal overlapping (**upper right**), 5′-terminal overlapping (**lower left**) and 3′-terminal overlapping (**lower right**).

4. Discussion

Several non-mutually exclusive theories have been proposed to explain the abundance of gene overlapping in viruses: (i) it has a role in gene regulation by providing an inherent mechanism for coordinated expression [7]; (ii) it is an effective mechanism for generating novel genes while keeping genome size minimized, by introducing a new reading frame on top of an existing one [32,33]; or (iii) as mutations in these regions affect more than one gene, gene overlapping amplifies the deleterious effect of mutations, thus quickly eliminating such mutations from the viral population, particularly in RNA viruses which have higher mutation rates [7,34,35]. Although there is general agreement on the role of gene overlapping in maintaining genomic compression [10,31,36,37], its effect on virus evolutionary rates remains more elusive [1,2]. This is particularly so for DNA viruses that despite having in general lower mutation rates than RNA viruses have in some cases larger proportion of their genes with at least one overlapping instance [2]. Here, we analyzed whether in the largest genus of plant DNA viruses, whose genome is enriched in gene overlapping instances, this feature modulates the rate of gene evolution.

Our comparative genomic analyses in species of the genus *Begomovirus* indicate that tree length (as a proxy of the rate of evolution) was generally smaller in OV than in NOV regions, with most overlapping instances following this rule. This agrees with the predictions of mathematical models [34,38,39]. Interestingly, these models also predict that the reduction in evolutionary rate is the consequence of correlations at overlapping sites, which are stronger in positions where a mutation would result in a nonsynonymous change in both overlapping genes than in positions where mutations are synonymous in one gene and nonsynonymous in the other [7,34,38]. This may explain why our results indicate that the reduction in the genetic diversity of OV regions is associated with decreased d_N, but not d_S although in most instances OV regions had lower values of both parameters:

gene overlapping would influence both synonymous and nonsynonymous substitution, but this effect would be stronger in nonsynonymous ones. There was significant negative (logarithmic) correlation between the length of overlap and the genetic diversity of each gene; that is, the longer the OV region, the lower the evolutionary rate. This agrees with theoretical models, which predict that evolutionary rate is expected to decline nonlinearly with increasing overlap [7]. This negative logarithmic association also indicates that an increased proportion of gene overlapping reduces begomovirus evolutionary rates up to a threshold, beyond which larger overlapping has no effect on tree length. Thus, long overlapping regions cannot be fully explained by their effect on evolutionary rates alone, and other selection pressures, such as genome compression or coordinated gene expression are likely to play a role.

Altogether, our results provide compelling evidence supporting the role of gene overlapping in reducing the rate of *Begomovirus* evolution. This observation is in accordance with previous reports for a variety of RNA viruses [1,40–42]. In most of these cases, the reduction of the rate of virus evolution associated with gene overlapping has been attributed to the need of these viruses to buffer excessive mutational load due to high mutation rates. To date, however, estimates of mutation rates in DNA viruses suggest that these are lower than for RNA viruses [11]. Two lines of evidence suggest that this might not be the case for begomoviruses. First, rough estimates of mutation frequency in TYLCCNV showed values around 1×10^{-4} [19], which is comparable to the variation reported for plant RNA viruses and higher than for other ssDNA viruses [11,43]. Second, it has been shown that some of the DNA polymerases involved in begomovirus replication are error-prone in conditions equivalent to those in which they amplify the viral genome [44,45]. Hence, begomoviruses could have evolved overlapping regions as a safety mechanism to control high mutation rates.

Evolutionary constraints imposed by gene overlapping are a double-edged sword. They restrict the fixation of deleterious mutations; but at the same time, they leave little room to increase virus fitness, as beneficial mutations in one gene are often deleterious in the other and are therefore purged [1,4]. Viruses are faced with the need to reconcile these two facets such that they limit the fixation of unfit mutations but allow generation of beneficial genetic diversity. To do so, it has been shown that viruses may use a "segregated" organization in which overlapped regions harbor functional domains of one gene or the other, but never both [4]. Thus, gene overlapping imposes a certain degree of evolutionary constraint, as mutations affect more than one gene at the same time. However, this is not as strong as if both genes would harbor functional domains in the overlapping region, or as relaxed as if both genes would not overlap. This strategy results in higher fitness peaks than in the absence of gene overlapping [4]. Interestingly, some evidence suggests that begomoviruses may use a similar strategy. For instance, AV1 functional domains involved in DNA shuttle into the nucleus or in vector transmission are located at the N-terminal region of the protein, overlapping with AV2 [46]; whereas hydrophobic domains involved in the silencing suppression activity of AV2 locate at the NOV region of this protein [47]. Similarly, in AC2 the domain responsible for repressing AV1 expression is in the NOV region of this gene [48], whereas the OV region of AC3 is rich in functional domains [47].

Despite the general trend toward a reduction of genetic diversity in OV compared with NOV regions, when each type of overlapping was analyzed separately, this effect remained significant only in instances with 5′- and 3′-terminal overlap, whereas nearly one-third of the instances with internal overlap showed the opposite trend. Different types of overlapping vary in the preponderance of the associated frameshifting [10]. However, in begomoviruses all overlapping instances have +1/−1 frameshift, which are identical in the extent to which they allow selective independence of the overlapping genes [7]. Alternatively, in our dataset we included mono- and bipartite begomoviruses, for which different functions have been attributed to the C4/AC4 proteins [30]. Different selective pressures on C4/AC4 depending on its function may impose different constraints on its evolution, modulating the buffering effects of gene overlapping on the accumulation of mutations on AC1. We do not favor this hypothesis as our results indicate that genomic structure has little effect on the role of gene overlapping as modulator of begomovirus evolution. Another possible explanation for the observed differences is that, as we restricted our analyses to a single virus genus and each type of overlap occurs in the same genes across species, differences between terminal and internal overlapping reflect particular characteristics of the genes involved. Indeed, internal overlapping instances involved the same two genes (AC1 and AC4) in all species. If, for instance, the AC1 gene is dominating the evolution of AC4, as has been shown for younger overlapping genes generated by overprinting over older ones [33,49], the resulting internal overlapping would have less effect in the evolution of AC1, in accordance with our results. In addition, note that AC1 is involved in virus replication, which is a key component of virus fitness, thus this gene is more likely to drive AC4 evolution rather than the other way around. In support of this hypothesis, it has been shown that AC1 is under strong negative selection, whereas AC4 is under positive selection [50]. Finally, at odds with the examples mentioned above, functional conserved domains are not segregated in AC1/AC4 [47,51], which would also support that the observed differences respond to gene-specific features.

An additional source of gene-specific heterogeneity in our dataset that could explain the differential effect of internal and terminal overlapping in begomovirus evolution is the presence of recombination. Large fragments of AC1 (including the region overlapping with AC4) are recombination hotspots, whereas AC2/AC3 and big portions of AV1 and AV2 are coldspots [26,52]. It has been hypothesized that recombination allows removing deleterious mutations with high efficiency, as reviewed by [53]. Hence, the limited effect of AC1/AC4 internal overlapping in virus evolutionary rates could be explained by a higher frequency of recombination in AC1, which in NOV regions would have similar consequences than gene overlapping. Although we cannot completely discard such a role of recombination, at least in our dataset several observations argue against it. First, the percentage of instances with over 10% of recombinant sequences was evenly distributed across types of overlapping (31–35%, Appendix A). If recombination in AC1/AC4 were to explain our results, we would have expected more frequent recombination in internal than in terminal overlapping. Rather, virus species identity seemed to explain most of the variation in recombination frequency, with three species (*Bhendi yellow vein mosaic virus*, *Chilli leaf curl virus* and *Okra enation leaf curl virus*) accounting for two thirds of the overlapping instances with excessive recombination. Second, when these instances (41/125) were removed from the analyses, we still observed higher t values in NOV than in OV regions in terminal (30/37 and 27/36 instances, for 5′-, and 3′-terminal overlapping, respectively), but not in internal (5/10 instances), overlapping. Hence, our conclusions hold regardless of the presence of extensive recombination.

Some cautionary comments on our results are called for, however. First, the number of instances among types of overlapping is not fully balanced, with lower numbers for internal than for both types of terminal overlap. Hence, the lack of a significant effect of OV regions in internal overlap could be due to reduced sample size. Second, because we restricted our analyses to a single virus genus, overlapping instances occur in the same genes across species, which may reduce the range of overlapping lengths included in the regression analyses with the subsequent reduction of statistical power. However, the range of terminal overlapping lengths included was enough to detect a significant correlation between percent of gene overlapping and genetic diversity. This range was much smaller for internal overlapping, which again may explain the lack of association between the two analyzed traits. Finally, we could include only 18 out of the 420 begomovirus species, as these were the only ones that fulfilled the criteria to be included in our analyses. Despite the small sample size, our results indicate strongly significant effects, which support the relevant role of gene overlapping in begomovirus evolution.

In sum, this work provides novel evidence of the selective constraints imposed by gene overlapping on the pace of begomovirus evolution. Whether this effect is general for DNA viruses would be an interesting avenue of future research.

Supplementary Materials: The following are available online at https://www.mdpi.com/article/10.3390/microorganisms10020366/s1, File S1: Sequence alignments used in this work.

Author Contributions: Conceptualization, I.P.; methodology, I.P.; formal analysis, I.M.-H.; data curation, I.M.-H.; writing—original draft preparation, I.P.; writing—review and editing, I.P.; funding acquisition, I.P. All authors have read and agreed to the published version of the manuscript.

Funding: This research was funded by Plan Nacional I + D + i, Ministerio de Economía y Competitividad (Agencia Estatal de Investigación), Spain [PID2019-109579RB-I00] to IP.

Institutional Review Board Statement: Not applicable.

Informed Consent Statement: Not applicable.

Data Availability Statement: Data available as Supplementary Material and as Appendix A.

Conflicts of Interest: The authors declare no conflict of interest. The funders had no role in the design of the study; in the collection, analyses, or interpretation of data; in the writing of the manuscript, or in the decision to publish the results.

Appendix A

Table A1. Statistical parameters of traits associated with the rate of evolutionary change in overlapping genes.

Species [1]	N [2]	Overlapping [3]		Gene Length [4]	t [5]		d_N/d_S	
		ORF	Length		NOV	OV	NOV	OV
Internal (11/17)								
African cassava mosaic virus (B)	27 (4%)	AC1-AC4	423	1077	0.53	0.41	0.39	0.11
Alternanthera yellow vein virus (M)	11 (8%)	AC1-AC4	291	1086	0.47	0.64	0.54	0.33
Bean golden mosaic virus (B)	121 (0%)	AC1-AC4	258	1086	0.44	0.86	0.34	0.06
Bhendi yellow vein mosaic virus (M)*	39 (98%)	AC1-AC4	294	1092	1.25	0.85	0.48	0.09
Bhendi yellow vein mosaic virus (M)	32 (99%)	AC1-AC4	303	1092	1.90	1.35	0.11	0.08
Chilli leaf curl virus (M)	16 (28%)	AC1-AC4	300	1086	0.86	0.83	0.22	0.20
Cotton leaf curl Gezira virus (M)	21 (9%)	AC1-AC4	294	1089	0.22	0.89	0.37	0.02
Cotton leaf curl Multan virus (M)	50 (2%)	AC1-AC4	303	1092	0.67	0.21	0.07	0.02
East African cassava mosaic Kenya virus (B)	50 (0%)	AC1-AC4	297	1065	0.27	0.31	0.05	0.07
East African cassava mosaic Malawi virus (B)	13 (0%)	AC1-AC4	234	1080	0.07	0.06	0.20	0.19
East African cassava mosaic virus (B)	153 (2%)	AC1-AC4	234	1080	1.55	1.15	0.41	0.36
East African cassava mosaic Zanzibar virus (B)	14 (29%)	AC1-AC4	258	1080	0.35	0.09	0.08	0.03
Okra enation leaf curl virus (M)	60 (92%)	AC1-AC4	308	1089	2.96	1.48	0.29	0.24
South African cassava mosaic virus (B)	125 (0%)	AC1-AC4	297	1080	0.55	0.79	0.09	0.14
Sweet potato leaf curl virus (M)	17 (41%)	AC1-AC4	258	1095	0.79	0.99	0.12	0.28
Tomato leaf curl New Delhi virus (B)	97 (3%)	AC1-AC4	303	1086	3.75	3.34	0.43	0.39
Tomato yellow leaf curl virus (M)	397 (68%)	AC1-AC4	294	1074	2.43	1.60	0.70	0.05
5′-terminal (42/54)								
African cassava mosaic virus (B)	32 (3%)	AC1-AC2	93	408	0.21	0.37	0.11	0.20
	31 (3%)	AC2-AC3	260	405	0.31	0.03	0.50	0.03
	26 (0%)	AV1-AV2	193	777	0.39	0.26	0.37	0.15
Alternanthera yellow vein virus (M)	11 (18%)	AC1-AC2	98	406	0.22	0.01	0.22	0.02
	13 (0%)	AC2-AC3	260	405	0.71	0.27	0.11	0.03
	10 (60%)	AV1-AV2	189	771	0.33	0.76	1.15	0.21
Bean golden mosaic virus (B)	158 (0%)	AC1-AC2	89	390	1.30	0.29	0.40	0.13
	158 (0%)	AC2-AC3	254	399	1.50	0.25	0.26	0.05
Bhendi yellow vein mosaic virus (M)*	51 (71%)	AC1-AC2	104	453	1.91	1.68	0.33	0.29
	51 (20%)	AC2-AC3	308	405	3.44	0.74	0.28	0.14
	50 (40%)	AV1-AV2	206	771	1.76	0.85	0.63	0.14
Bhendi yellow vein mosaic virus (M)	57 (54%)	AC1-AC2	104	453	2.42	2.82	1.01	0.58
	57 (2%)	AC2-AC3	308	405	1.80	0.56	1.18	0.56
	56 (16%)	AV1-AV2	206	771	1.63	0.61	0.38	0.21
Chilli leaf curl virus (M)	18 (33%)	AC1-AC2	98	405	0.99	0.30	0.17	0.02
	23 (17%)	AC2-AC3	260	405	3.08	0.79	0.47	0.25
	22 (18%)	AV1-AV2	197	771	1.22	1.33	0.18	0.24
Cotton leaf curl Gezira virus (M)	32 (6%)	AC1-AC2	101	405	0.01	0.33	0.02	0.52
	32 (0%)	AC2-AC3	257	402	0.47	0.04	0.42	0.07
	31 (0%)	AV1-AV2	209	777	0.35	0.37	0.16	0.24

Table A1. *Cont.*

Species [1]	N [2]	Overlapping [3]		Gene Length [4]	t [5]		d_N/d_S	
		ORF	Length		NOV	OV	NOV	OV
Cotton leaf curl Multan virus (M)	58 (2%)	AC1-AC2	104	453	0.59	0.65	0.12	0.14
	59 (2%)	AC2-AC3	308	405	1.08	0.80	1.11	0.78
	59 (15%)	AV1-AV2	206	771	1.04	0.75	0.44	0.37
East African cassava mosaic Kenya virus (B)	71 (0%)	AC1-AC2	77	408	0.40	0.29	0.25	0.08
	71 (0%)	AC2-AC3	260	405	0.34	0.22	0.18	0.06
	64 (0%)	AV1-AV2	197	774	0.23	0.18	0.55	0.10
East African cassava mosaic Malawi virus (B)	9 (0%)	AC1-AC2	92	408	0.00	0.00	-	-
	10 (0%)	AC2-AC3	260	405	0.06	0.03	0.57	0.14
	12 (0%)	AV1-AV2	191	777	0.05	0.08	0.17	0.22
East African cassava mosaic virus (B)	166 (2%)	AC1-AC2	92	405	1.79	0.97	2.47	0.27
	162 (0%)	AC2-AC3	260	405	1.38	0.60	0.67	0.05
	105 (0%)	AV1-AV2	197	775	0.79	0.31	0.58	0.09
East African cassava mosaic Zanzibar virus (B)	15 (27%)	AC1-AC2	92	408	0.23	0.01	0.50	0.15
	15 (0%)	AC2-AC3	260	405	0.11	0.10	1.76	0.42
	15 (0%)	AV1-AV2	197	775	0.20	0.07	0.63	0.27
Okra enation leaf curl virus (M)	67 (97%)	AC1-AC2	104	453	3.45	3.18	0.51	0.23
	68 (32%)	AC2-AC3	308	405	1.04	0.63	1.44	0.17
	41 (58%)	AV1-AV2	188	771	0.64	1.26	0.24	0.62
Pepper golden mosaic virus (B)	54 (57%)	AC1-AC2	59	390	1.49	0.66	0.47	0.12
	54 (0%)	AC2-AC3	254	399	0.94	0.89	0.34	0.94
Pepper huasteco yellow vein virus (B)	19 (5%)	AC1-AC2	80	417	1.01	0.01	0.38	0.28
	45 (0%)	AC2-AC3	254	399	0.74	0.42	1.99	0.32
South African cassava mosaic virus (B)	131 (0%)	AC1-AC2	92	408	0.56	0.77	0.47	1.52
	130 (0%)	AC2-AC3	260	405	0.67	0.48	0.40	0.64
	126 (4%)	AV1-AV2	191	777	0.73	0.64	1.17	0.65
Sweet potato leaf curl virus (M)	18 (39%)	AC1-AC2	92	452	0.54	1.68	0.39	0.33
	14 (7%)	AC2-AC3	278	435	2.96	0.73	1.78	0.25
	18 (9%)	AV1-AV2	176	765	0.52	0.17	0.87	0.10
Tomato leaf curl New Delhi virus (B)	88 (4%)	AC1-AC2	98	420	2.83	1.47	0.71	0.62
	113 (1%)	AC2-AC3	281	411	5.99	1.56	0.57	0.46
	115 (0%)	AV1-AV2	179	771	2.90	2.12	1.16	0.77
Tomato yellow leaf curl virus (M)	588 (9%)	AC1-AC2	92	408	7.68	5.22	0.96	0.43
	521 (10%)	AC2-AC3	260	405	6.37	3.08	0.95	0.64
	593 (10%)	AV1-AV2	191	777	3.47	3.16	0.89	0.42
$3'$-terminal (39/54)								
African cassava mosaic virus (B)	27 (3%)	AC1-AC2	93	1077	0.53	0.36	0.50	0.29
	32 (3%)	AC2-AC3	260	408	0.21	0.03	0.44	0.20
	33 (0%)	AV1-AV2	193	342	0.54	0.24	0.47	0.18
Alternanthera yellow vein virus (M)	11 (18%)	AC1-AC2	98	1086	0.47	0.01	0.48	0.02
	11 (0%)	AC2-AC3	260	406	0.22	0.27	0.45	0.57
	12 (60%)	AV1-AV2	189	348	1.21	0.76	0.57	0.13
Bean golden mosaic virus (B)	121 (0%)	AC1-AC2	89	1086	0.44	0.10	0.56	0.34
	158 (0%)	AC2-AC3	254	390	0.30	0.28	0.65	0.61
Bhendi yellow vein mosaic virus (M)*	39 (92%)	AC1-AC2	104	1092	1.25	0.91	0.58	0.38
	51 (20%)	AC2-AC3	308	453	1.91	0.91	0.27	0.03
	50 (40%)	AV1-AV2	206	366	1.34	0.46	0.13	0.10
Bhendi yellow vein mosaic virus (M)	23 (98%)	AC1-AC2	104	1092	1.90	0.88	0.16	0.70
	57 (2%)	AC2-AC3	308	453	2.42	0.57	0.21	0.51
	55 (16%)	AV1-AV2	206	366	0.89	0.32	0.78	0.54
Chilli leaf curl virus (M)	16 (38%)	AC1-AC2	98	1086	0.86	0.10	0.30	0.12
	18 (17%)	AC2-AC3	260	405	0.99	0.50	0.66	0.57
	16 (19%)	AV1-AV2	197	357	0.26	0.45	1.76	2.67
Cotton leaf curl Gezira virus (M)	21 (9%)	AC1-AC2	101	1089	0.22	0.33	0.07	0.33
	32 (0%)	AC2-AC3	257	405	0.01	0.03	0.24	0.47
	29 (0%)	AV1-AV2	209	369	0.08	0.50	0.07	0.63
Cotton leaf curl Multan virus (M)	50 (2%)	AC1-AC2	104	1092	0.67	0.38	0.27	0.03
	58 (2%)	AC2-AC3	308	453	0.59	0.75	0.11	0.14
	57 (16%)	AV1-AV2	206	366	0.45	0.46	0.24	0.26
East African cassava mosaic Kenya virus (B)	50 (0%)	AC1-AC2	77	1065	0.27	0.01	0.38	0.25
	71 (0%)	AC2-AC3	260	408	0.40	0.25	0.25	0.61
	56 (0%)	AV1-AV2	197	357	0.16	0.18	0.42	0.45
East African cassava mosaic Malawi virus (B)	13 (0%)	AC1-AC2	92	1080	0.07	0.00	0.16	0.00
	9 (0%)	AC2-AC3	260	408	0.00	0.00	-	-
	12 (0%)	AV1-AV2	191	351	0.06	0.08	0.34	0.39

Table A1. *Cont.*

Species [1]	N [2]	Overlapping [3]		Gene Length [4]	t [5]		d_N/d_S	
		ORF	Length		NOV	OV	NOV	OV
East African cassava mosaic virus (B)	153 (3%)	AC1-AC2	92	1080	1.55	0.62	0.43	0.25
	166 (0%)	AC2-AC3	260	405	1.79	0.62	0.43	0.35
	136 (0%)	AV1-AV2	197	357	0.75	0.32	0.36	0.22
East African cassava mosaic Zanzibar virus (B)	14 (29%)	AC1-AC2	92	1080	0.35	0.01	0.32	0.17
	15 (0%)	AC2-AC3	260	408	0.23	0.10	0.77	0.28
	13 (0%)	AV1-AV2	197	357	0.15	0.07	0.65	0.47
Okra enation leaf curl virus (M)	60 (92%)	AC1-AC2	104	1089	2.96	4.66	0.05	0.25
	67 (33%)	AC2-AC3	308	453	3.45	0.68	0.32	0.04
	45 (53%)	AV1-AV2	188	348	0.38	0.75	0.21	0.50
Pepper golden mosaic virus (B)	54 (57%)	AC1-AC2	59	1050	1.47	0.68	0.25	0.01
	54 (0%)	AC2-AC3	254	390	1.49	0.92	0.44	0.29
Pepper huasteco yellow vein virus (B)	44 (2%)	AC1-AC2	80	1050	0.81	0.12	0.26	0.11
	19 (0%)	AC2-AC3	254	417	1.01	0.21	1.40	1.03
South African cassava mosaic virus (B)	125 (0%)	AC1-AC2	92	1080	0.55	0.60	1.04	1.47
	131 (0%)	AC2-AC3	260	408	0.56	0.42	0.70	0.43
	128 (4%)	AV1-AV2	191	351	1.16	0.99	0.24	0.08
Sweet potato leaf curl virus (M)	17 (41%)	AC1-AC2	92	1095	0.79	1.61	0.57	1.50
	18 (6%)	AC2-AC3	278	452	0.54	1.21	0.70	1.16
	18 (9%)	AV1-AV2	176	345	0.62	0.16	0.18	0.06
Tomato leaf curl New Delhi virus (B)	97 (3%)	AC1-AC2	98	1086	3.75	1.86	0.16	0.13
	88 (1%)	AC2-AC3	281	420	2.83	0.88	0.12	0.84
	105 (0%)	AV1-AV2	179	339	2.78	1.58	0.71	0.52
Tomato yellow leaf curl virus (M)	397 (8%)	AC1-AC2	92	1074	2.43	2.23	0.40	1.52
	588 (9%)	AC2-AC3	260	408	7.67	4.27	0.20	0.87
	623 (9%)	AV1-AV2	191	351	3.01	2.27	0.40	0.18

[1] Number of genes with significant differences in tree length between OV and NOV regions over total number. Deviation from randomness was tested by Fisher's exact test ($p < 0.05$). M: Monopartite; B: Bipartite. Asterisks indicate sequences formerly classified as *Bendhi yellow vein India virus*; [2] Number of sequences used. Percentage of recombinant sequences is shown in parentheses; [3] Name of the genes involved in the overlapping instance, and length of the OV region; [4] Length of the largest gene (internal overlapping) of the gene with 5′-terminal overlapping (5′-terminal) and of the gene with 3′-terminal overlapping (3′-terminal); [5] Tree length of inferred phylogenies for OV and NOV regions of each overlapping instance.

References

1. Simon-Loriere, E.; Holmes, E.C.; Pagán, I. The effect of gene overlapping on the rate of RNA virus evolution. *Mol. Biol. Evol.* **2013**, *30*, 1916–1928. [CrossRef] [PubMed]
2. Schlub, T.E.; Holmes, E.C. Properties and abundance of overlapping genes in viruses. *Virus Evol.* **2020**, *6*, veaa009. [CrossRef] [PubMed]
3. Gogarten, J.P.; Townsend, J.P. Horizontal Gene Transfer, Genome Innovation and Evolution. *Nat. Rev. Microbiol.* **2005**, *3*, 679–687. [CrossRef] [PubMed]
4. Fernandes, J.D.; Faust, T.B.; Strauli, N.B.; Smith, C.; Crosby, D.C.; Nakamura, R.L.; Hernandez, R.D.; Frankel, A.D. Functional Segregation of Overlapping Genes in HIV. *Cell* **2016**, *167*, 1762–1773. [CrossRef] [PubMed]
5. Nelson, C.W.; Ardern, Z.; Goldberg, T.L.; Meng, C.; Kuo, C.-H.; Ludwig, C.; Kolokotronis, S.-O.; Wei, X. Dynamically evolving novel overlapping gene as a factor in the SARS-CoV-2 pandemic. *eLife* **2020**, *9*, e59633. [CrossRef] [PubMed]
6. Wright, B.W.; Ruan, J.; Molloy, M.P.; Jaschke, P.R. Genome modularization reveals overlapped gene topology is necessary for efficient viral reproduction. *ACS Synth. Biol.* **2020**, *9*, 3079–3090. [CrossRef]
7. Krakauer, D.C. Stability and evolution of overlapping genes. *Evolution* **2000**, *54*, 731–739. [CrossRef]
8. Belshaw, R.; Gardner, A.; Rambaut, A.; Pybus, O.G. Pacing a small cage: Mutation and RNA viruses. *Trends Ecol. Evol.* **2008**, *23*, 188–193. [CrossRef]
9. Brandes, N.; Linial, M. Gene overlapping and size constraints in the viral world. *Biol. Direct* **2016**, *11*, 26. [CrossRef]
10. Belshaw, R.; Pybus, O.G.; Rambaut, A. The evolution of genome compression and genomic novelty in RNA viruses. *Genome Res.* **2007**, *17*, 1496–1504. [CrossRef]
11. Sanjuán, R.; Domingo-Calap, P. Mechanisms of viral mutation. *Cell Mol. Life Sci.* **2016**, *73*, 4433–4448. [CrossRef]
12. Duffy, S.; Shackelton, L.A.; Holmes, E.C. Rates of evolutionary change in viruses: Patterns and determinants. *Nat. Rev. Genet.* **2008**, *9*, 267–276. [CrossRef] [PubMed]
13. Cui, J.; Schlub, T.E.; Holmes, E.C. An Allometric Relationship between the Genome Length and Virion Volume of Viruses. *J. Virol.* **2014**, *88*, 6403–6410. [CrossRef] [PubMed]
14. Wu, M.; Wei, H.; Tan, H.; Pan, S.; Liu, Q.; Bejarano, E.R.; Lozano-Durán, R. Plant DNA polymerases α and δ mediate replication of geminiviruses. *Nat. Commun.* **2021**, *12*, 2780. [CrossRef] [PubMed]

15. Fiallo-Olivé, E.; Lett, J.-M.; Martin, D.P.; Roumagnac, P.; Varsani, A.; Zerbini, F.M.; Navas-Castillo, J. ICTV Report Consortium. ICTV Virus Taxonomy Profile: *Geminiviridae* 2021. *J. Gen. Virol.* **2021**, *102*, 001696. [CrossRef] [PubMed]
16. Hanley-Bowdoin, L.; Bejarano, E.R.; Robertson, D.; Mansoor, S. Geminiviruses: Masters at redirecting and reprogramming plant processes. *Nat. Rev. Microbiol.* **2013**, *11*, 777–788. [CrossRef]
17. Zerbini, F.M.; Briddon, R.W.; Idris, A.; Martin, D.P.; Moriones, E.; Navas-Castillo, J.; Rivera-Bustamante, R.; Roumagnac, P.; Varsani, A.; ICTV Report Consortium. ICTV Virus Taxonomy Profile: Geminiviridae. *J. Gen. Virol.* **2017**, *98*, 131–133. [CrossRef]
18. Pagán, I.; García-Arenal, F. Population Genomics of Plant Viruses. In *Population Genomics: Microorganisms*; Polz, M.F., Rajora, O.P., Eds.; Springer International Publishing: New York, NY, USA, 2018.
19. Ge, L.; Zhang, J.; Zhou, X.; Li, H. Genetic Structure and Population Variability of Tomato Yellow Leaf Curl China Virus. *J. Virol.* **2007**, *81*, 5902–5907. [CrossRef]
20. Rodelo-Urrego, M.; García-Arenal, F.; Pagán, I. The effect of ecosystem biodiversity on virus genetic diversity depends on virus species: A study of chiltepin-infecting begomoviruses in Mexico. *Virus Evol.* **2015**, *1*, vev004. [CrossRef]
21. Edgar, R.C. MUSCLE: Multiple sequence alignment with high accuracy and high throughput. *Nucleic Acids Res.* **2004**, *32*, 1792–1797. [CrossRef]
22. Larsson, A. AliView: A fast and lightweight alignment viewer and editor for large data sets. *Bioinformatics* **2014**, *30*, 3276–3278. [CrossRef] [PubMed]
23. Kosakovsky Pond, S.L.; Frost, S.D. Datamonkey: Rapid detection of selective pressure on individual sites of codon alignments. *Bioinformatics* **2005**, *21*, 2531–2533. [CrossRef] [PubMed]
24. Kosakovsky Pond, S.L.; Frost, S.D.; Grossman, Z.; Gravenor, M.B.; Richman, D.D.; Leigh Brown, A.J. Adaptation to different human populations by HIV-1 revealed by codon-based analyses. *PLoS Comp. Biol.* **2006**, *2*, e62. [CrossRef] [PubMed]
25. Martin, D.P.; Varsani, A.; Roumagnac, P.; Botha, G.; Maslamoney, S.; Schwab, T.; Kelz, Z.; Kumar, V.; Murrell, B. RDP5: A computer program for analyzing recombination in, and removing signals of recombination from, nucleotide sequence datasets. *Virus Evol.* **2020**, *7*, veaa087. [CrossRef] [PubMed]
26. Lefeuvre, P.; Lett, J.-M.; Reynaud, B.; Martin, D.P. Avoidance of protein fold disruption in natural virus recombinants. *PLoS Pathog* **2007**, *3*, e181. [CrossRef] [PubMed]
27. Belgorodski, N.; Greiner, M.; Tolksdorf, K.; Schueller, K.; rriskDistributions: Fitting Distributions to Given Data or Known Quantiles. R Package v.2.1. Available online: https://CRAN.R-project.org/package=rriskDistributions (accessed on 16 December 2021).
28. Sokal, R.R.; Rohlf, F.J. *Biometry: The Principles and Practices of Statistics in Biological Research*; W.H. Freeman & Company: New York, NY, USA, 1995.
29. R Core Team. *R: A Language and Environment for Statistical Computing*; R Foundation for Statistical Computing: Vienna, Austria, 2021; Available online: https://www.R-project.org/ (accessed on 16 December 2021).
30. Luna, A.P.; Lozano-Durán, R. Geminivirus-Encoded Proteins: Not All Positional Homologs Are Made Equal. *Front. Microbiol.* **2020**, *11*, 878. [CrossRef]
31. Chirico, N.; Vianelli, A.; Belshaw, R. Why genes overlap in viruses. *Proc. Biol. Sci.* **2010**, *277*, 3809–3817. [CrossRef]
32. Rancurel, C.; Khosravi, M.; Dunker, A.K.; Romero, P.R.; Karlin, D. Overlapping Genes Produce Proteins with Unusual Sequence Properties and Offer Insight into De Novo Protein Creation. *J. Virol.* **2009**, *83*, 10719–10736. [CrossRef]
33. Sabath, N.; Wagner, A.; Karlin, D. Evolution of Viral Proteins Originated De Novo by Overprinting. *Mol. Biol. Evol.* **2012**, *29*, 3767–3780. [CrossRef]
34. Krakauer, D.C. Evolutionary principles of genomic compression. *Comments Theor. Biol.* **2002**, *7*, 215–236. [CrossRef]
35. Elena, S.F.; Carrasco, P.; Darós, J.A.; Sanjuán, R. Mechanisms of genetic robustness. *EMBO Rep.* **2006**, *7*, 168–173. [CrossRef] [PubMed]
36. Holmes, E.C. Error thresholds and the constraints to RNA virus evolution. *Trends Microbiol.* **2003**, *11*, 543–546. [CrossRef] [PubMed]
37. Lillo, F.; Krakauer, D.C. A statistical analysis of the three-fold evolution of genomic compression through frame overlaps in prokaryotes. *Biol. Direct* **2007**, *2*, 22. [CrossRef] [PubMed]
38. Krakauer, D.C.; Plotkin, J.B. Redundancy, antiredundancy, and the robustness of genomes. *Proc. Natl. Acad. Sci. USA* **2002**, *99*, 1405–1409. [CrossRef] [PubMed]
39. Peleg, O.; Kirzhner, V.; Trifonov, E.; Bolshoy, A. Overlapping messages and survivability. *J. Mol. Evol.* **2004**, *59*, 520–527. [CrossRef]
40. Hughes, A.L.; Westover, K.; da Silva, J.; O'Connor, D.H.; Watkins, D.I. Simultaneous positive and purifying selection on overlapping reading frames of the *tat* and *vpr* genes of simian immunodeficiency virus. *J. Virol.* **2001**, *75*, 7966–7972. [CrossRef]
41. Guyader, S.; Ducray, D.G. Sequence analysis of potato leafroll virus isolates reveals genetic stability, major evolutionary events and differential selection pressure between overlapping reading frame products. *J. Gen. Virol.* **2002**, *83*, 1799–1807. [CrossRef]
42. Pagán, I.; Holmes, E.C. Long-term evolution of the *Luteoviridae*: Time scale and mode of virus speciation. *J. Virol.* **2010**, *84*, 6177–6187. [CrossRef]
43. Schneider, W.L.; Roossinck, M.J. Genetic diversity in RNA viral quasispecies is controlled by host-virus interactions. *J. Virol.* **2001**, *75*, 6566–6571. [CrossRef]
44. Deem, A.; Keszthelyi, A.; Blackgrove, T.; Vayl, A.; Coffey, B.; Mathur, R.; Chabes, A.; Malkova, A. Break-induced replication is highly inaccurate. *PLoS Biol.* **2011**, *9*, e1000594. [CrossRef]

45. Hicks, W.M.; Kim, M.; Haber, J.E. Increased mutagenesis and unique mutation signature associated with mitotic gene conversion. *Science* **2010**, *329*, 82–85. [CrossRef] [PubMed]
46. Fondong, V.N. Geminivirus protein structure and function. *Mol. Plant. Pathol.* **2013**, *14*, 635–649. [CrossRef] [PubMed]
47. Luna, A.P.; Romero-Rodríguez, B.; Rosas-Díaz, T.; Cerero, L.; Rodríguez-Negrete, E.A.; Castillo, A.G.; Bejarano, E.R. Characterization of Curtovirus V2 protein, a functional homolog of Begomovirus V2. *Front. Plant Sci.* **2020**, *11*, 835. [CrossRef] [PubMed]
48. Lacatus, G.; Sunter, G. The Arabidopsis PEAPOD2 transcription factor interacts with geminivirus AL2 protein and the coat protein promoter. *Virology* **2009**, *392*, 196–202. [CrossRef]
49. Pagán, I.; Betancourt, M.; de Miguel, J.; Piñero, d.; Fraile, A.; García-Arenal, F. Genomic and biological characterization of chiltepín yellow mosaic virus, a new tymovirus infecting *Capsicum annuum* var. *aviculare in Mexico. Arch. Virol.* **2010**, *155*, 675–684. [CrossRef]
50. Deom, C.M.; Brewer, M.T.; Severns, P.M. Positive selection and intrinsic disorder are associated with multifunctional C4(AC4) proteins and geminivirus diversification. *Sci. Rep.* **2021**, *11*, 11150. [CrossRef]
51. Fondong, V.N. The ever-expanding role of C4/AC4 in Geminivirus infection: Punching above its weight? *Mol. Plant* **2019**, *12*, 145–147. [CrossRef]
52. Martin, D.P.; Lefeuvre, P.; Varsani, A.; Hoareau, M.; Semegni, J.-Y.; Dijoux, B.; Vincent, C.; Reynaud, B.; Lett, J.-M. Complex recombination patterns arising during geminivirus coinfections preserve and demarcate biologically important intra-genome interaction networks. *PLoS Pathog.* **2011**, *7*, e1002203. [CrossRef]
53. Simon-Loriere, E.; Holmes, E.C. Why do RNA viruses recombine? *Nat. Rev. Microbiol.* **2011**, *9*, 617–626. [CrossRef]

 *microorganisms*

Article

Improvement of Alternaria Leaf Blotch and Fruit Spot of Apple Control through the Management of Primary Inoculum

Jordi Cabrefiga *, Maria Victoria Salomon and Pere Vilardell

Sustainable Plant Protection, Intitute of Agrifood Research and Technology, La Tallada d'Empordà, 17134 Girona, Spain
* Correspondence: jordi.cabrefiga@irta.cat

Abstract: *Alternaria* spp. is the causal agent of apple leaf blotch and fruit spot, diseases of recent appearance in Spain. The overwinter inoculum of *Alternaria* spp. is the source of primary infections in apple, thus the aim of this work was to optimize the control of infection through two environmentally friendly inoculum-management strategies, the removal of winter fallen leaves and the treatment of leaves with the biological agent *Trichoderma asperellum* to inhibit or prevent inoculum development in commercial orchards. The results of commercial orchard trials showed that leaf aspiration and application of *T. asperellum* on the ground have efficacy to reduce fruit spot between 50 and 80% and leaf blotch of between 30 and 40% depending on the year. The efficacies on the reduction of leaf blotch were slightly lower than of fruit spot. Disease reduction has been related to a reduction of total spores released during the season. Results of dynamics of spore release indicate that factors influencing spore release were rainfall and temperature. In conclusion, the use of environmentally friendly strategies combined with standard fungicides, and with monitoring environmental conditions, might allow a reduction in the number of phytosanitary applications, thus achieving the goal of reducing their use.

Keywords: Alternaria leaf blotch; Alternaria fruit spot; *Trichoderma*; spore release; inoculum management

Citation: Cabrefiga, J.; Salomon, M.V.; Vilardell, P. Improvement of Alternaria Leaf Blotch and Fruit Spot of Apple Control through the Management of Primary Inoculum. *Microorganisms* **2023**, *11*, 101. https://doi.org/10.3390/microorganisms11010101

Academic Editor: Ana Palacio-Bielsa

Received: 17 November 2022
Revised: 24 December 2022
Accepted: 27 December 2022
Published: 30 December 2022

1. Introduction

Alternaria is a widespread genus of fungi that includes several species of saprophytes and pathogens of different plant species including a wide variety of crops during pre- and postharvest, thus causing production losses that have a significant economic impact on agriculture. In apples (*Malus domestica* Borkh.), some species of *Alternaria* have been reported as responsible for damage in leaves as well as in young and mature fruits causing leaf blotch, fruit spot, and moldy core diseases in susceptible varieties such as Fuji, Royal Gala, Red Delicious and Golden Delicious [1–3]. The first pathology was associated with the species *Alternaria mali* (Roberts) (syn. *Alternaria alternata* f. sp. *mali*, or *Alternaria alternata* apple pathotype) [4] and was first identified in 1924 in the USA [5]. Although multiple *Alternaria* species have been described as causing leaf blotch and fruit spot of apple, *A. alternata* and the *A. arborescens* species complexes have been commonly associated with them [3,6–8]. These studies indicate that the distribution of *Alternaria* species causing leaf blotch and fruit spot in apple is different among regions, though *A. mali* is still the most commonly cited pathogen worldwide [9]. In Spain, three species have been found to be associated with these diseases, including *A. arborescens*, *A. tenuissima* and *A. alternata* (this work, data not shown), as it has been described in Australia [10], in Italy [8], in Israel [3] and more recently in France [11].

The occurrence of Alternaria leaf blotch and fruit spot has been reported in all apple-producing regions of the world [11], being one of the most important diseases of apple in Southeast Asia, including Japan, South Korea and China [9], southeastern USA [9] and Australia [10]. In Europe, Alternaria leaf blotch and fruit spot were first reported in 1996 in Yugoslavia [12] and later described in some important apple growing areas such as France, where they started to be a relevant problem in 2016 [11], Netherlands where they were detected in a survey in 2014 [6], and Italy, where the problem dates back to 1999 [13]. In Spain, the first symptoms, affecting mainly Gala and Golden Delicious varieties, were observed in the northeastern part in 2006, although damages of concern appeared in 2014–2015 [14]. In this region, apple is an important fruit crop with a total production area of 2720 ha and a production of 101,745 T in 2021 [15]. Since the first focus, Alternaria leaf blotch and fruit spot have spread throughout the area to affect more than 20% of commercial orchards of sensitive varieties, including Golden Delicious, Gala and Pink Lady, especially.

The first symptoms appear on leaves at the end of May as small circular purplish-brown spots of 3 to 5 mm diameter bordered by a dark margin that evolve into irregular darker lesions. Spots increase in number and size during the season. When severe infection occurs in petioles, the leaves turn yellow and premature defoliation may occur and can result in 60–85% defoliation in susceptible cultivars [9,16], thus reducing vigor and affecting yield and quality even for the next season [11,17,18]. On fruits, symptoms generally start in late spring or early summer. Spots of 1 to 3 mm appear on the lenticels, sometimes surrounded by a reddish halo, often resulting in light tissue penetration. These spots are particularly visible during harvest time and postharvest storage and make the product unmarketable for fresh consumption or downgraded for juicing, resulting in significant losses to the grower [10]. It may cause soft rot, particularly when the skin has already been damaged by other means, especially mechanical wounds [16], or cracks around the apple calyx [19]. Fruit spot can cause losses between 10 and 40% of production depending on the year, farm and variety, with the Gala and Golden Delicious varieties being the most harmed.

The saprophytic phase of *Alternaria* spp. occurs during winter when the fungus overwinters as mycelium mainly in leaf residues on the orchard floor and also in twigs or dormant buds. During spring, spore production increases on dormant leaves under suitable conditions related to the increase of temperature combined with rainfall [20,21]. Factors such as water and wind favor the spread of conidia spores into the tree canopy, where they can colonize the growing leaves and fruits, producing primary infections with lesions of varying severity depending on the intensity [3,22,23]. Primary infection takes place one month after petal fall [10]. The disease progresses rapidly in optimum temperatures ranging from 25 to 31 °C and 5.1 h of wetting and symptoms can start appearing two days after infection [9]. The freshly emerging shoots are infected from about 20 days after bloom. Disease incidence and secondary infection increases during the growing season, where peak temperatures are combined with high rainfall and relative humidity. Disease progress on leaves and fruit continues to increase until the end of summer or beginning of autumn when defoliation occurs.

Due to the first symptoms generally appearing in late spring and developing during the summer until harvest, a high number of fungicide applications must be made even close to harvest, increasing the possibility of finding residues in the fruit. The chemicals recommended for control of apple leaf blotch and fruit spot are pyraclostrobin + boscalid, captan, mancozeb, fludioxonil and fluopyram + febuconazole. Moreover, the effectiveness of fungicide applications is not enough to reduce disease severity in some cases, especially under high inoculum pressure. Eradication of primary sources of inocula has shown to be a successful management option for other diseases [24,25] and may be a good option for the control of Alternaria leaf blotch and fruit spot of apple. Application of urea on fallen leaves, mulching, removal of weeds, discarding fallen apple litter from orchards, application of lime sulphur, and manual removal of leaf residues could reduce the sources of inocula in orchards [20]. Spores also reside on twigs and buds during the winter season. Protective spray of copper-based fungicide is recommended prior to development of new leaves during the end of autumn or early spring. Selective pruning of the canopy also reduces inocula present in twigs and buds in orchards [20].

Although, the disease cycle of leaf blotch and fruit spot of apple is poorly understood, it seems that the main source of overwintering inocula are the fallen leaves on the orchard floor. Thus, in the case of *A. mali*, the number of conidia detected was higher and had higher germination in leaves rather than in buds, suggesting the importance of leaves as an important source of primary inocula [26]. While, in *Alternaria* complex related with leaf blotch and fruit spot, the cumulative spore production is higher in leaf residue rather than in twigs and canopy leaves [27]. For this reason, an environmentally friendly strategy should be focused on reducing primary inocula production to increase control efficiency and to replace at least in part the synthetic phytosanitaries. According to this, the aim of this work was to optimize the control of *Alternaria* infection through two inoculum-management strategies, the removal of winter fallen leaves to reduce the availability of inocula, and the treatment of leaves with the biological agent *T. asperellum* to inhibit or prevent inocula development on fallen leaves. Trials were performed during three consecutive years in commercial orchards to determine if the disease control achieved by the standard fungicide sprays strategy was improved by the use of additional sanitation measures. This new strategy could mean a step forward on the way to much more sustainable apple production with the final goal of production of higher overall quality with the lowest environmental impact.

2. Materials and Methods

2.1. Field Trials Design and Conditions

Four trials were carried out in two apple orchards located in Catalunya in northeastern Spain during three consecutive years, specifically during 2019, 2020, and 2021. Orchards were selected that were naturally infected by Alternaria leaf blotch and fruit spot with a history of remarkable damage. The first orchard was of the Golden Reinders variety located in Garrigàs (42.19452, 2.97272), with trees grafted onto M9 and a 3.75 × 1.2 m plantation frame. The second orchard was of the Brookfield Gala variety located in Sant Pere Pescador (42.16517, 3.09509), with trees grafted onto M9 NAKB and a 3.8 × 1.0 m plantation frame. Both orchards had a central axis training system, irrigation by a drip system and were without anti-hail nets. During the assays, fertilization, pruning, herbicide and phytosanitary treatments were conducted following standards of integrated production used in commercial apple orchards of the region. For easy identification throughout the paper, trials have been codified (Table 1).

Table 1. Characteristics of trials performed for evaluating the effect of different sanitation treatments aimed at controlling Alternaria fruit spot of apple.

Trial	Year	Country	Orchard Location	Cultivar	Plot Size	Sanitation Strategy [1]
1	2019	Spain	Sant Pere Pescador	Gala	2000 m^2	ASP
2	2019	Spain	Garrigàs	Golden	2000 m^2	ASP
3	2020	Spain	Garrigàs	Golden	3000 m^2	ASP, TRI
4	2021	Spain	Garrigàs	Golden	3000 m^2	ASP, TRI

[1] ASP—leaf aspiration; TRI—application of *Trichoderma asperellum*.

Two primary inoculum-management methods, leaf aspiration (ASP) and application of *T. asperellum* on the ground (TRI) were tested in comparison with a control where no inoculum management was performed (CNT). ASP consisted of completely removing leaves from the ground from February to middle of March depending on the trial (Table 2). The fallen leaves were collected by raking and vacuuming with a tractor-driven vacuum machine (John Deere TC125 Turf Collection System). Collected leaves were burnt in order to inactivate the *Alternaria* inoculum. TRI consisted of a single application onto the ground surface with the T34 strain of *T. asperellum* (5×10^4 cfu/g) provided by Dr. María Isabel Trillas Gay from Biocontrol Technologies S.L (Barcelona, Spain). The application was made in the tree rows in April at a dose of 0.5 g/L at an application volume of 600 L/ha, when the temperature and humidity conditions were adequate. Depending on the trial, different strategies were applied (Table 2).

Table 2. Treatments performed in orchard trials for evaluating the effect of different sanitation methods aimed at controlling Alternaria fruit spot of apple. Treatments included are no inoculum-management strategy (CNT) and aspiration strategy (ASP).

Trial	Year	Treatment	Action Date	Fungicide Application Dates [1]	Disease Evaluations [2]
1	2019	CNT	-	10/06; 12/07; 29/07	11/07; 09/08
		ASP	06-March		
2	2019	CNT	-	10/06; 12/07; 29/07; 21/08	11/07; 09/08
		ASP	06-March		
3	2020	CNT	-	09/06; 26/06; 08/07; 22/07; 09/08	27/07; 03/09
		ASP	04-Feb		
		TRI	14-April		
4	2021	CNT	-	18/06; 07/07; 20/07; 02/08	29/07; 13/09
		ASP	23-March		
		TRI	03-May		

[1] Applications were done mainly with mancozeb 75% at 3.0 kg/ha to cover rain episodes. [2] Evaluations were performed according to the presence and the evolution of symptoms.

Each strategy was arranged in a single plot of approximately 1000 m^2, that was randomly selected in each orchard. The dominant wind direction was considered at the time of distributing strategy plots, to avoid interference of sources of inocula. Each plot consisted of three rows of 80 m long in trial 1 and 90 m long in the other three trials. The external rows were considered buffer zones while the center row was where the evaluations were performed. Four replicates of ten trees were randomly selected and labeled in the central row for disease and spore release assessment.

Disease control during the growing season was based on fungicide applications. Fungicides were applied in all the plots independently of the inoculum-management strategy. Applications were made to preventively cover rain periods from June to August. The type of fungicides used were different in function for the risk period. Thus, the first rain in early June was covered with boscalid 25.2% + piraclostrobin 12.8% at 80 mL/hl (Bellis, BASF Española S.L.U, Barcelona, Spain) while the rest of the rains were covered with mancozeb 75% at 0.3 kg/hl (Vondozeb GD, UPL Iberia, S.A, Barcelona, Spain). Applications between 800 and 1000 L/ha were performed with a 3000-litre commercial Multi-fan sprayer (Teyme, Lleida, Spain). The number of applications was different in every trial and was dependent of the number of rainfalls during the risk period, from June to August (Table 2).

2.2. Spore Release

In order to determine the effect of leaf aspiration on the overwintering inocula, seasonal spore release was determined in the aspiration and control plots. Two spore traps consisting of microscope slides (2.6 cm wide and 7.6 cm long) painted with silicone solution (Lanzoni s.r.l., Bologna, Italy) and joined with a clothespin to 1 m bamboo were placed in each replicate of each plot. Thus, a total of 8 spore traps were installed per plot. These spore traps were placed in the field with the slides 75 cm above the ground and facing the main wind direction of the area. Sampling was carried out continuously for 2 years from 1 January 2020 to 31 December 2021, with slides changing every week. *Alternaria* spores are easily recognizable due to their typical morphology and were counted by direct microscopic observation (Carl Zeiss, Jena, Germany), using the methodology proposed by the Spanish Aerobiology Network, based on the observation of two longitudinal sections of a microscopic slide at a magnification of 400 [28]. Each section corresponds to a total area of 380 mm^2. Means of the number of spores per cm^2 were calculated and plotted in relation to the number of days. In addition, the area under the spore release curve (AUSRC) was calculated by the method of trapezoid integration [29].

2.3. Disease Assessment

The presence of lesions was evaluated in each replicate on 10 leaves of 20 shoots and on 100 fruits chosen at random and distributed on both sides of the trees. The incidence was assessed as the number of infected leaves or fruits over the total. For severity evaluation, each leaf or fruit was assessed according to the following semi-quantitative categorical severity index (SI): 0—no symptoms observed; (1)—up to 10% of the affected leaf surface or 1 lesion per fruit; (2)—10–50% of the affected leaf surface or 2–3 lesions per fruit; (3)—more than 50% of the affected leaf surface or more than 3 lesions per fruit. Then the following formula was used: $S = \sum_{i \to 1}^{i} \left(\frac{SI_i}{n \times 3} \right) \times 100$ (1), where S is the severity (0–100); SI is the disease severity index for each leaf or fruit; i is the number of infected fruits or leaves, n is the number of leaves or fruits; and 3 is the maximum level of severity.

2.4. Weather Conditions

In order to relate climatic factors to spore release dynamics in the orchards, weather data including daily minimum and maximum temperatures, relative humidity, wetness and rainfall were obtained from an automatic weather station placed in the experimental orchards. Temperature and relative humidity were measured every 10 min and wetness and rainfall every 20 s. Mean temperature, relative humidity, duration of wetness, and total rainfall were recorded every hour.

2.5. Data Analysis

All data are presented as mean $\pm$ standard deviation (SD). To statistically analyze results, Student's t test and one way analysis of variance (ANOVA) were applied and significant differences ($p < 0.05$) among the treatments were determined using a post-hoc Tukey HSD test when necessary. Analyses were performed using the statistical package JMP (v16, SAS Institute Inc., Cary, NC, USA).

3. Results

3.1. Effect of Treatments on Spore Release

Climatic conditions in the two years, where assessment of spore release was conducted, were very different taking into account the cumulative rainfall and the number of rainy days. The year 2020 was the rainiest, with a total of 639 mm of cumulative rainfall and 159 rainy days, while in 2021 the total cumulative rainfall was only 437.4 mm with 100 rainy days. The amount of rainfall and the rainy days affected the spore release; thus, the cumulative release of spores was higher during the year 2020 than in 2021 (Figure 1). Spore release started when the median temperature was above 12.5 °C and was associated with rain episodes. The period of maximum spore release, independent of the year, was between May and June, accumulating to 31.66% and 56.62% of the total of released spores in 2020 and 2021, respectively. During this period, the differences between released spores in the control plots and the aspiration plots were highest. After this period, the accumulation of spores slowed down during the months of July and August, with a slight increase in September, and finally stopped from October until December. In the case of the year 2020, an increase in spore release was observed due to the exceptional rainfall conditions between September and November. When the total amount of released spores was compared between control plots and aspiration plots on the basis of the AUSRC, significant reductions were observed in both years, with a similar reduction, 28.18% in 2020 and 29.07% in 2021 (Table 3). Thus, although the AUSRC was higher in 2020 because the spore release was greater throughout the season, the effect of spores release was the same in both years independent of the total amount of spores released.

When the release of spores was analyzed month by month, differences between the control and aspiration plots were very clear in some month, such as May and June, where the number of spores released in the aspiration plot decreased by 39.22% and by 34.58% in 2020, and by 70.38% and by 59.29% in 2021, respectively. This reduction was very important and occurred during the period of maximum risk of infection. In addition, in most months, the release was lower in the aspiration plots, with some exceptions that coincide with periods of low spore release. Results also show low spore release in cold months, including January, February, March and December, where the amount of spores was residual (Figure 2).

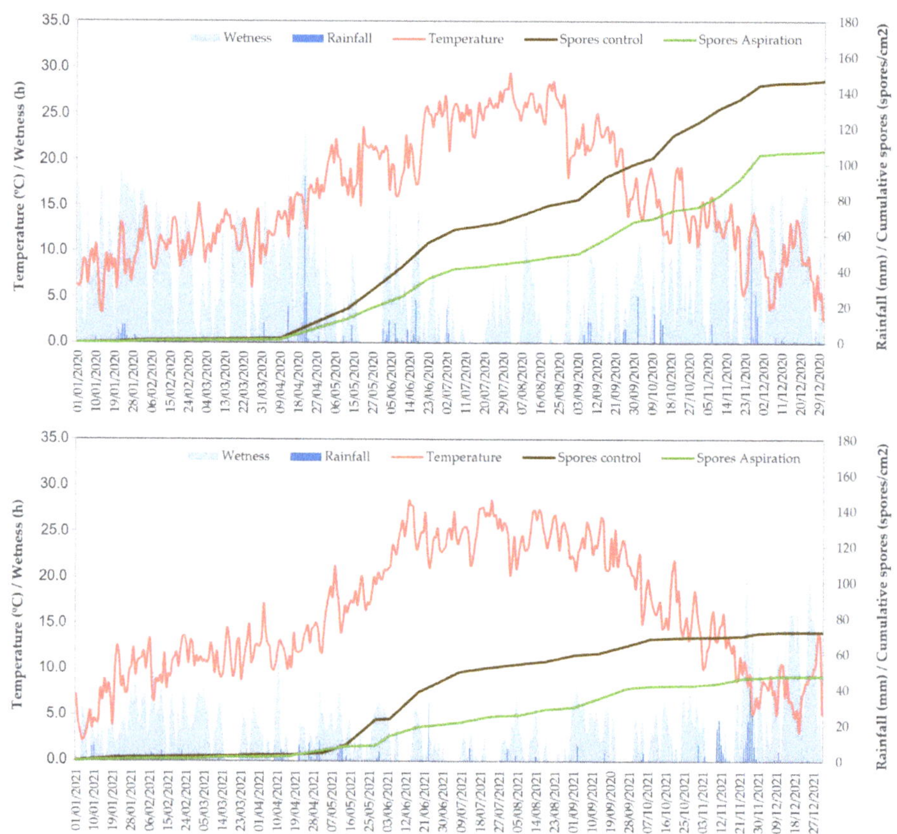

Figure 1. Dynamics of spore release and mean temperature, daily wetness duration and rainfall in trials 3 and 4 performed on 2020 and 2021, respectively.

Table 3. Comparison of area under the spore release curve (AUSRC) obtained in the control plots and the aspiration plots in 2020 and 2021 by Student's *t* test.

Trial	Year	AUSRC–CNT [1]	AUSRC-ASP [1]	F-Ratio	*p*-Value > F
3	2020	131.09	94.14	8.1021	0.0291
4	2021	63.40	44.97	6.2126	0.0470

[1] CNT—control where no inoculum management was performed; ASP, leaf aspiration.

3.2. Effects of Treatments on Alternaria Leaf Blotch and Fruit Spot

The effects of different inoculum-management strategies on the development of Alternaria leaf blotch and fruit spot were tested in trials 1, 2, 3 and 4 (Figure 3). The level of natural infection in the different trials was different depending on the year and between varieties. In 2020, the incidence of Alternaria leaf blotch and fruit spot in golden varieties were too high with values around 65% and 40%, respectively. While in the rest of the years, 2019 and 2021, the incidence was moderate, with values in the control around 20% for leaf blotch and 8% for fruit spot. In contrast, the incidence in the Gala variety was too low, with incidences of 6% and 1% of leaf blotch and fruit spot, respectively. In all cases, severity was quite lower than incidence, indicating that infections were moderate, with few leaf surfaces affected and few spots per fruit. When the evolution of symptoms was analyzed in comparison with climatic conditions, the appearance of new spots was strongly related to episodes of rain (data not shown).

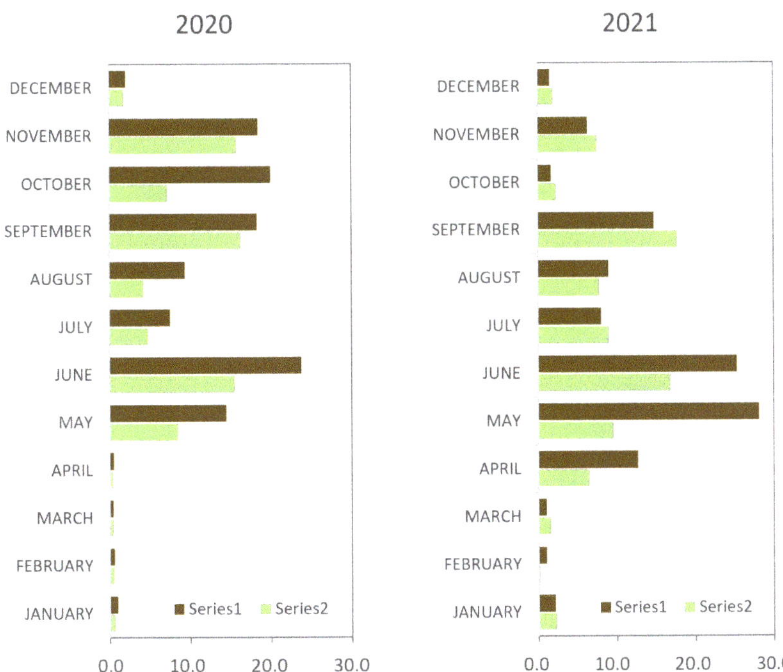

Figure 2. Spore release month by month in the control and in the aspiration plots in 2020 and 2021.

In trials 1 and 2, only aspiration of fallen leaves was tested in comparison with the control where no inoculum management was performed. In both assays, the leaf aspiration significantly reduced the incidence and severity of leaf blotch and fruit spot. The efficacy on reduction of fruit spot was higher with values above 80%, while the efficacy on the reduction of the leaf blotch was lower, with values around 40% in trial 1 and 30% in trial 2. In trials 3 and 4, both strategies were tested with good results, and a significant reduction of the incidence and severity of leaf blotch and fruit spot were observed in comparison with the control. Although, efficacies were lower than in trials 1 and 2. Again, the efficacies on the reduction of fruit spot were higher with values around 60% in trial 3 and 50% in trial 4. In contrast, the efficacies on the reduction of leaf blotch incidence and severity were slightly lower, with values around 30% in both assays. No differences were observed between the two inoculum-reduction strategies in any of the trials, showing in all cases similar efficacies.

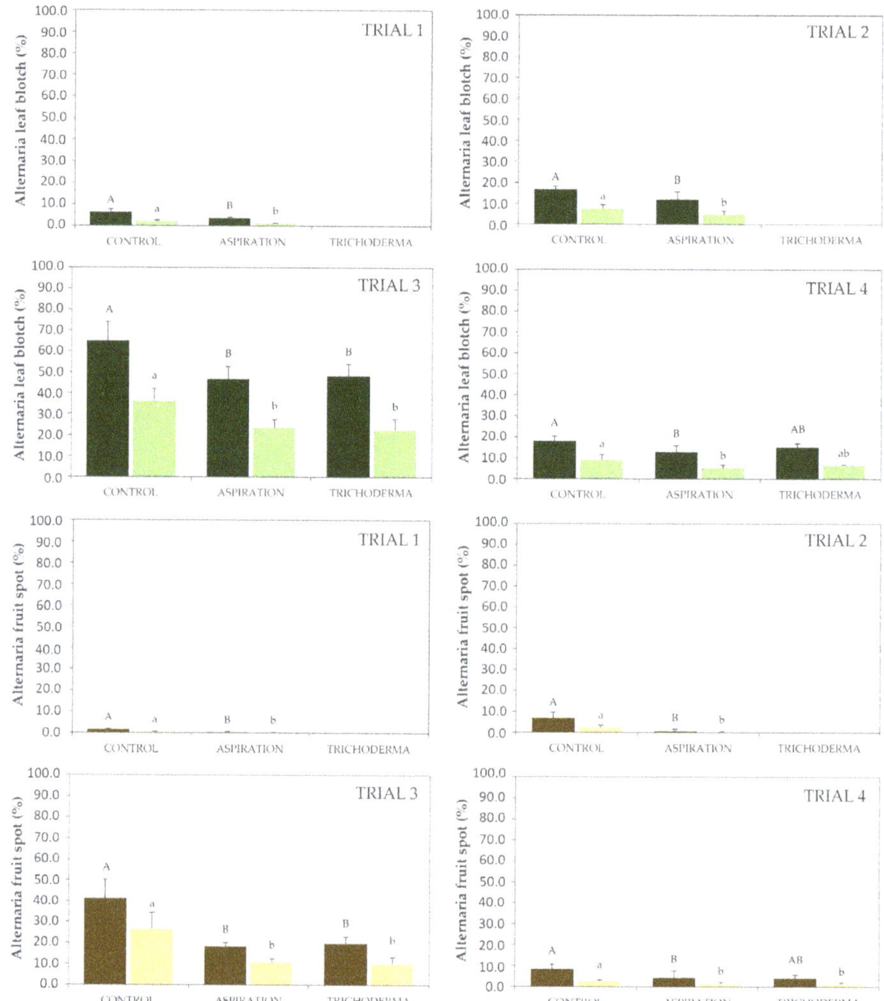

Figure 3. Incidence (dark plots) and severity (clear plots) of Alternaria leaf blotch (green series) and fruit spot (brown series) in commercial orchards where fallen leaves were collected by aspiration during winter or treated with *T. asperellum* during spring in comparison with controls without any inoculum management. Data correspond to the evaluations of 5 August 2019 for trial 1, 9 August 2019 for trial 2, 27 July 2020 for trial 3, and 29 July 2021 for trial 4. Data are presented as the mean of four replicates with the standard deviations (vertical bars). Different letters (capital letters for incidence and lowercase letters for severity) show significant differences between the treatments according to Student's *t* test ($p < 0.05$) in trials 1 and 2, or Tukey's test ($p < 0.05$, ANOVA, LSD) in trials 3 and 4.

4. Discussion

In recent years, Alternaria fruit spot has become one of the most important problems in apple production in the area of Girona in Spain. Its emergence has led to an increase of fungicide treatments and even so the production losses have been from 10 to 40% depending on the year, the orchard and the variety, with Gala and Golden being the most problematic. The disease has been expanding since 2009, arriving in 2017 to cause problems in around 20% of commercial apple orchards in Girona. Specifically, the explosion of this disease has been accompanied by an increase of between 20 and 30% in the number of treatments with fungicides. This increase is very significant, and furthermore these are concentrated very close to the harvest and have a direct effect on the residues on the fruit. This circumstance represents a serious obstacle to the aim of increasing the sustainability of exploitations, and obtaining fruit free of residue for commercial purposes. Moreover, in most years, the increase of fungicide applications has not represented a significant reduction in damage, in accordance with similar experience reported in Australia when the effectiveness of fungicide applications was often erratic in reducing disease severity and varied among regions [10]. Failure to control diseases may be due to high inoculum pressure in the orchards, but also due to the products used, which are probably not the most suitable, nor their positioning. Increasing knowledge of the identification of inoculum sources and also of the periods of maximum risk of infection is essential to attain satisfactory control of Alternaria fruit spot. According to spore release, results showed that release starts when temperatures are above 12.5 °C and ends when the temperatures drops below this value. Moreover, release is related to rain episodes indicating that rain splash helps in the release and projection of spores. In addition, when comparing released spores between 2020 and 2021, the difference was clear, being much higher in 2020 when the amount of rainfall, number of rainy days, and total wetness duration were higher. This evidence indicates that these climatic events can be considered important factors in increasing the incidence and severity during infection periods. These results are in agreement with findings for Alternaria brown spot of mandarin caused by *A. alternata* [30] and for *A. mali* in apple [17]. In this way, a comprehensive use of this information can be very useful to define risk episodes and the degree of severity, and therefore, the first step in the development of a forecasting model to predict infection episodes and help in the correct positioning of fungicides in a preventive strategy. Use of decision-support systems based on epidemic models have been demonstrated to be a very effective strategy to control fungal diseases, such as apple scab [31] and downy mildew and powdery mildew in grapevine [32], and it is a strategy to be explored and developed in the future.

When comparing the released spores in the control plots and in the aspiration plots, the amount of spores was lower in the aspiration plots, indicating that leaves are a reservoir of some disposable *Alternaria* spores, and probably one of the main inoculum sources as indicated by different authors [26,27]. Thus, the hypothesis is that *Alternaria* spores maturate in the leaf residue during autumn and winter, and probably early spring, generating the first spores that will be released during May and June. After this release, new spores are produced in infected leaves producing more spores that will be released during summer. This is supported by the fact that differences between spore release in the standard blocks and aspirated blocks are mainly in the period of May–June. After this period, light differences in spore release were observed between blocks. Historically this period of May–June has been related to the maximum risk of infection by *Alternaria*, thus limiting the release of spores during this period is important for a suitable control of infection. These results indicate that improving the management of the orchard to reduce the inocula of the pathogen through good management of leaves can help in the reduction of spore release and probably in disease control. This is in agreement with some reports that show good efficacy in the control of similar pathogens, like brown spot in pear caused by *Stemphylium vesicarium* (Wallr.) E.G. Simmons [33] or apple scab caused by *Venturia inaequalis* (Cooke) G. Winter in apple [34,35] using similar strategies focused primary on inoculum management. Therefore, it is necessary to use new control strategies based on reduction of the primary inocula to complement fungicide strategies. In this way, the findings from 3 year trials in commercial orchards showed that leaf litter management by aspiration and by *Trichoderma* application allowed a significant reduction in Alternaria leaf blotch and fruit spot development. The incidence and severity of the diseases in both leaves and fruits decreased significantly, with reductions between 50 and 80% of fruit spot and between 30 and 40% of leaf blotch depending on the year. These results support the idea that reducing the source of inocula by removing fallen leaves is a suitable strategy that can be used complementally to fungicides or biological control agent application in this area. Other studies have reported similar results in controlling other fungus diseases. Management of litter by leaf shredding and removal reduced the incidence and severity of apple scab in a French orchard by 50–80% and by about 90% in North Carolina [25,36], and successfully reduced apple scab ascospore production [34,37]. In pear it was reported that leaf shredding or removal were the most effective methods of reducing overwintering inocula of *Pleospora allii* (Rabenh.) Ces. & De Not and *S. vesicarium*, the causal agent of brown spot [33,38]. On the other hand, application of *Trichoderma* significantly decreased the incidence and severity of leaf blotch and fruit spot by around 50% and 20–30%, respectively. In agreement with this result, this biological control agent has shown the ability to reduce both overwintering inocula and conidia production of *S. vesicarium* and *P. allii* when it was applied to a pear orchard [39,40]. Moreover, *Trichoderma* inhibits the causal agent of apple ring rot *Botryosphaeria berengeriana* (De Notaris) and reduces the incidence on fruit [41], the causal agent of apple valsa canker *Valsa ceratosperma* (Tode) G.C. Adams & Rossman [42], and *Colletotrichum gloeosporioides* (Penz.) Penz. and Sacc, the cause of pre-harvest fruit drop in citrus [43]. In some of these studies, the mycoparasitic activity of *Trichoderma* was evidenced as a mechanism of action [41,42]. Although *Trichoderma* species are widely used in agriculture as biocontrol agents, their use to control *Alternaria* overwintering inocula in apples needs further studies due to its great potential. Considering that both tested strategies were applied in combination with phytosanitary treatments following the standards of integrated production used in commercial apple orchards of the region, it is suggested that the control of inocula source through alternative environmentally friendly strategies may act as an important complementary factor to fungicides, thus achieving the goal of reducing their use. This has been shown in previous studies performed in pear and apple orchards, in which also an alternative management was added to conventional fungicides [34,44]. Moreover, the consistent use of this strategy of inoculum management can be a good way to clean problematic orchards, gradually reducing the inoculum pressure. This can be an important aspect to take into account in organic orchards, where the limited number of fungicides make the efficient control of Alternaria fruit spot difficult.

5. Conclusions

The findings of this study point in the direction that sanitation of orchards through the elimination or the treatment with *Trichoderma* of the falling leaves reduces spore release mainly during the period between May and June. Moreover, this reduction is directly related to Alternaria leaf blotch and fruit rot reduction suggesting that this strategy based on the reduction of the primary inocula can be useful to complement a fungicide strategy.

Author Contributions: Conceptualization, J.C. and P.V.; Methodology, J.C. and P.V.; Formal analysis, J.C.; Investigation, J.C. and P.V.; Resources, M.V.S. and J.C.; Writing—original draft preparation, M.V.S. and J.C.; Writing—review and editing, M.V.S., J.C. and P.V.; Funding acquisition, J.C. and P.V. All authors have read and agreed to the published version of the manuscript.

Funding: This research has been funded by innovative pilot projects in agricultural productivity and sustainability by Operational Groups of the Generalitat de Catalunya and the European agricultural fund for rural development (EAFRD) with grant number 56.21.023.2019.P4.

Institutional Review Board Statement: Not applicable.

Informed Consent Statement: Not applicable.

Data Availability Statement: Data is contained within the article.

Acknowledgments: We thank cooperating fruit growers for allowing us to use their orchards for studies, and Cesca Alcalá and Anaís Marsal for technical assistance.

Conflicts of Interest: The authors declare no conflict of interest.

References

1. Gao, L.L.; Zhang, Q.; Sun, X.Y.; Jiang, L.; Zhang, R.; Sun, G.Y.; Zha, Y.L.; Biggs, A.R. Etiology of moldy core, core browning, and core rot of Fuji apple in China. *Plant Dis.* **2013**, *97*, 510–516. [CrossRef] [PubMed]
2. Johnson, R.D.; Johnson, L.; Kohmoto, K.; Otani, H.; Lane, C.R.; Kodama, M. A Polymerase chain reaction-based method to specifically detect *Alternaria alternata* apple pathotype (*A. mali*), the causal agent of Alternaria blotch of apple. *Phytopathology* **2000**, *90*, 973–976. [CrossRef] [PubMed]
3. Gur, L.; Reuveni, M.; Cohen, Y. Occurrence and etiology of Alternaria leaf blotch and fruit spot of apple caused by *Alternaria alternata* f. sp. mali on cv. Pink Lady in Israel. *Eur. J. Plant Pathol.* **2017**, *147*, 695–708. [CrossRef]
4. Grove, G.G.; Eastwell, K.C.; Jones, A.L.; Sutton, T.B. Diseases of apple. In *Apples: Botany, Production and Uses*; Feree, D.C., Warrington, I.J., Eds.; CABI Publishing: Cambridge, MA, USA, 2003; Chapter 18; pp. 459–488. [CrossRef]
5. Roberts, J.W. Morphological characters of *Alternaria mali*. *J. Agric. Res.* **1924**, *25*, 699–708.
6. Wenneker, M.; Pham, K.T.K.; Woudenberg, J.H.C.; Thomma, B.P.H.J. First report of *Alternaria arborescens* species complex causing leaf blotch and associated premature leaf drop of 'Golden Delicious' apple trees in the Netherlands. *Plant Dis.* **2018**, *102*, 1654. [CrossRef]
7. Toome-Heller, M.; Baskarathevan, J.; Burnip, G.; Alexander, B. First Report of apple leaf blotch caused by *Alternaria arborescens* complex in New Zealand. *N. Z. J. Crop Hortic. Sci.* **2018**, *46*, 354–359. [CrossRef]
8. Rotondo, F.; Collina, M.; Brunelli, A.; Pryor, B.M. Comparison of *Alternaria* spp. collected in Italy from apple with *A. mali* and other AM-toxin producing strains. *Phytopathology* **2012**, *102*, 1130–1142. [CrossRef]
9. Filajdic, N.; Sutton, T.B. Influence of temperature and wetness duration on infection of apple leaves and virulence of different isolates of *Alternaria mali*. *Phytopathology* **1992**, *82*, 1279–1283. [CrossRef]
10. Harteveld, D.O.C.; Akinsanmi, O.A.; Drenth, A. Multiple *Alternaria* species groups are associated with leaf blotch and fruit spot diseases of apple in Australia. *Plant Pathol.* **2013**, *62*, 289–297. [CrossRef]
11. Fontaine, K.; Fourrier-Jeandel, C.; Armitage, A.D.; Boutigny, A.L.; Crépet, M.; Caffier, V.; Gnide, D.C.; Shiller, J.; Cam, B.L.; Giraud, M.; et al. Identification and pathogenicity of *Alternaria* species associated with leaf blotch disease and premature defoliation in French apple orchards. *PeerJ* **2021**, *9*, e12496. [CrossRef]
12. Bulajic, A. First report of *Alternaria mali* on apples in Yugoslavia. *Plant Dis.* **1996**, *80*, 709. [CrossRef]
13. Marschall, K.; Bertagnoll, M. *Alternaria alternata*, causal agent of lenticel rot and leaf necrosis on apple in Italy [*Malus pumila* Mill.; South Tyrol]. In Proceedings of the Atti delle Giornate Fitopatologiche, Bologna, Italy, 27–29 March 2006.
14. Vilardell, P. Una nueva enfermedad causada por *Alternaria* afecta a plantaciones de manzana en Girona. *Vida Rural.* **2018**, *448*, 30–33.
15. Generalitat de Catalunya de Departament d'Acció Climàtica, Alimentació i Agenda Rural. *Estadístiques Defin. De Conreus.* Available online: https://agricultura.gencat.cat/ca/departament/estadistiques/agricultura/estadistiques-definitives-conreus (accessed on 19 December 2022).
16. Jung, K.-H. Growth inhibition effect of pyroligneous acid on pathogenic fungus, *Alternaria mali*, the agent of Alternaria blotch of apple. *Biotechnol. Bioprocess Eng.* **2007**, *12*, 318–322. [CrossRef]
17. Filajdic, N.; Sutton, T.B.; Walgenbach, J.F.; Unrath, C.R. The influence of European red mites on intensity of Alternaria blotch of apple and fruit quality and yield. *Plant Dis.* **1995**, *79*, 683–690. [CrossRef]
18. Cooke, T.; Persley, D.; House, S. (Eds.) *Diseases of Fruit Crops in Australia*; CSIRO Publishing: Collingwood, Australia, 2019; p. 279. ISBN 9780643069718. [CrossRef]
19. Stern, R.; Ben-Arie, R.; Ginzberg, I. Reducing the incidence of calyx cracking in 'Pink Lady' apple using a combination of cytokinin 6-benzyladenine and gibberellins (GA4+7). *J. Hortic. Sci. Biotechnol.* **2013**, *88*, 147–153. [CrossRef]
20. Madhu, G.S.; Sajad, U.N.; Mir, J.I.; Raja, W.H.; Sheikh, M.A.; Sharma, M.A.; Singh, D.B. Alternaria leaf and fruit spot in apple: Symptoms, cause and management. *Eur. J. Biotechnol. Biosci.* **2020**, *8*, 24–26.
21. Rozo, M.E.B.; Pinto, V.F.; Pose, G. Especies de *Alternaria* asociadas a cultivos de manzana y pera en la región del Alto Valle del Río Negro, Argentina. *Cult. Cient.* **2019**, *17*, 18–31. [CrossRef]
22. Linares, C.D.; Belmonte, J.; Canela, M.; de la Guardia, C.D.; Alba-Sanchez, F.; Sabariego, S.; Alonso-Pérez, S. Dispersal patterns of *Alternaria* conidia in Spain. *Agric. For. Meteorol.* **2010**, *150*, 1491–1500. [CrossRef]
23. Harteveld, D.O.C.; Akinsanmi, O.A.; Chandra, K.; Drenth, A. Timing of infection and development of *Alternaria* diseases in the canopy of apple trees. *Plant Dis.* **2014**, *98*, 401–408. [CrossRef]
24. Kim, Y.C.; Lee, J.H.; Bae, Y.-S.; Sohn, B.-K.; Park, S.K. Development of effective environmentally-friendly approaches to control Alternaria blight and anthracnose diseases of Korean ginseng. *Eur. J. Plant Pathol.* **2010**, *127*, 443–450. [CrossRef]
25. Gomez, C.; Brun, L.; Chauffour, D.; Vallée, D.D.L. Effect of leaf litter management on scab development in an organic apple orchard. *Agric. Ecosyst. Environ.* **2007**, *118*, 249–255. [CrossRef]
26. Filajdic, N.; Sutton, T.B. Overwintering of *Alternaria mali*, the causal agent of Alternaria blotch of apple. *Plant Dis.* **1995**, *79*, 695–698. [CrossRef]
27. Harteveld, D.O.C.; Akinsanmi, O.A.; Dullahide, S.; Drenth, A. Sources and seasonal dynamics of *Alternaria* inoculum associated with leaf blotch and fruit spot of apples. *Crop Prot.* **2014**, *59*, 35–42. [CrossRef]
28. Almaguer, M.; Díaz, L.; Fernández-González, M.; Valdéz, E. Allergenic fungal spores and hyphal fragments in the aerosol of Havana, Cuba. *Aerobiologia* **2020**, *36*, 441–448. [CrossRef]

29. Scherm, H.; Savelle, A.T.; Boozer, R.T.; Foshee, W.G. Seasonal dynamics of conidial production potential of *Fusicladium carpophilum* on twig lesions in southeastern peach orchards. *Plant Dis.* **2008**, *92*, 47–50. [CrossRef]

30. Bassimba, D.D.M.; Mira, J.L.; Vicent, A. Inoculum sources, infection periods, and effects of environmental factors on Alternaria brown spot of mandarin in mediterranean climate conditions. *Plant Dis.* **2014**, *98*, 409–417. [CrossRef]

31. Jamar, L.; Cavelier, M.; Lateur, M. Primary scab control using a "during-infection" spray timing and the effect on fruit quality and yield in organic apple production. *Biotechnol. Agron. Soc. Environ.* **2010**, *14*, 423–439.

32. Caffi, T.; Rossi, V.; Bugiani, R. Evaluation of a warning system for controlling primary infections of grapevine downy mildew. *Plant Dis.* **2010**, *94*, 709–716. [CrossRef]

33. Llorente, I.; Vilardell, A.; Vilardell, P.; Pattori, E.; Bugiani, R.; Rossi, V.; Montesinos, E. Control of brown spot of pear by reducing the overwintering inoculum through sanitation. *Eur. J. Plant Pathol.* **2010**, *128*, 127–141. [CrossRef]

34. Holb, I.J. Effect of six sanitation treatments on leaf litter density, ascospore production of *Venturia inaequalis* and scab incidence in integrated and organic apple orchards. *Eur. J. Plant Pathol.* **2006**, *115*, 293–307. [CrossRef]

35. Beckerman, J.; Abbott, C. Comparative Studies on the effect of adjuvants with urea to reduce the overwintering inoculum of *Venturia inaequalis*. *Plant Dis.* **2019**, *103*, 531–537. [CrossRef] [PubMed]

36. Sutton, D.K.; MacHardy, W.E.; Lord, W.G. Effects of shredding or treating apple leaf litter with urea on ascospore dose of *Venturia inaequalis* and disease buildup. *Plant Dis.* **2000**, *84*, 1319–1326. [CrossRef] [PubMed]

37. Vincent, C.; Rancourt, B.; Carisse, O. Apple leaf shredding as a non-chemical tool to manage apple scab and spotted tentiform leafminer. *Agric. Ecosyst. Environ.* **2004**, *104*, 595–604. [CrossRef]

38. Llorente, I.; Vilardell, A.; Montesinos, E. Infection potential of *Pleospora allii* and evaluation of methods for reduction of the overwintering inoculum of brown spot of pear. *Plant Dis.* **2006**, *90*, 1511–1516. [CrossRef] [PubMed]

39. Moragrega, C.; Carmona, A.; Llorente, I. Biocontrol of *Stemphylium vesicarium* and *Pleospora allii* on pear by *Bacillus subtilis* and *Trichoderma* spp.: Preventative and curative effects on inoculum production. *Agronomy* **2021**, *11*, 1455. [CrossRef]

40. Rossi, V.; Pattori, E. Inoculum reduction of *Stemphylium vesicarium*, the causal agent of brown spot of pear, through application of *Trichoderma*-based products. *Biol. Control.* **2009**, *49*, 52–57. [CrossRef]

41. Kexiang, G.; Xiaoguang, L.; Yonghong, L.; Tianbo, Z.; Shuliang, W. Potential of *Trichoderma harzianum* and *T. Atroviride* to Control *Botryosphaeria berengeriana* f. sp. *piricola*, the cause of apple ring rot. *J. Phytopathol.* **2002**, *150*, 271–276. [CrossRef]

42. Valetti, L.; Lima, N.B.; Cazón, L.I.; Crociara, C.; Ortega, L.; Pastor, S. Mycoparasitic *Trichoderma* isolates as a biocontrol agent against *Valsa ceratosperma*, the causal agent of apple valsa canker. *Eur. J. Plant Pathol.* **2022**, *163*, 923–935. [CrossRef]

43. Vu, T.X.; Tran, T.B.; Hoang, C.Q.; Nguyen, H.T.; Do, L.M.; Dinh, M.T.; Thai, D.H.; Tran, T.V. Potential of *Trichoderma* asperellum as a biocontrol agent against citrus diseases caused by *Penicillium digitatum* and *Colletotrichum gloeosporioides*. *Int. J. Agric. Technol.* **2021**, *1*, 2005–2020.

44. Llorente, I.; Vilardell, A.; Vilardell, P.; Montesinos, E. Evaluation of new methods in integrated control of brown spot of pear (*Stemphylium vesicarium*, teleomorph *Pleospora allii*). *Acta Hortic.* **2008**, *800*, 825–832. [CrossRef]

 *microorganisms*

Review

Evaluation of Control Strategies for *Xylella fastidiosa* in the Balearic Islands

Bàrbara Quetglas [1], Diego Olmo [2], Alicia Nieto [2], David Borràs [2], Francesc Adrover [2,3], Ana Pedrosa [1,3], Marina Montesinos [1], Juan de Dios García [3], Marta López [3], Andreu Juan [3] and Eduardo Moralejo [1,*]

[1] TRAGSA Group, 07009 Palma de Mallorca, Spain
[2] Institute of Agro-Food and Fisheries Research and Training of the Balearic Islands (IRFAP), 07009 Palma de Mallorca, Spain
[3] Agriculture Service, Ministry of Agriculture, Fisheries and Food of the Balearic Government, 07006 Palma de Mallorca, Spain
* Correspondence: emoralejor@gmail.com

Abstract: The emergence of *Xylella fastidiosa* (Xf) in the Balearic Islands in October 2016 was a major phytosanitary challenge with international implications. Immediately after its detection, eradication and containment measures included in Decision 2015/789 were implemented. Surveys intensified during 2017, which soon revealed that the pathogen was widely distributed on the islands and eradication measures were no longer feasible. In this review, we analyzed the control measures carried out by the Balearic Government in compliance with European legislation, as well as the implementation of its control action plan. At the same time, we contrasted them with the results of scientific research accumulated since 2017 on the epidemiological situation. The case of Xf in the Balearic Islands is paradigmatic since it concentrates on a small territory with one of the widest genetic diversities of Xf affecting crops and forest ecosystems. We also outline the difficulties of anticipating unexpected epidemiological situations in the legislation on harmful exotic organisms on which little biological information is available. Because Xf has become naturalized in the islands, coexistence alternatives based on scientific knowledge are proposed to reorient control strategies towards the main goal of minimizing damage to crops and the landscape.

Keywords: risk assessment; phytosanitary measures; invasive pathogens; vector-borne disease; *Philaenus spumarius*

Citation: Quetglas, B.; Olmo, D.; Nieto, A.; Borràs, D.; Adrover, F.; Pedrosa, A.; Montesinos, M.; de Dios García, J.; López, M.; Juan, A.; et al. Evaluation of Control Strategies for *Xylella fastidiosa* in the Balearic Islands. *Microorganisms* **2022**, *10*, 2393. https://doi.org/10.3390/microorganisms10122393

Academic Editor: Ana Palacio-Bielsa

Received: 14 November 2022
Accepted: 29 November 2022
Published: 2 December 2022

Publisher's Note: MDPI stays neutral with regard to jurisdictional claims in published maps and institutional affiliations.

1. Introduction

No other activity of humanity has previously been impacted the negative side effects of globalization like agriculture and forestry [1]. Since the beginning of the Modern Age, the economic damage caused by invasive pathogens and pests to crops and forests has had a considerable impact [2,3]. Even today, between 20% and 40% of harvests in main crops are lost directly or indirectly due to pathogens and pests [4]. To reduce the risk of introducing alien pathogens while minimally interfering with international trade, phytosanitary regulations have been accorded among countries through the International Plant Protection Convention (IPPC). In Europe, although plant health legislation is implemented with high standards and scientific consensus [3], it has failed to prevent the entry of harmful alien pathogens, such as *Phytophthora ramorum*, *Hymenoscyphus fraxineus*, and *Xylella fastidiosa* [5–7].

The bacterium *Xylella fastidiosa* (Xf) native to the American continent is one of the most feared phytopathogens in the world due to the damage it causes to crops of great economic value such as grapevine, citrus, coffee, almond, and olive trees, among others [8]. Xf is transmitted non-specifically by insects belonging to sharpshooter leafhoppers (Hemiptera: Cicadellidae, Cicadellinae) and spittlebugs (Hemiptera: Cercopoidae) [9]. Once introduced into the xylem vessels, the bacterium colonizes the vascular system of the plant, compromising the water supply in susceptible hosts. Xf's bad reputation stems mainly from the fact there are no known curative treatments for infected plants [10]. Although experimental injection of antibiotics (e.g., streptomycin) into infected plants reduces disease symptoms, the infection is not eliminated [11,12] and reappears if treatment is stopped. Furthermore, antibiotics are banned in the EU as a treatment against bacterial plant diseases [13]. Several minerals and compounds in addition to microbial endophytes as biocontrol agents have been tested in America and Italy, showing some protection against Xf infections in grapevines [14], citrus [15], and olive trees [16]; however, their large-scale application is still economically expensive [13]. For this reason, efforts are also being made to find alternative control systems to those directed solely at the bacterium as an economically sustainable strategy over time.

Other factors adding difficulties to control strategies are the diversity and wide host range of the pathogen [17]. Three main subspecies with allopatric origins, *pauca*, *fastidiosa* and *multiplex*, are known from South, Central, and North America, respectively [8]. Within each subspecies, diverse genetic lineages with different host ranges have evolved [14]. Genetic recombination among subspecies seems to provide the main source of genetic variation, which may lead to host jumps [17,18]. To date, Xf as a taxonomic unit is known to infect over 638 plant species [19]. However, each of the 90 known sequence-type (ST) profiles has smaller host ranges, sometimes with overlapping hosts [20,21].

Although there have been some previous unconfirmed reports [22], Xf was first detected in Europe on olive trees in Apulia, Italy [7]. The olive quick decline syndrome (OQDS) induced by Xf has destroyed millions of olive trees in Apulia since 2013, causing a significant economic and landscape impact [23]. In 2015, Xf was detected on the island of Corsica and shortly after in the Provence-Alpes-Côte d'Azur in France [24], followed by the Balearic Islands in October 2016 [25]. Currently, the pathogen is also established in Alicante (Spain) [26] and Israel [27], and new outbreaks have emerged in Tuscany [28] and Portugal [29]. The origin of these introductions and their recent evolutionary history have been investigated [30]. This is a necessary first step to search for control strategies, as will be explained later.

In the Balearic Islands, several factors have converged to considerably delay Xf detection: (i) Xf strains are not excessively virulent on the main hosts, almond and wild olive trees; (ii) symptom development coincides with the summer drought peak; (iii) the presence of other wood or root pathogens is already established causing basal mortality rates enhanced by drought; (iv) there is a lack of renewal of an ageing almond tree population together the abandonment of land care due to low yields; and finally, (v) and perhaps most importantly, Xf was not expected.

Since Xf first detection in the Balearic Islands in 2016, phytosanitary measures, included in Decision (EU) 2015/789 and the updated Regulation (EU) 2020/1201, have been implemented, while additional measures have been enacted by regional and Spanish authorities. At the same time, research programs were initiated to find out the chronology of the introductions and the incidence of the pathogen in crops. To date, Xf in the islands shows one of the largest genetic diversities in Europe, while their impact on various crops and forest species allows for an evaluation of the implementation of quarantine measures in a varied context. We believe that the pathogen's situation in the Balearic Islands is illustrative of the difficulties of legislating on introduced organisms with a complex and poorly understood biology and wide host range. Paradoxically, Xf would likely have entered Mallorca in 1993, around 22 years before the legislation to be applied to control the spread of the pathogen in Europe was launched and eight years before being in force the Council Directive 2000/29/EC on protective measures against the introduction and spread of organisms harmful to plants inEurope. In this review, we expose the chronology of the events and how, as the research on the origin and phylogenetic relationships of the Xf populations of the Balearic Islands progresses, the perception of the effectiveness of phytosanitary measures is changing. Our purpose is to focus exclusively to control strategies in the Balearic Islands, as excellent general reviews of control attempts and management of Xf have been recently published [31]

2. Action Plans to Combat *Xylella fastidiosa* in the Balearic Islands

On November 25, 2016, an outbreak of *Xylella fastidiosa* (Wells et al. [32]) was officially declared in the Balearic Islands, and thereby a containment plan was adopted to eradicate and control it after the Resolution of the Ministry of the Environment, Agriculture and Fisheries of the Balearic Islands Government. A demarcated area was established around the outbreak focus in a garden center in Porto Cristo, Mallorca, and an infected and buffer zone was defined (Article 4, Decision 2015/789). The action plan for the implementation of phytosanitary measures was coordinated by the phytosanitary authority, the General Directorate of the Agriculture Department of the of Agriculture, Livestock and Rural Development of the Government of the Balearic Islands in collaboration with the other local administrations and the General Directorate of Natural Spaces and Biodiversity. In addition, a management and coordination group together with a scientific group were designated by the official body. A mixed commission between the Balearic Government and the Civil Guard was formed to coordinate the controls at airports, ports, and roads and prevent the spread of the Xf outside the Balearic territory and between islands.

Mandatory surveys began with the entry into force of the decision and were intensified following the first detection of Xf. The Official Plant Health Laboratory of the Balearic Islands (hereinafter LOSVIB) was in charge of analyzing plant samples collected in the demarcated zone and beyond at the end of 2016. The result analysis successively revealed that the pathogen was widely established in Mallorca, Ibiza, and Menorca. Consequently, the buffer zone had to be reviewed at each new outbreak communication, soon extending to almost the entire surface of the islands in June 2017 (Figure 1). At the same time, the local media echoed the situation of the new pathogen, causing some concern in public opinion [33].

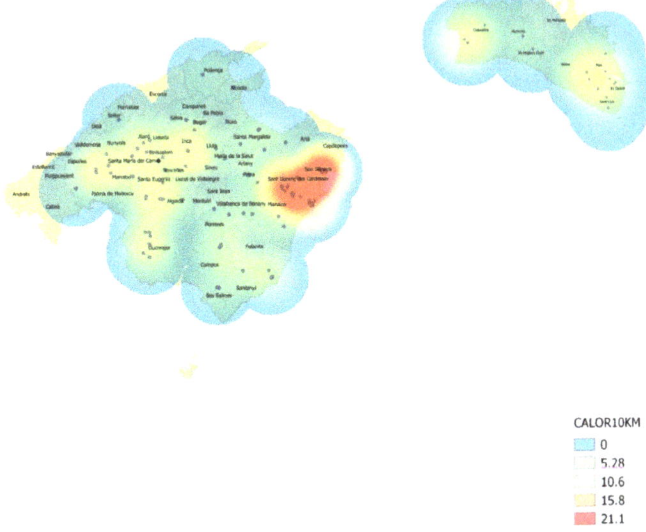
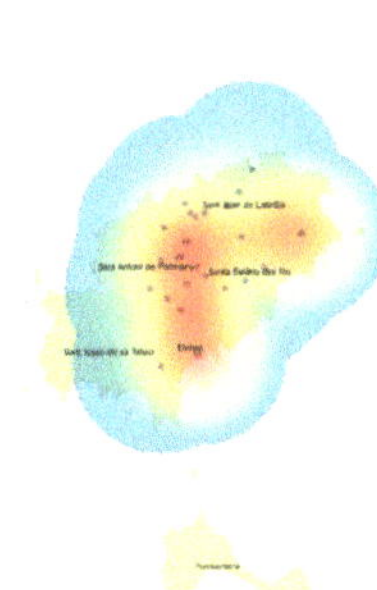
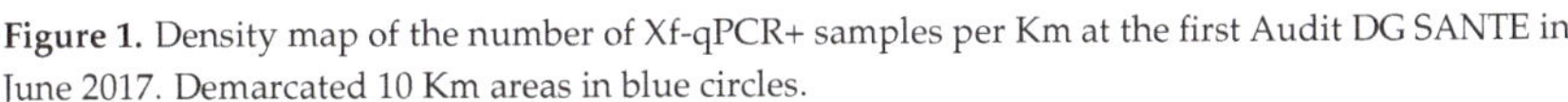

Figure 1. Density map of the number of Xf-qPCR+ samples per Km at the first Audit DG SANTE in June 2017. Demarcated 10 Km areas in blue circles.

From October 2016 to the end of 2022, significant resources have been dedicated to understanding the situation of Xf in the field. The LOSVIB has analyzed more than 18,016 plant samples collected from a total of 382 plant species. In the 32 notifications to the SG SANTE, 1360 Xf-qPCR positives and 37 species as host plants have been declared (Figure 2, Table 1). All samples were tested according to the EPPO Diagnostic Protocol (PM 7/24(4), [29]) both on symptomatic and asymptomatic plants, as well as on insect vectors. All new qPCR-Xf-positive hosts were confirmed by the National Reference Laboratory for phytopathogenic bacteria in Valencia, Spain, and their genetic profiles were determined by MLST in the IAS-CSIC laboratory in Córdoba, Spain. To date, three subspecies and four sequence types (ST1, ST7, ST80, and ST 81) have been identified on the islands [30], and the genomes of several isolates have been sequenced [34,35]. More details on the situation of Xf in the Balearic Islands can be found in Olmo et al. [36].

Table 1. Host plants for *Xylella fastidiosa* recorded in the Balearic Islands in 2022.

Host	Family	Islands [1]	Strain [2]
Acacia saligna	Leguminosae	Ma ǀ Ib	ST81 ǀ ST80
Calicotome spinosa	Leguminosae	Ma	ST1
Cistus albidus	Cistaceae	Ma ǀ Me ǀ Ib	ST81 ǀ ST81 ǀ ST80
Cistus monspeliensis	Cistaceae	Ma	ST1
Clematis cirrhosa	Ranunculaceae	Me	ST81
Elaeagnus angustifolia	Elaeagnaceae	Ib	Not determined
Ficus carica	Moraceae	Ma ǀ Me	ST81 ǀ ST81
Fraxinus angustifolia	Oleaceae	Ma	ST81
Genista hirsuta	Leguminosae	Ib	Not determined
Genista lucida	Leguminosae	Ma	ST1
Genista valdes-bermejoi	Leguminosae	Ma	ST81
Helichrysum stoechas	Compositae	Ma ǀ Me	Not determined ǀ ST81
Juglans regia	Juglandaceae	Ma	ST1
Lavandula angustifolia	Labiatae	Ma ǀ Ib	ST81 ǀ ST80
Lavandula dentata	Labiatae	Ma ǀ Ib	ST81 ǀ ST80
Nerium oleander	Apocynaceae	Ma ǀ Ib	ST81 ǀ Not determined
Olea europaea var. europaea	Oleaceae	Ma ǀ Me ǀ Ib	ST81 ǀ ST81 ǀ ST80
Olea europaea var. sylvestris	Oleaceae	Ma ǀ Me ǀ Ib	ST81 ǀ ST81 ǀ ST80
Phagnalon saxatile	Compositae	Ma	Not determined
Phillyrea angustifolia	Oleaceae	Ma	ST81
Phlomis italica	Labiatae	Me	Not determined
Polygala myrtifolia	Polygalaceae	Ma ǀ Ib	ST1, ST7, ST81 ǀ ST80
Prunus avium	Rosaceae	Ma	ST1
Prunus domestica	Rosaceae	Ma	ST81
Prunus dulcis	Rosaceae	Ma ǀ Me ǀ Ib	ST1, ST7, ST81 ǀ ST81 ǀ ST80
Rhamnus alaternus	Rhamnaceae	Ma ǀ Me	ST1, ST81 ǀ ST81
Rosmarinus officinalis	Labiatae	Ma ǀ Me ǀ Ib	ST81 ǀ ST81 ǀ ST80
Ruta chalepensis	Rutaceae	Ma	ST1
Santolina chamaecyparissus	Compositae	Ma ǀ Me	Not determined ǀ ST81
Santolina magonica	Compositae	Me	ST81
Salvia officinalis	Labiatae	Ma	ST81
Spartium junceum	Leguminosae	Ma	ST81
Teucrium capitatum	Labiatae	Ma	ST1
Thymus vulgaris	Labiatae	Ib	Not determined
Ulex parviflorus	Leguminosae	Ib	ST80
Vitex agnus-castus	Verbenaceae	Me	Not determined
Vitis vinifera	Vitaceae	Ma	ST1

[1] Ma= Mallorca; Ib= Ibiza; Me= Menorca. [2] ST= sequence types found in the Balearic Islands.

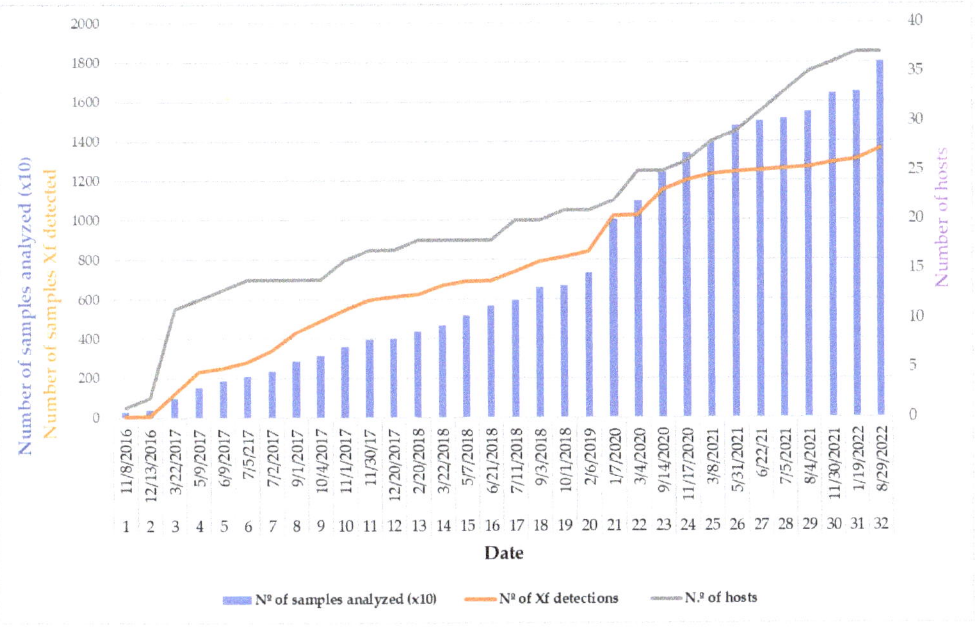

Figure 2. Progression of the number of total samples, Xf-qPCR-positive samples, and hosts over time in the Balearic Islands.

As surveys intensified, awareness grew that the strict application of Decision 2015/789 was not possible in the Balearic Islands. These concerns of the official body were discussed and exposed in the commission audits of June 2017. Shortly afterwards, as a result, the Balearic Islands were declared an infected area in the Execution Decision (EU) 2017/2352 of the commission of 14 December 2017, which amends Execution Decision (EU) 2015/789. Since then, containment measures could be applied instead of eradication measures in Corsica and the Balearic Islands. In both territories, the bacterium was widely established and could no longer be eradicated. Elimination was limited to infected and positive plants within a 50 m radius. In the new Regulation (EU) 2020/1201 in force that repeals Decision 2015/789, these areas are included in Annex III, with the containment measures focused on minimizing the amount of bacterial inoculum and keeping the vector population in check, the lowest possible levels. The containment measures are applied in the infected areas and are aimed at the elimination and destruction exclusively of those infected plants detected within the framework of the annual control that is carried out in certain places within the vicinity of plant complexes with cultural and social interest.

2.1. Elimination of Plants in Infected Areas

All plant samples that enter the LOSVIB are registered in an official database that includes relevant information on the host, location, date, etc., as well as the coordinates where the sample was collected. Decision (EU) 2015/789 required all member states to eliminate all host plants, regardless of their phytosanitary status, within a 100 m radius around the Xf-positive plant. In the amendment of the decision in December 2017, given that the strict application of Article 6 would mean wiping out the entire agricultural landscape, the phytosanitary status of the Balearic Islands was modified by declaring the entire archipelago as an infected zone. In the new status, the containment plan approved allows only the affected plants to be removed, ensuring that no new plant infections occur within a radius of 50 m.

In the currently in-force Regulation (EU) 2020/1201, all qPCR-positive plants detected in an infected zone are eliminated, except plants for scientific purposes or with particular cultural or social value. Since October 2016, a total of 13,757 plants were eliminated on all the islands, 10,113 plants in Mallorca, 329 in Menorca, 3314 in Ibiza, and 1 in Formentera. Of these eliminations, only 1276 were positive samples for Xf, with a total of 37 host species in all the Balearic Islands (Table 1). The distribution by islands is as follows: 773 positive samples were from Mallorca, 369 from Ibiza, and 218 from Menorca, obtaining together a total of 1360 positive samples, and thus approximately 94% of the positives were eliminated.

2.2. Measures against Vectors of the Specified Pest in Containment Zones

In compliance with Article 7 of Decision (EU) 2015/789, appropriate agricultural practices for vector management in the demarcated area have been promoted by the official body. Training courses have been held for farmers on the management of vectors in crops. In 2018, a phytosanitary supply campaign was started to combat Xf. Among the containment measures, farmers have been subsided for the use of the following authorized insecticides: Kaolin 95% WP p/p; Deltamethrin 10% (EC) p/v; and Lambda-Cyhalothrin 10% (CS) p/v on olive trees; Deltamethrin 10% (EC) p/v, Lambda-Cyhalothrin 10% (CS), Pyrethrin 4% (EC) p/v, and Azadirachtin 3,2% (EC) p/v on grapevines; and Deltamethrin 10% (EC) p/v and Lambda-Cyhalothrin 10% (CS) on almond trees. Between 2019 and 2021, a total of 2675 ha of almond plantations, 1888 ha of olive trees, and 1440 ha of vineyards have been treated with insecticides to control vector populations.

2.3. Annual Surveillance of Infected Areas

Until the amendment of Decision (EU) 2015/789, with the entry into force of Regulation (EU) 2020/1201, censuses and two annual samplings of host plants and vectors were carried out within a 100 m radius of the positive sample point, including one during the vector's flight season. Since 2018, sampling has been reduced to an annual one with a 50 m radius. In both cases, Xf-positive plants within the radius were eliminated. Until 2020, 1563 samples were analyzed. Since 2020, the risk-based estimate of the system sensitivity tool RIBESS+ has been applied to sample the specified plants in places that have particular cultural, social, or scientific values.

A total of 3001 samples were recorded during the 2018–2021 period that were within the infected zone, and of these, 76% of the infected radios were inspected. Given that the average density of hosts is 100 for every 50 m of radius and the occupied surfaces are more than 673 ha, the measure supposes analyzing the unfeasible quantity of more than 60,000 samples per year. Within the areas of interest with social and cultural value, careful monitoring of the germplasm banks has been carried out on 499 crop varieties in four fields in Mallorca, 117 varieties in Ibiza in one field, and 448 in Menorca in one field.

2.4. Authorization Regarding the Planting of Specified Plants in Infected Areas

A strict application of Article 5 of the EU Decision 2015/789 that prohibits the planting of hosts in infected areas would have condemned almond tree plantations and vineyards in Mallorca and olive trees in Ibiza to extinction. Since 2017, massive sampling and field studies have been carried out in which information has been collected on the epidemiological situation of different crop varieties. At the same time, inoculation tests have been conducted in an insect-proof net tunnel exposed to environmental temperature to determine the most resistant or susceptible varieties. In addition, germplasm banks of crop varieties have been intensively monitored and sampled during the past years. This information has been used to legislate the planting authorization of certain varieties through the Resolution of 14 February 2018, of the General Directorate of Agricultural Production Health, which approves the request of the Autonomous Community of the Balearic Islands to the planting of certain host plants of Xf in infected areas.

New almond plantations have been allowed, except for the following list of varieties: Marcona, Garrigues, Bord de Santa Maria, Bord de Selva, Bord des Raiguer, Corona, Filau, Lluca, Menut, Mollar, Morro de vaca, Pere Gelabert, Pintadeta, Rutlo, Trinxets, Desmai Victoria, Viveta, and Vivot. For olive tree (*Olea europaea* var. *europaea*), the following varieties are allowed: Empeltre, Mallorquina, Arbequina, Picual, Arbosana, Koroneiki, Hojiblanca, Cornachuela, Cornicabra, Morruda, Sikitita, and Frantoio. For winemaking (*Vitis vinifera* L.), the varieties authorized are Cabernet Sauvignon, Callet, Chardonnay, Escursac, Fogoneu, Garnacha Blanca, Garnacha Negra, Giró Ros, Gorgollasa, Macabeo/viura, Malvasía aromatica/Malvasía de Banyalbufar, Manto Negro, Merlot, Moll/Prensal Blanc/Prensal, Monastrell, Muscat of Alexandria, Muscat Petit Grans, Parellada, Petit Verdot, Pinot noir, Riesling, Sauvignon Blanc, Syrah, Tempranillo, and Viognier.

In addition, the launch of New Generation funds is promoting a restructuring plan for rainfed fruit plantations in the Balearic Islands for the period 2021–2027. The objectives of the plan are the recovery of part of the lost area, the improvement of the efficiency in the production and transformation, and the increase in the commercialization of derived or elaborated products raising the agricultural income associated with the agriculture sector.

2.5. Movement of Plant Material out of the Balearic Islands and Between Islands, Phytosanitary Passports, and Border Controls

Articles 4 and 9 of Decision 2015/789 banned the export of plants from demarcated areas. These measures were reinforced with the Ministerial Order APM/21/2017 of specific prevention measures; the Resolution of the Minister of the Environment, Agriculture and Fisheries, which prohibits the departure from the territory of the island of Ibiza to the rest of the Balearic Islands; and the declaration of public utility in the fight against Xf through decree 65/2019. It prohibits the exit from the territory of the Balearic Islands for all plants for planting, except seeds, belonging to the genera or species listed in Annex I of the Execution Decision (EU) 2015/789, which is maintained by Regulation (EU) 2020/1201.

Fortunately, the Mediterranean Sea provides an effective natural barrier to the spread of Xf outside the islands, as indicated in the absence of the pathogen on the island of Formentera, only a 3 km distance from Ibiza [36]. Given the importance of the movement of tourists in the Balearic Islands, priority has been given to controlling the departure of plants from airports and ports. The exit of unauthorized plants is prohibited, as well as circulation between islands. Controls are carried out by the state security forces and authorized customs inspection agents in ports and airports in collaboration with the Plant Health Service of the Balearic Islands (Decree 65/2019). To support the control measures, informative posters and controlled containers for passengers have been placed at the airport control points (Figure 3). Since 2017, the Civil Guard has carried out more than 65,432 passenger inspections, and 417 inspections have been carried out in ports and airports in customs areas, registering 129 incidents with the interception of infected plant material.

Plant production for export outside the islands is negligible. Between 2017 and 2022, the official body only received one request for authorization to export Xf host plants outside of the Balearic Islands. In 2020, a bonsai production nursery was authorized after verifying compliance with articles 19 and 20 of regulation 2020/1201. In total, 204 authorized plants, mainly *Olea europaea* var. *sylvestris* and *Myrtus communis*, have been exported to mainland Spain and other European countries.

Most ornamental plants that are sold in small retailers and garden centers (plant operators) in the Balearic Islands are imported. Inspections of plant phytosanitary passports have been conducted on the 143 registered professional plant operators. More than 180 health inspections have been carried out in nurseries and garden centers, 66 in the perimeter of these facilities. Between 2017 and 2021, a total of 127 prohibited plants for sale were seized, 17 positive plants in operator facilities and 10 in the surroundings of a total of 1561 samples.

Figure 3. Informative, leaflet brochures, and posters of *Xylella fastidiosa* at the airport of Palma de Mallorca.

To carry out the controls in the authorized border control posts (BCP), there is a plant health inspector assigned to the functional area of agriculture and fishing of the Government sub-delegation in Palma. So far, there is no evidence of entries or exits of host plants or specified plants, according to the definition established in Article 1, Decision (EU) 2015/789, and then in Regulation (EU) 2020/1201. In the authorized BCPs, there have been no interceptions of any type of plant material intended for planting as established in the community legislation.

2.6. Awareness Campaigns

A main goal of the phytosanitary authorities has been to disseminate information campaigns for main stakeholders and the general public in the detection and control of the pathogen. The list of host plants for Xf is kept updated to help the authorities in charge of carrying out controls on the entry and exit of plant material. About 14 training days have been organized for managers of the national airport management agency (AENA), Port Authority and Ports of the Balearic Islands, and a total of 358 campaigns at points of entry (ports and airports), in which information and dissemination, as well as supply of containers, informative leaflets, and brochures, have been provided. In addition, the Agriculture Service has carried out 379 training campaigns for professional users (garden centers and nurseries) and delivered brochures to publicize the presence of Xf in the Balearic Islands. Moreover, a plant health bulletin board is published monthly, and Xf regular public information is available on the plant health website: Pub-https://www.caib. es/sites/xf (accessed on 15 November 2022) and contact telephone numbers 900 102 186 and an informative email on Xf: sanitatvegetal@dgagric.caib.es

On the other hand, the Forest Health Service has prepared an informative video to show the biology of the bacterium and its vector, as well as the symptoms, signs, and damage they generate, published an edition of the "Visual Guide to symptoms of Xf in forest species of the Balearic Islands". Periodic updating of information is available on the forest health website: http://sanidadforestal.caib.es. (accessed on 15 November 2022) Consultations about Xf can be done through the contact telephone number (+34 971176666), through the Environmental Information Point (PIA): 900151617 (free), or by email at sanidadforestal@gmail.com. In addition, there have been periodic training on Xf for the Forestry Technical Group and also informative meetings with environmental agents.

3. Research on the Reconstruction of the Introduction and Spread of *Xylella fastidiosa* in the Balearic Islands

As mentioned in Article 6 of Decision 2015/789, the Member States should carry out the appropriate investigations to identify the origin of the transferred infected plants. Since 2017, research has been developed to establish the origin and date of the possible entries of the different Xf genotypes on the different islands. From the beginning, there were suspicions that Xf could be behind the great mortality of almond trees that had previously been attributed to the interaction of a complex of wood fungi with drought [37]. In the spring of 2017, it was already known that there were qPCR-positive almond trees scattered throughout the island. MLST analyses carried out in Cordoba, Spain, on leaf samples from Xf-qPCR-positive almond trees pointed out a close genetic relatedness to Xf subsp. *fastidiosa* and *multiplex* that were causing almond leaf scorch disease (ALSD) in California. Even in 2010, the possibility that Xf could be involved in the death of the almond trees had been considered. For all these reasons, in the spring of 2017, we worked on the hypothesis that ALSD could be the cause of the death of almond trees and because the first cases of mass mortality began to be seen in 2007, the bacteria must have been introduced in the early 2000s or earlier.

3.1. Disease Incidence

Knowing the disease incidence caused by Xf on different crop diseases was pivotal for several reasons. First, it allowed us to quantify the outbreaks in terms of population rather than scattered infected plant units verified in the laboratory by qPCR. Secondly, it enabled us to properly infer the spatiotemporal spread of the ALSD epidemic, affecting the most important crop in Mallorca. At the same time that the number of samples detected by qPCR was periodically reported as outbreaks notifications to the commission (Figure 1), field studies were carried out to visually estimate the incidence of ALSD throughout the island. A total of 126 almond orchards distributed throughout Mallorca were inspected to estimate the incidence of ALSD in 2017. By counting the trees that showed disease symptoms previously attributed to fungal trunk pathogens as an advanced stage of ALSD, it was visually determined that the incidence of the disease affected approximately 79.5% of the almond trees [38]. To investigate the disease progress, we used the Google Street View panoramic-image repository to approximate the ALSD incidence in 2012. Around 249 orchards distributed throughout the island were visually examined, and the average incidence of ALSD was estimated at 53.4% [38].

On the other hand, there was a certain urgency in knowing the incidence of Pierce's disease established for the first time in Europe. In the summer of 2018, extensive sampling was carried out in vineyards in Mallorca to verify that the incidence was very heterogeneous. It was found that the disease incidence ranged from less than 1% to 99%, depending mainly on the management of the vegetative cover in spring, the treatments received with insecticides, the age of the plantation, and the varieties planted [39].

Although no specific studies have been carried out to assess the incidence of Xf in wild olive trees, an abundant and widespread species in the Balearic Islands, a very conservative estimate of 10% would indicate hundreds of thousands of infected trees. All these approximations of Xf incidence on crops and forest trees at the end of 2017, transformed into units of infected plants, suggested a range between 1 and 3 million Xf-infected plants in Mallorca alone (*cf.* Figure 1). These differences in the way of describing and interpreting an epidemiological outbreak in a territory deserve further reflection and thus are discussed later in this perspective.

3.2. Phylogenetic Analysis

Our second research priority was to establish the geographical origin and the approximate date of introduction of Xf in the Balearic Islands. To do this, we needed to obtain a sufficient number of isolates to sequence their genomes and compare them with other genomes published in GenBank. In an international collaborative study using several Xf isolates of subsp. *multiplex* from Mallorca and Menorca and other isolates collected in Europe and America, it was possible to show that the Balearic isolates were introduced from California [30]. To determine the probable date of introduction, we faced the limitation of the sampling time (2016 to 2019), which was quite scarce to estimate the substitution rate with some confidence and thus calculate the molecular clock. This problem could be solved by anchoring the minimum node date for the Balearic Islands clade to the year of the oldest ring in which Xf DNA sequences were detected in the growth rings of almond trees. Xf DNA sequences of subsp. *fastidiosa* were detected in rings corresponding to 1998 in trees with ALSD symptoms, while Xf. subsp. *multiplex* was found in rings of 2000. This allowed us to properly use the priors in Bayesian inference to estimate with some confidence that the introduction of both subspecies from California occurred around 1990–1997 [38]. By revealing the date and origin of the introduction of the pathogen, another dimension was given to the epidemiological situation of Xf in the Balearic Islands, which cast doubt on the effectiveness of the current phytosanitary measures.

3.3. Vector Transmission

To explain how Xf had spread within the islands and to control its transmission, we needed to identify the insect vectors involved, their prevalence, and their ecology. Several research projects were funded by the Balearic Government, the Ministry of Agriculture of the Government of Spain, and the European Food Safety Authority (EFSA) [40]. Early surveys indicated a predominance of *P. spumarius* and much lower proportions of the species *Neophilaenus campestris* in fields of all three islands. In addition, a high density of *P. spumarius* nymphs was observed in different plots between February and April. However, the captures of adults in the vegetative cover and crop canopy were considerably lower. The first detections of infective adults of *P. spumarius* occur in May and accumulate during the summer in variable proportions according to the different studies and the detection methods used [40]. In general, there is a seasonal pattern of the vector populations similar to that observed in Apulia (Italy) in the OQSD [41].

In trials conducted in an insect-proof tunnel, the transmission of Pierce's disease and ALSD by *P. spumarius* has been demonstrated, after an acquisition access period (APP) and inoculation access period (IAP) of 72–96 h [42]. Cross-transmissions between vineyards and almond trees have also been successfully carried out, and infections of the *multiplex* subspecies of wild olive trees have been transmitted to almond trees. Furthermore, on one occasion, Pierce's disease was transmitted from an infected to a healthy grapevine through *N. campestris*.

3.4. Climatic Conditions

In a recent study carried out in Mallorca, risk maps for Pierce's disease have been developed for the main wine-producing areas of the world on the basis of epidemiological models [43]. These specific models for the pathosystem *Vitis vinifera*-Xf ST1- *P. spumarius* suggest that the disease could only have become established in Europe on Mediterranean islands, such as the Balearic Islands, or very specific areas in Mediterranean coastlands. The current distribution of Xf in Europe does not seem to be a coincidence, but rather it would indicate where the pathogen could establish itself in a situation where the movement of infected plants in Europe during the last 20 years would have been the norm. These models are important because they reveal that other strains of the pathogen could easily become established in the Balearic Islands or other Mediterranean islands, and they provide an idea of the stochasticity of the invasion process in the initial stage after entry.

3.5. Field Observations and Inoculation Experiments

The four factors that intervene in the development of a disease, namely, the hosts, the climate, the insect vectors, and the genetic diversity of the pathogen, have been addressed in these six years of research. On the other hand, studies have been carried out in the field and in insect-proof tunnels on the susceptibility of the different varieties of almond, vine, and olive trees to better understand the diseases and adequately guide the policies aimed at controlling Xf in the Balearic Islands. As a general rule, most almond trees and grapevine varieties are affected by Xf, whereas olive trees are resistant to *fastidiosa* and *multiplex* subspecies. Because grapevines are affected only by subsp. *fastidiosa*, these observations are more conclusive. All grapevine varieties inoculated or monitored in the field are in some way susceptible to Xf. In the inoculation tests, there were significant differences between varieties in the development of symptoms and Xf infection, but part of these differences were due to rootstock–variety interactions. Varietal response to Xf in the inoculations were similar to those observed in the field, although it was not possible to correlate the different incidences with the severities observed in the inoculations due to the intervention of other non-controllable factors such as crop management and its impact on the vector population.

Germplasm bank varietal collections of almond trees naturally exposed to the pathogen have provided quality experimental data to categorize the susceptibility of well-identified crop varieties. The results of the observations in these germplasm banks have been very similar to those in the field. These data have been used to recommend the planting of some tolerant varieties and prohibit the most susceptible ones.

4. Discussion

The implementation of phytosanitary measures represents the first combat front for eradicating and containing harmful organisms. If these are taken at the right time and with diligence, they can be very effective in their ultimate purpose. However, this is rarely the case; in particular, pathogenic microorganisms are generally much more difficult to detect than insect pests. In addition, symptoms produced by harmful microorganisms can be non-specific or confused with other root or vascular pathogens, or even with plant physiological disorders. Overall, there is often a significant period elapsing between the introduction event and the detection of the harmful organism, sometimes making posterior eradication or even containment unfeasible. The exposed case of Xf in the Balearic Islands is a good example of this. Between the introduction and its detection, around 23 years passed. Much of the discussion is situated in this context.

In early 2017, the official body soon recognized the infeasibility of the eradication measures established by Decision 2015/789, given the situation of Xf in the Balearic Islands. Technical and political efforts were devoted to convincing the European plant health authorities of the need to adapt the decision for the islands' specific case. With the entry into force of the modification of the decision and later included in the regulation, the infeasibility of eradication in infected areas is recognized. Instead, it emphasizes the need to intensify surveillance to detect a possible spread of the pathogen. In addition, sampling methods are harmonized in all the delimited areas through the use of the statistical and risk-based sampling tool RIBESS+ developed by the EFSA. Although the benefits of this statistical tool as a surveillance guide are beyond doubt, in areas where Xf has become naturalized, such as the three largest Balearic Islands, the aim of the survey has to be necessarily different to those of detecting, delimiting, and monitoring Xf in buffer zones. Research has shown that there are millions of infected plants in the field and wild borders, so removing only qPCR-positive plants (1360 qPCR+) represents much less than 0.1% of the infected plants. Nowadays, the pathogen and the vector integrate the landscape, so reducing the amount of inoculum below significant levels would imply the elimination of millions of plants and the massive use of insecticides. In support of this argument, a recent model for the ALSD epidemic in Mallorca has estimated that the mean value of the basic reproductive number (R_0) < 1 occurred approximately in 2011. This indicates that the number of new infections has been decreasing year after year and the trend is directed to the extinction of the epidemic if there is no replacement of susceptible plants [43]. All this suggests that instead of directing the effort to indiscriminately detect new plants infected with qPCR, the focus should be on reducing the inoculum in new plantations and in areas where the incidence is still low, such as in the mountains or endangered ecosystems far away from crops.

Except in mountain areas, the extreme southwest of the island of Mallorca, the north of Menorca, and the little island of Formentera, Xf is not spreading in the sense of increasing an epidemic front. As explained above, the pathogen has been established for decades on the islands and is everywhere there are susceptible hosts. In some crops such as the almond tree, the spread on a regional scale is even decreasing, since the incidence is very high, and few trees remain uninfected. The idea of monitoring Xf spread makes little sense in the case of the Balearic Islands. It is not that the usefulness of the RIBESS+ statistical tool is questioned, but we believe that the aim of the surveys has to be redefined to target other risks, having in mind that the pathogen has been around for a long time. Our suggestion would be to focus on selective surveys aimed at investigating specific risks, such as monitoring the introduction of new strains of Xf or the introduction of strains between islands as well as new potential insect potential vectors, which seems more reasonable.

This connects with another widespread but probably misconceived idea about the emergence of new genotypes. Although much remains to be learned, homologous recombination in Xf appears to be a major driver in host shifts [17]. Jumps to new hosts could be more abrupt than gradual, that is, the pathogenicity and virulence of the genotypes would be established from the first contact with the new host and not progressively due to the accumulation of point mutations in the pathogen's genome. For example, we assume that the ST81 of subsp. *multiplex* must have been virulent from the first contact with wild olive trees in Mallorca, but instead, it rarely infects nearby olive trees. We do not expect, though we cannot discard, that Xf colonies living as commensals on olive trees will progressively become more aggressive through the accumulating mutations or incorporation of genetic material from other species. In other words, new host–Xf strain contacts would be under extreme episodic selection. On the contrary, in the case of ALSD and Pierce's disease in Mallorca, these strains and diseases have been transferred from California to Mallorca. They are not new contacts and, therefore, have been under stabilizing selection for some time [44]. After 27 years since their introduction, it is not expected that there will be a sudden outbreak in a new host without recombination events, the introduction of a new pathogen strain, or the intervention of extreme environmental factors.

We believe that when Xf becomes naturalized, as has occurred in Corsica or the Balearic Islands, the best strategy to detect new variants of the pathogen is to focus on vectors, as has been recently done in Corsica [45]. The main reason is that if genetic recombination occurs, the fittest new genotype will accumulate in the newly formed soft tissue of the infected plant where vectors feed more frequently. Colony bacterial adhesion to the walls of insect mouthparts would act as a sink where foreign DNA could be detected through adequate molecular methods. Therefore, the likelihood of detecting the new variant would be expected to be greater in the vector mouthparts than in a tree with hundreds of twigs. The other reason that needs more theoretical and experimental support is that genetic recombination between Xf strains might occur more frequently in vectors than in hosts.

Varietal resistance/tolerance to Xf has been identified as a strategy to control the pathogen in crops [46]. It is believed that greater genetic diversity in the host population protects against possible epidemic outbreaks; however, this well-established paradigm was not fulfilled in the case of ALSD in the Balearic Islands where both fastidiosa and multiplex subspecies spread within and among almond orchards (>100 almond tree varieties) of the island of Mallorca for approximately three decades. On a regional scale, the decision to at least prohibit new plantations of the most susceptible varieties in order to reduce the inoculum within new plantations seems accurate, as has been done in the Balearic Islands. In Majorcan viticulture, the average size of the vineyard ranges from 1 to 10 hectares. Rows of different varieties are usually found in the same plot (Figure 4), so monoculture of more tolerant varieties to the bacteria does not seem to be a viable or realistic strategy for controlling Pierce's disease, given the large number of wineries in a small territory. Instead, the use of more tolerant almond varieties would make more sense as it is a more industrial crop than vines and does not depend as much on organoleptic qualities.

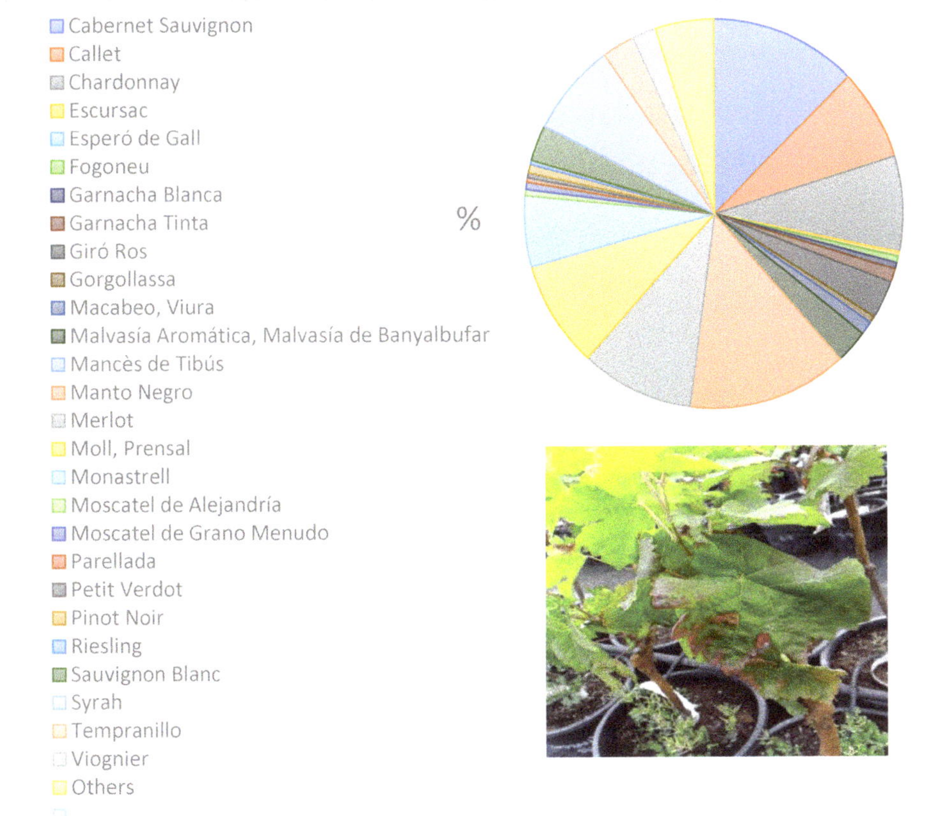

Figure 4. Diversity and surface-area proportions of grapevine varieties planted in Mallorca. All these varieties have been shown to be susceptible *to Xylella fastidiosa* in the field and in inoculation experiments. Symptoms of a grapevine (var. Tempranillo) seven weeks after inoculation (bottom).

5. Future Perspectives

There is no choice but to live with Xf in the Balearic Islands. Knowing how to live together, however, should not only become a positivist slogan but should also recognize that there are dangerous diseases established whose control can be achieved at a reasonable cost. Simple actions, such as the control of nursery material, weed control, and the elimination of infected plants in new vineyards or almond plantations make cultivation of main crops possible with little economic loss.

It is also important to anticipate new possible risks. The climate of the Balearic Islands and other Mediterranean islands seems to be suitable for all Xf subspecies [47]. Therefore, we believe that efforts should focus on preventing new entries of the pathogen or potential vectors through international plant trade. Finally, this review aims to share the experience of Xf control measures in the Balearic Islands with other areas where the pathogen has not yet been introduced or detected. We have exposed the facts and the action plans applied for the control of bacterium and vectors in the Balearic Islands, as well as the advances in the knowledge of the origin, phylogenetic relationship, and dating of the introduction of the two Xf strains in Majorca. We believe that the experience of Xf in the Balearic Islands invites us to consider a more epidemiological approach in the legislation of harmful organisms, in which it contemplates the possibility of unexpected events that force the adoption of control measures proportional to the real epidemiological situation of that territory.

Author Contributions: Conceptualization, B.Q., D.O. and E.M.; investigation, B.Q., M.M., A.N., A.P. and D.B.; resources, J.d.D.G., A.N., F.A., D.B, M.M, A.P. and M.L.; data curation, M.M., D.O., J.d.D.G. and B.Q.; writing—original draft preparation, E.M.; writing—review and editing, E.M., B.Q. and D.O.; visualization, J.d.D.G. and D.O.; supervision, M.L., E.M. and A.J.; project administration, M.M., D.O., M.L. and A.J.; funding acquisition, D.O. and A.J. All authors have read and agreed to the published version of the manuscript.

Funding: This research was partially funded by project E-RTA2017-00004-04 (Desarrollo de estrategias de erradicación, contención y control de X. fastidiosa en España) from 'Programa Estatal de I + D + I Orientada a los Retos de la Sociedad of the Spanish Government' and FEDER), and from the Organización Profesional del Aceite de Oliva, as well as by the agreement Diseño e Implementación de Estrategias de Control contra *Xylella fastidiosa* (EXPED. contract. 4384/18) from Govern de les Illes Balears.

Institutional Review Board Statement: Not applicable.

Informed Consent Statement: Not applicable.

Data Availability Statement: Not applicable.

Conflicts of Interest: The authors declare no conflict of interest.

References

1. Anderson, P.K.; Cunningham, A.A.; Patel, N.G.; Morales, F.J.; Epstein, P.R.; Daszak, P. Emerging infectious diseases of plants: Pathogen pollution, climate change and agrotechnology drivers. *Trends Ecol. Evol.* **2004**, *19*, 535–544. [CrossRef] [PubMed]
2. Borkar, S.G. *History of Plant Pathology*; CRC Press: Boca Raton, FL, USA, 2017; ISBN 1-351-37194-0.
3. Brasier, C.M. The biosecurity threat to the UK and global environment from international trade in plants. *Plant Pathol.* **2008**, *57*, 792–808. [CrossRef]
4. Savary, S.; Ficke, A.; Aubertot, J.-N.; Hollier, C. Crop losses due to diseases and their implications for global food production losses and food security. *Food Secur.* **2012**, *4*, 519–537. [CrossRef]
5. Werres, S.; Marwitz, R.; In't Veld, W.A.M.; De Cock, A.W.; Bonants, P.J.; De Weerdt, M.; Themann, K.; Ilieva, E.; Baayen, R.P. *Phytophthora ramorum* sp. nov., a new pathogen on rhododendron and viburnum. *Mycol. Res.* **2001**, *105*, 1155–1165. [CrossRef]
6. McMullan, M.; Rafiqi, M.; Kaithakottil, G.; Clavijo, B.J.; Bilham, L.; Orton, E.; Percival-Alwyn, L.; Ward, B.J.; Edwards, A.; Saunders, D.G. The ash dieback invasion of Europe was founded by two genetically divergent individuals. *Nat. Ecol. Evol.* **2018**, *2*, 1000–1008. [CrossRef] [PubMed]
7. Saponari, M.; Boscia, D.; Nigro, F.; Martelli, G.P. Identification of DNA sequences related to *Xylella fastidiosa* in oleander, almond and olive trees exhibiting leaf scorch symptoms in Apulia (southern Italy). *J. Plant Pathol.* **2013**, *95*, 668.
8. Sicard, A.; Zeilinger, A.R.; Vanhove, M.; Schartel, T.E.; Beal, D.J.; Daugherty, M.P.; Almeida, R.P. *Xylella fastidiosa*: Insights into an emerging plant pathogen. *Annu. Rev. Phytopathol.* **2018**, *56*, 181–202. [CrossRef] [PubMed]
9. Redak, R.A.; Purcell, A.H.; Lopes, J.R.; Blua, M.J.; Mizell Iii, R.F.; Andersen, P.C. The biology of xylem fluid-feeding insect vectors of *Xylella fastidiosa* and their relation to disease epidemiology. *Annu. Rev. Entomol.* **2004**, *49*, 243–270. [CrossRef] [PubMed]
10. EFSA; Carrasco Cabrera, L.; Medina Pastor, P. The 2019 European union report on pesticide residues in food. *EFSA J.* **2021**, *19*, e06491.
11. Hopkins, D.L. Effect of tetracycline antibiotics on Pierce's disease of grapevine in Florida [caused probably by bacterium]. *Proc. Annu. Meet. Fla. State Hortic. Soc. USA* **1980**, *92*, 284–285.
12. Kirkpatrick, B.; Anderson, P.; Civerolo, E.; Jones, D.; Purcell, A.; Smith, R.; Vargas, C.; Weber, E. Evaluation of bactericides and modes of delivery for managing Pierce's disease. *Pierce's Dis. Res. Symp.* **2003**, *103*, 101–103.
13. EFSA; Bragard, C.; Dehnen-Schmutz, K.; Di Serio, F.; Gonthier, P.; Jacques, M.-A.; Jaques Miret, J.A.; Justesen, A.F.; MacLeod, A.; Magnusson, C.S. Effectiveness of in planta control measures for *Xylella fastidiosa*. *EFSA J.* **2019**, *17*, e05666.
14. Baccari, C.; Antonova, E.; Lindow, S. Biological control of Pierce's disease of grape by an endophytic bacterium. *Phytopathology* **2019**, *109*, 248–256. [CrossRef] [PubMed]
15. Muranaka, L.S.; Giorgiano, T.E.; Takita, M.A.; Forim, M.R.; Silva, L.F.; Coletta-Filho, H.D.; Machado, M.A.; de Souza, A.A. N-acetylcysteine in agriculture, a novel use for an old molecule: Focus on controlling the plant–pathogen *Xylella fastidiosa*. *PLoS ONE* **2013**, *8*, e72937. [CrossRef] [PubMed]

16. Scortichini, M.; Chen, J.; De Caroli, M.; Dalessandro, G.; Pucci, N.; Modesti, V.; L'aurora, A.; Petriccione, M.; Zampella, L.; Mastrobuoni, F. A zinc, copper and citric acid biocomplex shows promise for control of *Xylella fastidiosa* subsp. pauca in olive trees in Apulia region (southern Italy). *Phytopathol. Mediterr.* **2018**, *57*, 48–72.

17. Vanhove, M.; Retchless, A.C.; Sicard, A.; Rieux, A.; Coletta-Filho, H.D.; De La Fuente, L.; Stenger, D.C.; Almeida, R.P. Genomic diversity and recombination among *Xylella fastidiosa* subspecies. *Appl. Environ. Microbiol.* **2019**, *85*, e02972-18. [CrossRef] [PubMed]

18. Landa, B.B.; Saponari, M.; Feitosa-Junior, O.R.; Giampetruzzi, A.; Vieira, F.J.; Mor, E.; Robatzek, S. *Xylella fastidiosa*'s relationships: The bacterium, the host plants, and the plant microbiome. *New Phytol.* **2022**, *234*, 1598–1605. [CrossRef]

19. EFSA; Delbianco, A.; Gibin, D.; Pasinato, L.; Morelli, M. Update of the *Xylella* spp. host plant database–systematic literature search up to 30 June 2021. *EFSA J.* **2022**, *20*, e07039.

20. Jolley, K.A.; Bray, J.E.; Maiden, M.C. Open-access bacterial population genomics: BIGSdb software, the PubMLST. Org Website and Their Applications. *Wellcome Open Res.* **2018**, *3*, 124. [CrossRef]

21. Nunney, L.; Azad, H.; Stouthamer, R. An experimental test of the host-plant range of nonrecombinant strains of north American *Xylella fastidiosa* subsp. multiplex. *Phytopathology* **2019**, *109*, 294–300. [CrossRef] [PubMed]

22. Berisha, B.; Chen, Y.D.; Zhang, G.Y.; Xu, B.Y.; Chen, T.A. Isolation of Peirce's disease bacteria from grapevines in Europe. *Eur. J. Plant Pathol.* **1998**, *104*, 427–433. [CrossRef]

23. Saponari, M.; Giampetruzzi, A.; Loconsole, G.; Boscia, D.; Saldarelli, P. *Xylella fastidiosa* in olive in Apulia: Where we stand. *Phytopathology* **2019**, *109*, 175–186. [CrossRef] [PubMed]

24. Denancé, N.; Legendre, B.; Briand, M.; Olivier, V.; De Boisseson, C.; Poliakoff, F.; Jacques, M.-A. Several subspecies and sequence types are associated with the emergence of *Xylella fastidiosa* in natural settings in France. *Plant Pathol.* **2017**, *66*, 1054–1064. [CrossRef]

25. Olmo, D.; Nieto, A.; Adrover, F.; Urbano, A.; Beidas, O.; Juan, A.; Marco-Noales, E.; López, M.M.; Navarro, I.; Monterde, A. First detection of *Xylella fastidiosa* infecting cherry (*Prunus avium*) and *Polygala myrtifolia* plants, in Mallorca island, Spain. *Plant Dis.* **2017**, *101*, 1820. [CrossRef]

26. Marco-Noales, E.; Barbé, S.; Monterde, A.; Navarro-Herrero, I.; Ferrer, A.; Dalmau, V.; Aure, C.M.; Domingo-Calap, M.L.; Landa, B.B.; Roselló, M. Evidence that *Xylella fastidiosa* is the causal agent of almond leaf scorch disease in Alicante, mainland Spain (Iberian peninsula). *Plant Dis.* **2021**, *105*, 3349–3352. [CrossRef] [PubMed]

27. Zecharia, N.; Krasnov, H.; Vanunu, M.; Castillo, A.I.; Haberman, A.; Dror, O.; Vakel, L.; Almeida, R.; Blank, L.; Shtienberg, D. *Xylella fastidiosa* outbreak in Israel: Population genetics, host range and temporal and spatial distribution analysis. *Phytopathology* **2022**. *Online ahead of print* . [CrossRef]

28. Saponari, M.; D'Attoma, G.; Abou Kubaa, R.; Loconsole, G.; Altamura, G.; Zicca, S.; Rizzo, D.; Boscia, D. A new variant of *Xylella fastidiosa* subspecies *multiplex* detected in different host plants in the recently emerged outbreak in the region of Tuscany, Italy. *Eur. J. Plant Pathol.* **2019**, *154*, 1195–1200. [CrossRef]

29. Carvalho-Luis, C.; Rodrigues, J.M.; Martins, L.M. Dispersion of the bacterium *Xylella fastidiosa* in Portugal. *J. Agric. Sci. Technol. A* **2022**, *12*, 35–41. [CrossRef]

30. Landa, B.B.; Castillo, A.I.; Giampetruzzi, A.; Kahn, A.; Román-Écija, M.; Velasco-Amo, M.P.; Navas-Cortés, J.A.; Marco-Noales, E.; Barbé, S.; Moralejo, E. Emergence of a plant pathogen in Europe associated with multiple intercontinental introductions. *Appl. Environ. Microbiol.* **2020**, *86*, e01521-19. [CrossRef]

31. Morelli, M.; García-Madero, J.M.; Jos, Á.; Saldarelli, P.; Dongiovanni, C.; Kovacova, M.; Saponari, M.; Baños Arjona, A.; Hackl, E.; Webb, S. *Xylella fastidiosa* in olive: A review of control attempts and current management. *Microorganisms* **2021**, *9*, 1771. [CrossRef]

32. Wells, J.M.; Raju, B.C.; Hung, H.-Y.; Weisburg, W.G.; Mandelco-Paul, L.; Brenner, D.J. *Xylella fastidiosa* Gen. Nov., Sp. Nov: Gram-negative, xylem-limited, fastidious plant bacteria related to *Xanthomonas* spp. *Int. J. Syst. Evol. Microbiol.* **1987**, *37*, 136–143. [CrossRef]

33. Reisman, E. Plants, pathogens, and the politics of care: *Xylella fastidiosa* and the intra-active breakdown of Mallorca's almond ecology. *Cult. Anthropol.* **2021**, *36*, 400–427. [CrossRef]

34. Landa, B.B.; Velasco-Amo, M.P.; Marco-Noales, E.; Olmo, D.; López, M.M.; Navarro, I.; Monterde, A.; Barbé, S.; Montes-Borrego, M.; Román-Écija, M. Draft genome sequence of *Xylella fastidiosa* subsp. fastidiosa strain IVIA5235, isolated fromPrunus avium in Mallorca island, Spain. *Microbiol. Resour. Announc.* **2018**, *7*, e01222-18. [PubMed]

35. Gomila, M.; Moralejo, E.; Busquets, A.; Segui, G.; Olmo, D.; Nieto, A.; Juan, A.; Lalucat, J. Draft genome resources of two strains of *Xylella fastidiosa* XYL1732/17 and XYL2055/17 isolated from Mallorca vineyards. *Phytopathology* **2019**, *109*, 222–224. [CrossRef] [PubMed]

36. Olmo, D.; Nieto, A.; Borràs, D.; Montesinos, M.; Adrover, F.; Pascual, A.; Gost, P.A.; Quetglas, B.; Urbano, A.; García, J.d.D.; et al. Landscape epidemiology of *Xylella fastidiosa* in the Balearic Islands. *Agronomy* **2021**, *11*, 473. [CrossRef]

37. Gramaje, D.; Agustí-Brisach, C.; Pérez-Sierra, A.; Moralejo, E.; Olmo, D.; Mostert, L.; Damm, U.; Armengol, J. Fungal trunk pathogens associated with wood decay of almond trees on Mallorca (Spain). *Persoonia-Mol. Phylogeny Evol. Fungi* **2012**, *28*, 1–13. [CrossRef] [PubMed]

38. Moralejo, E.; Gomila, M.; Montesinos, M.; Borràs, D.; Pascual, A.; Nieto, A.; Adrover, F.; Gost, P.A.; Seguí, G.; Busquets, A. Phylogenetic inference enables reconstruction of a long-overlooked outbreak of almond leaf scorch disease (*Xylella fastidiosa*) in Europe. *Commun. Biol.* **2020**, *3*, 560. [CrossRef]

39. Moralejo, E.; Borràs, D.; Gomila, M.; Montesinos, M.; Adrover, F.; Juan, A.; Nieto, A.; Olmo, D.; Seguí, G.; Landa, B.B. Insights into the epidemiology of Pierce's disease in vineyards of Mallorca, Spain. *Plant Pathol.* **2019**, *68*, 1458–1471. [CrossRef]

40. López-Mercadal, J.; DelgaDo, S.; MeRCaDal, P.; Seguí, G.; Lalucat, J.; BusQuets, A.; Gomila, M.; LesteR, K.; KeNyoN, D.M.; Ruiz-Pérez, M. Collection of data and information in Balearic Islands on biology of vectors and potential vectors of *Xylella fastidiosa* (GP/EFSA/ALPHA/017/01). *EFSA Support. Publ.* **2021**, *18*, 6925E. [CrossRef]

41. Cornara, D.; Saponari, M.; Zeilinger, A.R.; de Stradis, A.; Boscia, D.; Loconsole, G.; Bosco, D.; Martelli, G.P.; Almeida, R.P.; Porcelli, F. Spittlebugs as vectors of *Xylella fastidiosa* in olive orchards in Italy. *J. Pest Sci.* **2017**, *90*, 521–530. [CrossRef]

42. Borràs, D. Studies on the competence of potential Xylella fastidiosa vectors in the Balearic Islands (Spain). In Proceedings of the 3rd European Conference on Xylella fastidiosa and XF-ACTORS Final Meeting (xylella21), Online, 26–30 April 2021. [CrossRef]

43. Gimenez-Romero, A.; Galvan, J.; Montesinos, M.; Bauza, J.; Godefroid, M.; Fereres, A.; Ramasco, J.J.; Matias, M.A.; Moralejo, E. Global predictions for the risk of establishment of Pierce's disease of grapevines. *bioRxiv* **2022**. [CrossRef]

44. Brasier, C. Rapid Evolution of Introduced Tree Pathogens via Episodic Selection and Horizontal Gene Transfer. In *Proceedings of the Fourth International Workshop on the Genetics of Host-Parasite Interactions in Forestry: Disease and Insect Resistance in Forest Trees*; Gen. Tech. Rep. PSW-GTR-240; Sniezko, R.A., Yanchuk, A.D., Kliejunas, J.T., Palmieri, K.M., Alexander, J.M., Frankel, S.J., Eds.; Pacific Southwest Research Station, Forest Service, US Department of Agriculture: Albany, CA, USA, 2012; Volume 240, pp. 133–142.

45. Cruaud, A.; Gonzalez, A.-A.; Godefroid, M.; Nidelet, S.; Streito, J.-C.; Thuillier, J.-M.; Rossi, J.-P.; Santoni, S.; Rasplus, J.-Y. Using insects to detect, monitor and predict the distribution of *Xylella fastidiosa*: A case study in Corsica. *Sci. Rep.* **2018**, *8*, 15628. [CrossRef] [PubMed]

46. Sabella, E.; Luvisi, A.; Aprile, A.; Negro, C.; Vergine, M.; Nicolì, F.; Miceli, A.; De Bellis, L. *Xylella fastidiosa* induces differential expression of lignification related-genes and lignin accumulation in tolerant olive trees cv. Leccino. *J. Plant Physiol.* **2018**, *220*, 60–68. [CrossRef] [PubMed]

47. Godefroid, M.; Cruaud, A.; Streito, J.-C.; Rasplus, J.-Y.; Rossi, J.-P. *Xylella fastidiosa*: Climate suitability of European continent. *Sci. Rep.* **2019**, *9*, 8844. [CrossRef] [PubMed]

 *microorganisms*

Review

Synthetic Peptides against Plant Pathogenic Bacteria

Esther Badosa [1,*], Marta Planas [2], Lidia Feliu [2], Laura Montesinos [1], Anna Bonaterra [1] and Emilio Montesinos [1]

1   Laboratory of Plant Pathology, Institute of Food and Agricultural Technology-CIDSAV-Xarta, University of Girona, Campus Montilivi, 17071 Girona, Spain
2   LIPPSO, Department of Chemistry, University of Girona, Campus Montilivi, 17071 Girona, Spain
*   Correspondence: esther.badosa@udg.edu

Abstract: The control of plant diseases caused by bacteria that seriously compromise crop productivity around the world is still one of the most important challenges in food security. Integrated approaches for disease control generally lack plant protection products with high efficacy and low environmental and health adverse effects. Functional peptides, either from natural sources or synthetic, are considered as novel candidates to develop biopesticides. Synthetic peptides can be obtained based on the structure of natural compounds or de novo designed, considering the features of antimicrobial peptides. The advantage of this approach is that analogues can be conveniently prepared, enabling the identification of sequences with improved biological properties. Several peptide libraries have been designed and synthetized, and the best sequences showed strong bactericidal activity against important plant pathogenic bacteria, with a good profile of biodegradability and low toxicity. Among these sequences, there are bacteriolytic or antibiofilm peptides that work against the target bacteria, plant defense elicitor peptides, and multifunctional peptides that display several of these properties. Here, we report the research performed by our groups during the last twenty years, as well as our ongoing work. We also highlight those peptides that can be used as candidates to develop novel biopesticides, and the main challenges and prospects.

Keywords: synthetic peptides; pathogenic bacteria; biopesticides; lytic peptides; plant defense

Citation: Badosa, E.; Planas, M.; Feliu, L.; Montesinos, L.; Bonaterra, A.; Montesinos, E. Synthetic Peptides against Plant Pathogenic Bacteria. *Microorganisms* **2022**, *10*, 1784. https://doi.org/10.3390/microorganisms10091784

Academic Editor: Ana Palacio-Bielsa

Received: 31 July 2022
Accepted: 1 September 2022
Published: 3 September 2022

Publisher's Note: MDPI stays neutral with regard to jurisdictional claims in published maps and institutional affiliations.

1. Introduction

Losses in crop yield caused by plant diseases account for between 16 and 18 % of global agricultural productivity [1], which result in a substantial economic impact and endanger food security. Plant diseases caused by bacteria are the main factors limiting crop production. Some especially relevant plant pathogenic bacteria are *Erwinia amylovora*, *Xanthomonas arboricola* pv. vesicatoria and *Pseudomonas syringae* pv. syringae, the causal agents of fire blight in the Rosaceous family, the bacterial leaf spot on tomatoes and peppers and the floral bud necrosis, respectively [2,3]. The current effective approaches to control bacterial diseases are mostly based on the use of copper compounds or antibiotics, with the latter being allowed in several countries, but not in Europe [4,5]. However, their extensive use has adverse effects on the environment and human health, as well as on animal health, and evokes antibiotic resistance of bacterial pathogens [6]. Even so, the difficulty of control increases in the case of fastidious phytopathogenic bacteria, such as *Xylella fastidosa* and *Candidatus* Liberibacter asiaticus, for which there are still no effective methods for disease control in infected plants, due to the lack of appropriate bactericides and the difficulty of accessing the vascular vessels, where these pathogens propagate. Current control is based on preventive measures to limit their spread, such as eradication, pathogen-free propagation of plant material, exclusion, and vector control [7]. Taking the regulatory framework of pesticides (International Plant Protection Convention) into account, there is a strong need for alternative, environmentally compatible products to control phytopathogens. Antimicrobial peptides (AMPs) have been identified as good candidates for plant disease control, and show strong potential to manage diseases caused by plant pathogenic bacteria, fungi and phytoplasmas [8,9]. In fact, the main properties of AMP include the following: (1) high antimicrobial activity with minimal inhibitory concentration (MIC) at the micromolar level, being, in general, similar to that of conventional antibiotics and significantly lower than that of copper compounds, (2) quick response in the target pathogen, (3) higher biodegradability than conventional pesticides, (4) low cytotoxicity, and (5) multiple modes of action [10]. The main mechanism of action of AMPs is the electrostatic interaction with the bacterial membrane, leading to its disruption that results in bacterial death. It has also been described that some AMPs can interact with intracellular targets, such as nucleic acids and proteins [11,12]. Apart from being antibacterial, AMPs can act as pathogen-associated molecular patterns (PAMPs) or plant elicitor peptides (Peps), stimulating the plant immune system [13–15].

There is currently an abundance of updated information on antimicrobial peptides of different origins, with numerous recent reviews providing an overview of the progress on natural peptides of plant [16–19], animal [20–23] or microbial origin [24], as well as synthetic peptides based on the structure of natural sequences [25]. Research on the applications of these peptides, such as in agriculture [19,26] or human health [25,27–29], has also been extensively covered in previous reviews.

During the last twenty years, our research groups from the University of Girona have been extensively working to design and develop AMPs, with the aim to identify potential candidates to combat phytopathogenic bacteria of great economic importance. In particular, we have been interested in finding AMPs that are useful to control diseases caused by *Erwinia amylovora* (Ea) in pears, *Xanthomonas axonopodis* pv. vesicatoria (Xav) in tomatoes and pepper plants, *Xanthomonas fragariae* (Xfr) in strawberries, *Xanthomonas axonopdis* pv. pruni (Xap) in stone fruit, *Pseudomonas syringae* pv. syringae (Pss), pv. tomato (Pst) or pv. actinidiae (Psa) in Rosaceous plants, tomato or kiwi plants, respectively.

More recently, our efforts have been focused on the control of the quarantine pathogens *X. fastidiosa* (Xf), affecting at least 638 plant species [30] and *Candidatus* Liberibacter asiaticus, a fastidious, phloem-limited bacterium that affects citrus [7]. *X. fastidiosa* causes Pierce's disease in grapevines, almond leaf scorch, citrus variegated chlorosis and olive decay syndrome [31] and *Candidatus* Liberibacter asiaticus is the main causative agent of Huanglongbing (HLB) [32,33]. Within this context, up to now, we have developed linear and cyclic peptides, which have been designed from the structure of natural peptides or based on the common features of AMPs. Our main objective is to find short sequences with an optimal biological profile in terms of high antibacterial activity, low toxicity and high stability to protease degradation.

The aim of this review is to contextualize our research lines developed in the framework of several national and European projects, which have resulted in the identification of good candidates to be used in integrated plant bacterial disease management strategies for economically important crops. Finally, a brief summary of the current research that focused on the control of *X. fastidiosa* and *Candidatus* Liberibacter asiaticus is reported.

2. Peptides with Distinct Activity: Design and Identification of Leads

2.1. Antimicrobial Peptides (AMPs)

In this section, we describe the linear and cyclic peptides developed to find suitable leads as agents to control phytopathogens. With the aim of improving their biological activity profile, we have incorporated several modifications in the structure of these leads. All the research in this regard is described herein.

2.1.1. Linear Peptides

The linear peptides were first designed based on the structure of the 11-residue peptide **Pep3** (WKLFKKILKVL-NH$_2$), derived from the hybrid peptide cecropin A(1-7)-melittin (2-9), which was reported to display antimicrobial activity against several phytopathogens [34,35]. Taking into account the structure of **Pep3**, 22 analogues were designed by reducing its length, changing the C-terminal amide group with a carboxylic acid, derivatizing the N-terminus and replacing the residues at positions 1 and 10 (Figure 1A, Table 1). The best peptide **BP76** resulted from the replacement of Trp^1 and Val^{10} in **Pep3** with Lys and Phe, respectively, and showed high activity against Ea, Xav and Pss (MIC values between 2.5 and 5.0 µM) and low hemolysis (34% at 150 µM) [36]. The ideal Edmunson wheel projection of **BP76** was the basis for the design of a 125-member peptide library (CECMEL11) by incorporating amino acids with various degrees of hydrophobicity and hydrophilicity at positions 1 and 10 and also by varying the N-terminal derivatization [37]. This library allowed the identification of **BP100** (KKLFKKILKYL-NH_2), which, apart from having a good biological activity profile in vitro (MIC values between 2.5 and 7.5 µM, 22% hemolysis at 150 µM), was also more effective than **BP76** in vivo to treat Ea infections in detached apple and pear flowers and, interestingly, showed low oral acute toxicity in mice (LD_{50} > 2000 mg/kg of body weight) [38].

With the aim of further improving the biological activity of **BP100**, four sets of analogues were designed by introducing different unnatural amino acids (D-, triazolyl or biaryl amino acids) or an acyl chain in its sequence (Figure 1B, Table 1) [39]. First, 31 peptides, containing 1 to 11 D-amino acids, were prepared [40]. The incorporation of a D-amino acid is a widely used strategy to increase proteolytic stability and decrease hemolysis, while maintaining the antimicrobial activity. From this set, highlighted **BP143** (KKLf^{4}KKILKYL-NH_2) with a D-Phe^4 (MIC values between 2.5 and 7.5 µM, 4% hemolysis at 150 µM), which displayed high activity in planta assays, and it was as effective as streptomycin for the control of bacterial blight of pepper and pear, and fire blight of pear.

A second set of **BP100** analogues incorporated a 1,2,3-triazole ring (Figure 1B, Table 1). Eleven of these analogues were designed by derivatizing the side chain of each Lys in **BP100** with a triazolyl moiety and three derivatives were obtained by replacing the benzene ring of Phe^4 with this heterocycle [41]. It is well-known that this ring is stable to hydrolysis and redox conditions, as well as to metabolic degradation. In addition, the introduction of this heterocycle into a peptidic chain is straightforward. Peptidotriazoles with significant activity were identified. In particular, **BP250**, resulting from the replacement of the benzene ring of Phe^4 with a benzyl triazole, displayed high antibacterial activity (MIC values between 1.6 and 12.5 µM) and no hemolysis (0% at 150 µM). The lower hemolytic activity of **BP250** compared to **BP100** could be attributed to the lower hydrophobic character of the triazole moiety compared to the benzene ring.

A third study, focused on the derivatives from the CECMEL11 library bearing a His or a 5-arylhistidine residue instead of Phe^4, showed that this modification significantly reduced the hemolysis due to the hydrophilic character of the imidazole ring (Figure 1B, Table 1) [42]. Thus, **BP275** and **BP279** that incorporated a His^4 or a $His(5-Ph)^4$, respectively, were similarly active compared to the corresponding CECMEL11 derivatives and were considerably less hemolytic.

The fourth set consisted of lipopeptides derived from **BP100** by acylating its N-terminus or by incorporating an acyl lysine residue at each position of this sequence [43]. Butanoyl, hexanoyl and lauroyl groups were selected as acyl chains. The introduction of a fatty acid into an antimicrobial peptide has been demonstrated to improve its biological activity because it facilitates the insertion into the membrane bilayer. The best lipopeptides were **BP387** (MIC values between 1.6 and 6.2 µM, 11% hemolysis at 150 µM) and **BP389** (MIC values between 0.8 and 12.5 µM, 16% hemolysis at 150 µM), which contain a butanoyl group at position 8 and 10, respectively. Interestingly, these sequences were more active than **BP100** against the *Xanthomonas* species. From this set, 18 lipopeptides were chosen to reduce their hemolysis by incorporating a D-amino acid at position 4 [44]. This study led to the identification of **BP475** (MIC values between 1.6 and 6.2 µM, 0% hemolysis at 150 µM), bearing a D-Phe^4 and a butanoyl group at position 10. Structural characterization of this lipopeptide by NMR revealed that it adopts an α-helix from residues 6 to 10, pointing out that the presence of D-Phe^4 disrupts this secondary structure. Similar to **BP100** [45], this

C-terminal α-helix may favor the insertion of this lipopeptide into the bacterial membrane bilayer, enhancing the activity.

All the above peptides were synthesized following solid-phase procedures, which is a very convenient methodology for the preparation of short peptide sequences. However, the production of peptides can also be achieved through biotechnology techniques using bacteria, yeasts or plants as biofactories. Towards this aim, peptides derived from **BP100** were designed based on the structural requirements for their expression in rice plants and to facilitate the downstream process (Figure 1C, Table 1). Thus, to achieve the minimum expressability length, **BP100** was conjugated to other units of **BP100** and to fragments of natural antimicrobial peptides, such as cecropin A, magainin and melittin [46]. One or two AGPA hinges were incorporated as a stabilitzation/distortion moiety. In addition, a signal for peptide retention in the endoplasmic reticulum (KDEL), a residue that acts as a protease recognition site (Ser or Gly) and an epitope tag for peptide detection or purification (tag54) were included. It was observed that the presence of a KDEL unit or of tag54 did not affect the biological activity. The best peptide conjugates were **BP209**, **BP210** and **BP211** (MIC values between <1.25 and 5.0 µM; 1–17% hemolysis at 150 µM), resulting from the conjugation of **BP100** to a magainin fragment through an AGPA hinge. Among the peptides including the KDEL retention signal sequence, **BP178** that incorporated **BP100** and magainin (1-10) was the more active against Ea, Xav and Pss, being also low hemolytic, and it was chosen for its expression in plants. The *BP178* gene was efficiently expressed in transgenic rice seed endosperm, where the peptide stably accumulated at 21 µg/g of seed. Interestingly, the seedlings of transgenic lines showed enhanced protection to bacterial phytopathogens [47].

2.1.2. Cyclic Peptides

Cyclic peptides were de novo designed based on the general features described for antimicrobial peptides. Thus, cyclic peptides that contained 4 to 10 residues and incorporated hydrophilic (Lys) and hydrophobic (Leu, Phe and D-Phe) amino acids at alternating positions were prepared (Figure 2, Table 1). In addition, a Gln was incorporated at position 10 to facilitate the solid-phase synthesis. The general formula of these peptides was c(X_n-Y-X_m-Gln), where X was Lys or Leu, Y was L-Phe or D-Phe, m = n = 1 or m = 3 and n = 0 to 5 [48]. The most active cyclic peptide was **BPC16** (c(KLKLKFKLKQ)), which displayed MIC values between 6.2 and 12.5 µM against Pss and Xav and it was not active against Ea.

To improve the biological activity of BPC16, two combinatorial libraries were designed (Figure 2, Table 1) [49]. The first one comprised 56 cyclic decapeptides and contained Phe and Gln at positions 6 and 10, respectively, and all possible combinations of five Lys and three Leu at the other positions. From this library, the first cyclic peptides reported with activity against Ea were identified (MIC values between 12.5 and 25 µM). The second library of 16 sequences obeyed the general structure c(X^1XXX^{4}LysPheLysLysLeuGln), where X was Lys or Leu. From this library, the peptides BPC194 (c(KKLKKFKKLQ)) and BPC198 (c(KLKKKFKKLQ)) stood out for their good biological activity. BPC194 was more active against Ea (MIC values between 6.2 and 12.5 µM) than the best cyclic peptides from the first library and, in addition, both peptides were poorly hemolytic (10–13% at 150 µM). In a later study, Phe^6 in these libraries was replaced by a Trp. The results showed that, in general, the sequences obtained were more active against Xav, Pss and Ea than the ones with Phe and showed similar hemolysis [50]. This study allowed the identification of cyclic peptides BPC086W (c(LKKKLWKKLQ)) and BPC108W (c(LKKKKWLLKQ)), with MIC values between 0.8 and 12.5 µM against these bacteria and hemolysis ≤8% at 125 µM. The active conformation of BPC194 and BPC198 was predicted by molecular dynamics simulations [51]. These conformations were the basis for the design of 15 new analogues. The best three sequences were as active as the parent peptides. Moreover, the orientation of the hydrophilic pair interactions in the active conformation of these three peptides was analogous to that in BPC194 and BPC198, but the position of the hydrophobic residues was different.

LINEAR PEPTIDES

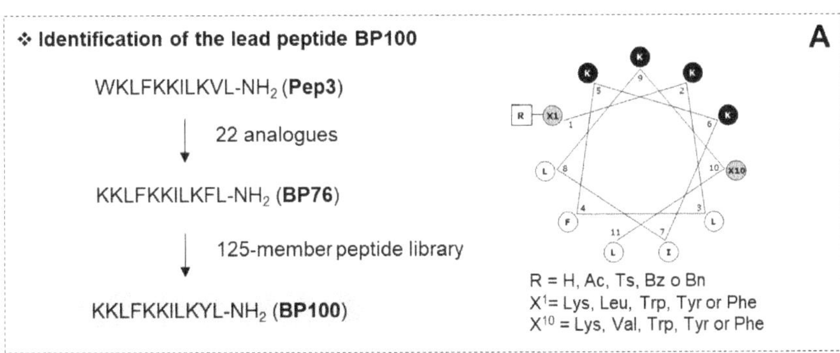

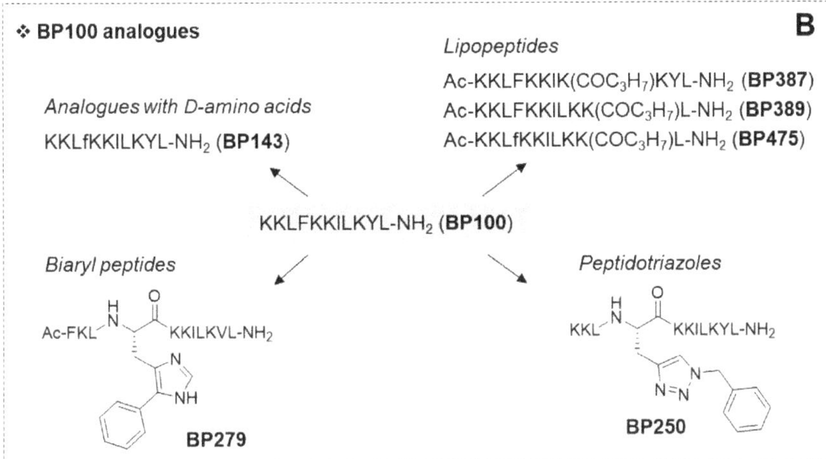

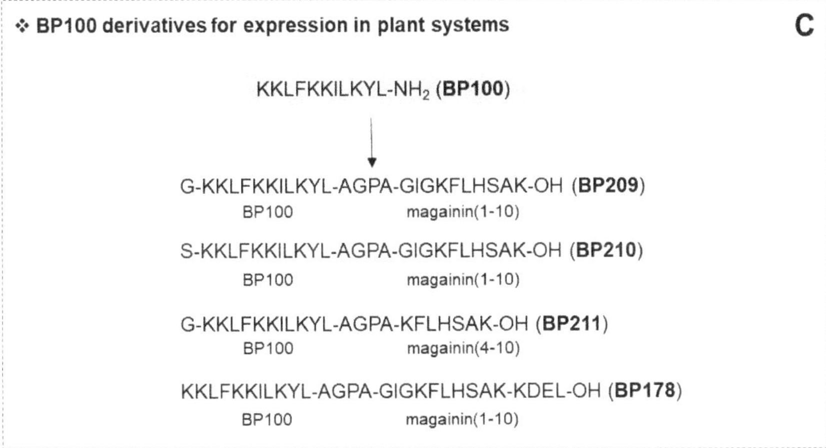

Figure 1. Linear peptides. Identification of the lead peptide **BP100** (**A**) [36,37], design of analogues (**B**) [39–44], and derivatives for expression in plant systems (**C**) [46], as described in Section 2.1.

From the lead peptide BPC194, two sets of analogues were designed, bearing a 1,2,3-triazole ring and/or an acyl chain (Figure 2B, Table 1). The triazolyl derivatives resulted from replacing Leu^3 with Ala, Glu, Lys or a Nle that incorporated a 1,2,3-triazolyl substituent at the side chain. Except for the Glu derivative, the other modifications were also included at position 5 in BPC194 instead of Lys [52,53]. This work led to cyclic peptidotriazoles BPC548 and BPC550, bearing a substituted triazolylnorleucine at position 3, with high antibacterial activity (MIC values between 3.1 and 25 μM) and low toxicity (7–12% at 150 μM).

The cyclic lipopeptides derived from BPC194 contained a fatty acid group at the N^ε-amino group of a Lys residue (Figure 2B, Table 1) [54,55]. The length and the position of the acyl chain was evaluated, as well as the incorporation of one or two D-amino acids and of a His. In particular, different Lys residues in BPC194 were derivatized with acyl groups of different lengths, ranging from 4 to 18 carbon atoms. In addition, Phe^6 or the acylated Lys were replaced with the corresponding D-enantiomer. Phe^6 was also replaced with a His. From the 51 cyclic lipopeptides prepared, those with the highest activity contained the acyl group at Lys^1, Lys^2 or Lys^5. It was observed that the presence of a D-amino acid maintained the activity and reduced the hemolysis. The incorporation of a His also resulted in a decrease in the hemolytic activity. Among the most active sequences, BPC702 was highlighted. This cyclic lipopeptide incorporated D-Lys^5 acylated with a butanoyl group, and displayed MIC values < 12.5 μM against Xav and Pss and low hemolysis (1% at 150 μM).

Table 1. Antimicrobial activity (minimum inhibitory concentration, MIC) and hemolysis of selected linear and cyclic peptides.

Peptide	Sequence [a]	MIC Intervals (μM)			Hemolysis (%) [c]
		Ea [b]	Pss [b]	Xav [b]	
Linear					
BP76	KKLFKKILKFL-NH$_2$	2.5–5.0	2.5–5.0	2.5–5.0	34
BP100	KKLFKKILKYL-NH$_2$	2.5–5.0	2.5–5.0	5.0–7.5	22
BP143	KKLfKKILKYL-NH$_2$	2.5–5.0	2.5–5.0	5.0–7.5	4
BP250	KKLA(Tr-Bn)KKILKYL-NH$_2$	3.1–6.2	1.6–3.1	6.2–12.5	0
BP275	Ts-FKLHKKILKVL-NH$_2$	12.5–25	3.1–6.2	3.1–6.2	4
BP279	Ac-FKLH(5-Ph)KKILKVL-NH$_2$	12.5–25	12.5–25	3.1–6.2	25
BP387	Ac-KKLFKKIK(COC$_3$H$_7$)KYL-NH$_2$	3.1–6.2	3.1–6.2	1.6–3.1	11
BP389	Ac-KKLFKKILKK(COC$_3$H$_7$)L-NH$_2$	3.1–6.2	6.2–12.5	0.8–1.6	16
BP475	Ac-KKLfKKILKK(COC$_3$H$_7$)L-NH$_2$	3.1–6.2	3.1–6.2	1.6–3.1	0
BP209	G-KKLFKKILKYL-AGPA-GIGKFLHSAK-OH	1.2–2.5	2.5–5	<1.2	13
BP210	S-KKLFKKILKYL-AGPA-GIGKFLHSAK-OH	1.2–2.5	2.5–5	<1.2	17
BP211	G-KKLFKKILKYL-AGPA-KFLHSAK-OH	1.2–2.5	2.5–5	<1.2	1
BP178	KKLFKKILKYL-AGPA-GIGKFLHSAK-KDEL-OH	2.5–5.0	2.5–5.0	2.5–5.0	3
Cyclic					
BPC16	c(KLKLKFKLKQ)	>100	12.5–25	6.2–12.5	17
BPC194	c(KKLKKFKKLQ)	6.2–12.5	3.1–6.2	3.1–6.2	13
BPC198	c(KLKKKFKKLQ)	12.5–25	3.1–6.2	3.1–6.2	10
BPC086W	c(LKKKLWKKLQ)	6.2–12.5	3.1–6.2	0.8–1.6	8
BPC108W	c(LKKKKWLLKQ)	6.2–12.5	1.6–3.1	1.6–3.1	4
BPC548	c(KK-Nle(Tr-Ph-Me)-KKFKKLQ)	12.5–25	3.1–6.2	3.1–6.2	12
BPC550	c(KK-Nle(Tr-Ph-OMe)-KKFKKLQ)	12.5–25	3.1–6.2	3.1–6.2	7
BPC702	c(KKLKk(COC$_3$H$_7$)FKKLQ)	25–50	6.2–12.5	6.2–12.5	1

[a] Lower case letters correspond to D-amino acids. Tr-Bn, triazole bearing a benzyl group. Ts, tosyl. Ac, acetyl. Ph, phenyl. COC$_3$H$_7$, butanoyl. Nle(Tr-Ph-Me), norleucine residue incorporating a triazole ring with a tolyl group at the side chain. Nle(Tr-Ph-OMe), norleucine residue incorporating a triazole ring with an anisole group at the side chain. [b] Ea, Xav and Pss stand for *Erwinia amylovora; Xanthomonas axonopodis* pv. vesicatoria and *Pseudomonas syringae* pv. syringae, respectively. [c] Percent hemolysis at 150 μM.

CYCLIC PEPTIDES

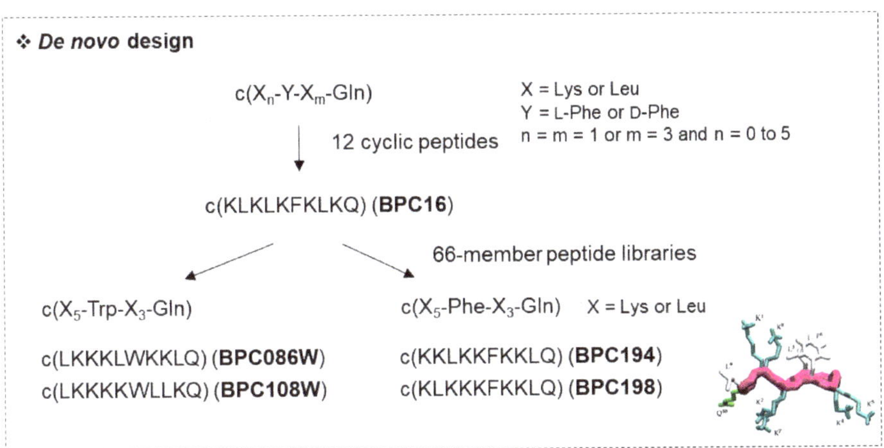

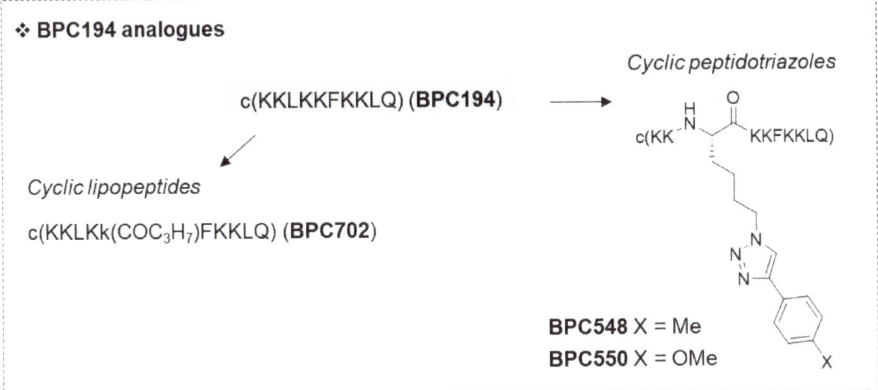

Figure 2. Cyclic peptides. De novo design of cyclic peptides [48–51], identification of leads and design of **BP194** analogues [49,52–55], as described in Section 2.1.

2.2. Plant Defense Elicitor Peptides

Antimicrobial peptides not only work directly against pathogens, but they can also act as plant defense elicitors and inducers of resistance to infections (IR). Two types of IR have been described in plants, including one that is basically pathogen-induced (systemic resistance, SAR) and another one that is caused by the colonization of plant roots by beneficial microorganisms (systemically induced resistance, ISR). This IR can be induced by molecular patterns and other molecules, including peptides [56].

With the aim of finding antimicrobial peptides that are able to trigger plant defense responses, we undertook several studies in which the following peptide elicitors were used as positive controls: **PIP-1** (YGIHTH-NH₂) [57], **Pep-13** (VWNQPVRGFKVYE-OK) [58] and **flg15** (RINSAKDDAAGLQIA-OH), which corresponds to 15 amino acids of the N-terminal conserved domain of the well-known bacterial flagellin [59,60]. The sequences tested for their capacity to induce plant defenses included cyclic peptides, linear peptides from the CECMEL11 library, lipopeptides derived from **BP100**, cyclic lipopeptides derived from **BPC194**, linear and cyclic peptidotriazoles, as well as peptide conjugates derived from **BP100** [10,43,61,62]. The best results were obtained for the linear peptide **BP13**, the cyclic peptide **BPC200W**, the linear lipopeptide **BP378**, and the peptide conjugates **flg15-BP16** and **BP178** (Table 2). Notably, these two peptide conjugates also display high antibacterial activity; therefore, they can be considered as bifunctional compounds.

2.3. Bifunctional Peptides

The interesting biological activity profiles exhibited by the above peptide conjugates **flg15-BP16** and **BP178** prompted us to further explore this type of compounds. In fact, the conjugation of two antimicrobial peptides in a single sequence provides new peptides with improved activity compared to the individual monomers. However, reports on peptide conjugates, resulting from the combination of two monomers with distinct activity, are scarce. Within this context, we decided to design peptide conjugates that contained an antimicrobial peptide (**BP16, BP100, BP143, KSLW, BP387, BP475**) at the N- or C-terminus of a plant defense elicitor peptide (**flg15, BP13, Pep13, PIP-1**) [62,63]. Analysis of the in vitro and in planta activity provided two bifunctional peptide conjugates, **flg15-BP475** and **flg15-BP16**, that were able to reduce Ea infections in pear plants and Xav infections in tomato, respectively, through a dual mechanism (Table 2). On the one hand, they showed high antibacterial activity and, on the other hand, they triggered plant defense responses.

3. Synthesis of Peptides

Linear peptides were synthesized following solid-phase procedures using the standard 9-fluorenylmethoxycarbonyl (Fmoc)/*tert*-butyl (*t*Bu) protocol. Cyclic peptides were prepared following the orthogonal tridimensional Fmoc/*t*Bu/allyl strategy. Fmoc-Rink-MBHA or ChemMatrix resins were used as solid support. The corresponding amino acids were anchored to the solid support using common coupling reagents, such as N,N-diisopropyl carbodiimide (DIC) and ethyl 2-cyano-2-(hydroxyamino)acetate (Oxyma). The Fmoc group was removed by treatment with piperidine-DMF and the allyl group with $Pd(PPh_3)_4$. Cyclization was carried out with the peptide sequence linked to the solid support using [ethylcyano(hydroxyimino)acetato-O^2]tri-1-pyrrolidinylphosphonium hexafluorophosphate (PyOxim)/Oxyma/N,N-diisopropylethylamine (DIPEA).

The peptides that incorporated a 1,2,3-triazole ring were synthesized through the reaction of an alkynyl peptidyl resin with an azide or by treatment of an azido peptidyl resin with an alkyne. The linear biaryl peptides were prepared via a Suzuki–Miyaura reaction by treating a bromopeptidyl resin with a boronic acid in solution. Alternatively, the biaryl bond was formed through the reaction of a boronopeptidyl resin, prepared by borylation of the corresponding bromopeptidyl resin, with an aryl iodide or bromide in solution.

The preparation of lipopeptides involved the synthesis of a sequence that incorporated a lysine residue with the side chain protected with 1-(4,4-dimethyl-2,6-dioxocyclohex-1-ylidene)-3-methylbutyl (ivDde). This group can be selectively removed in mild conditions that neither affect the other protecting groups nor the linkage of the sequence with the solid support. The acyl chain was incorporated using the same conditions as those of the amino acid coupling.

Once the synthesis of the sequence was completed, peptides were cleaved from the support, purified on a Combi*Flash* Rf200 automated flash chromatography system using a RediSep Rf Gold reversed-phase column packed with high-performance C_{18} derivatized silica (Teledyne ISCO, Lincoln, NE, USA). They were analyzed under standard analytical HPLC conditions with a 1260 Infinity II liquid chromatography instrument (Agilent, Santa Clara, CA, USA), using a Kromasil 100 C_{18} (4.6 mm × 40 mm, 3 μm) column. Peptides were characterized by electrospray-ionization mass spectrometry (ESI-MS) with an Esquire 6000 ESI ion trap LC/MS instrument (Bruker Daltonics, Billerica, MA, USA), equipped with an electrospray ion source. The instrument was operated in the positive ESI(+) ion mode. High-resolution mass spectrometry (HRMS) data were recorded on a Bruker MicroTof-QIITM instrument (Bruker Daltonics, Billerica, MA, USA), using ESI ionization sources. The instrument was also operated in the positive ion mode.

4. Activity of Peptides against Plant Pathogenic Bacteria

The use of peptides as control methods against plant pathogenic bacteria is mainly focused on the direct effect of peptides against the pathogens. Nevertheless, nowadays, there is evidence that some peptides can act in an indirect manner through the induction of the plant defense response. More information can be found in previous reviews [11,56]. Our research has taken into account both approaches, including the antibacterial activity and the ability to elicit the plant defense response.

4.1. In Vitro Activity

To determine the in vitro antibacterial activity of the peptides, the biological characteristics of the pathogens have to be taken into account, including their ability to grow on culture plates and their nutritional needs. Almost all of the phytopathogenic bacteria that we studied grew well in nutritional broth and agar plates and did not have complex nutritional requirements. Therefore, although some of bacteria of the genus *Pseudomonas*, *Erwinia*, and *Xanthomonas* are considered quarantine phytopathogens in some zones (found in the entire EU territory but not in some countries or zones, EU 2019/2072 of 28 November 2019), conventional methods, such as viable plate counting, could be used. Depending on the type of antimicrobial activity to be assessed after exposure of the pathogens to the peptides, several methodologies could be employed, such as those that determine the metabolic activity of cells or the amount of the free DNA in the sample due to pore formation (Figure 3). The most widely used methodology is the growth inhibition test in nutritional broth by measuring the optical density (Figure 3A). However, it has to be noted that this test does not allow to determine if the peptide has bacteriostatic or bactericidal activity. To determine the killing activity, contact tests should be used, and the effect can be measured as culturable cells recovered or metabolic activity (e.g., resarzurin). The bacteriolytic activity can be determined using the dye SYTOX green, which binds to free DNA and is able to enter the cells with damaged membranes due to the formation of pores by the peptide (Figure 3B).

4.2. Defense Elicitor Activity of Peptides in Plants

To determine the defense elicitor activity of the peptides, we started analyzing the capacity of selected peptides to induce extracellular pH changes and hydrogen peroxide production in tobacco cell cultures that are considered indicators of eliciting activity [61]. However, an in planta screening platform was preferred, so, we developed a platform in which tomato plant was used as a model plant and peptides were applied through a spray. The expression of defense-related genes was assessed by performing a retrotranscription of the RNA extracted from the plant coupled to a quantitative PCR, using actine as the endogenous gene for the normalization of the results and the $\Delta\Delta Ct$ method to perform the relative quantification of the treatment compared to the non-treated control [49]. Different types of peptides were tested, such as linear, cyclic and conjugates [43,61,62,64,65]. The results were promising and allowed the identification of peptide conjugates, incorporating an antimicrobial peptide and a plant defense elicitor peptide, that displayed bifunctional activity.

Taking into account the promising results of the peptide conjugates, in particular those of **BP178**, experiments were carried out taking into account the whole genome of the plants, including microarrays in tomato plants and RNAseq in *Prunus dulcis* using the Illumina platform. A microarray analysis of tomato plants (GeneChipTM tomato gene 1.0 ST array (Affymetrix)), treated with **BP178**, was performed. **flg15**, ethylene (ET), salicylic acid (SA) and jasmonic acid (JA) were used as the reference compounds [10]. It could be concluded that **BP178** elicits a similar gene expression pattern related to defenses than **flg15** and that it is related to several defense pathways (ET, SA, JA) (Figure 3C). Nevertheless, using tomato plant as a model system does not mean that the observed response will be the same in other plants of interest. Therefore, it is better to use the desired host of study whenever possible. In that sense, **BP178** was applied by endotherapy to almond plants, which were analyzed using RNAseq to obtain a holistic view of the effect of the peptide in the almond transcriptome. The results confirmed the observations obtained in tomato plants regarding the ability of **BP178** to elicit a defense response [62].

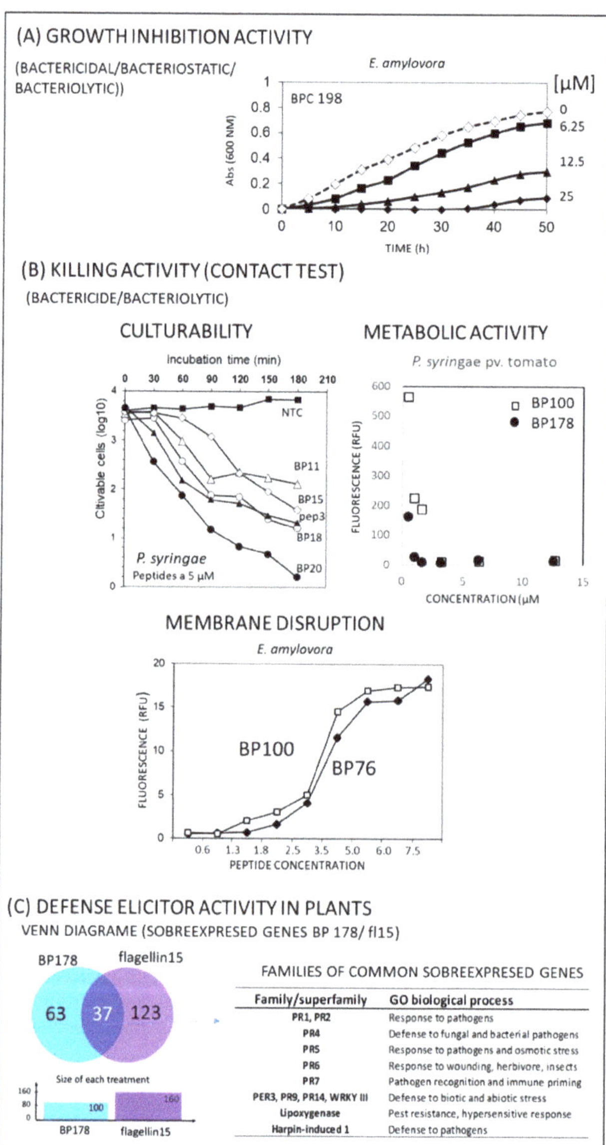

Family/superfamily	GO biological process
PR1, PR2	Response to pathogens
PR4	Defense to fungal and bacterial pathogens
PR5	Response to pathogens and osmotic stress
PR6	Response to wounding, herbivore, insects
PR7	Pathogen recognition and immune priming
PER3, PR9, PR14, WRKY III	Defense to biotic and abiotic stress
Lipoxygenase	Pest resistance, hypersensitive response
Harpin-induced 1	Defense to pathogens

Figure 3. Examples of the results obtained for different peptides to determine their antibacterial activity and their defense elicitor activity in plants. (**A**) Results based on growth inhibition experiments (data from Monroc et al. [49]). (**B**) Results based on killing assays using contact tests (data from Ferre et al. and Montesinos et al. [10,36] and non-published results). (**C**) Venn diagram of the overexpressed genes in tomato plants after treatment with **BP178** and **flg15** using microarray analysis, showing 37 common overexpressed genes, and the table shows the family of common overexpressed genes and their Gene Ontology biological process (data from Montesinos et al. [10]).

4.3. Other Activities

Other activities that need to be taken into account for the development of a peptide-based product are the phytotoxicity, the protease susceptibility and the hemolytic activity (Figure 4). To assess the phytotoxicity, peptides are placed into the mesophyll of tobacco plants and the diameter of the lesion they cause is measured. This assay shows the direct activity of peptides in plant cells, which must be taken into account when selecting the peptides to be tested in planta.

Another parameter to consider is the susceptibility of peptides to protease degradation, since they will be exposed to proteases once applied to the plants. If the peptides are not resistant enough, their half-life time will be low. With that in mind, the sequence of the lead peptides has been modified through the incorporation of unnatural amino acids, as mentioned above. For example, **BP143**, which bears a D-Phe^4, displayed 18% degradation after being exposed to proteinase K for 1 h compared to its parent peptide **BP100**, which showed 75% degradation [40].

Finally, hemolytic activity is the capacity to lyse erythrocytes, which may give an idea of the peptide toxicity. Therefore, it is important for peptides to display low hemolysis, since it will indicate that they are safe for the manipulators in future field applications. Many peptide modifications have allowed us to obtain peptides with similar or higher antimicrobial activity and, at the same time, with lower hemolytic activity. This is the case of **BP143**, in which the incorporation of a D-Phe^4 in **BP100** led to a reduction in the rate of hemolysis from 54% to 5% [40].

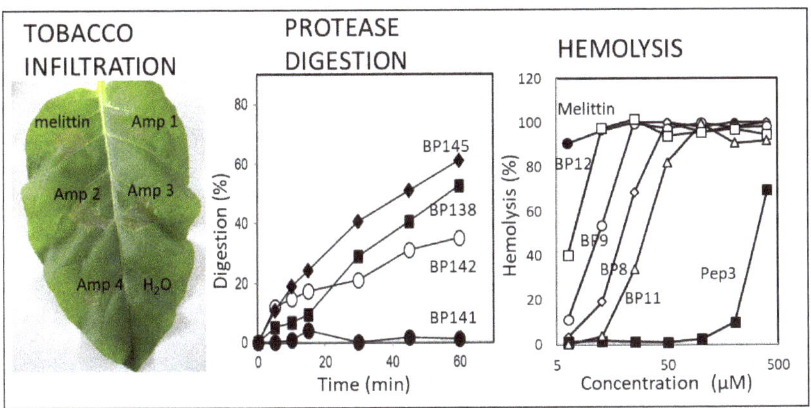

Figure 4. Examples of results obtained for different peptides related to phytotoxicity on tobacco leaves to the susceptibility to protease degradation using proteinase K and to the hemolytic activity using blood erythrocytes. Data from Ferre et al. and Güell et al. [36,40].

The results of tobacco leaf infiltration and the hemolysis assays are indicative of the potential toxicity of peptides. However, toxicity tests on whole plants or animals have to be performed before further development stages, because the main method of interaction of peptides with humans and animals will be dermal, by inhalation or oral ingestion. Therefore, it is essential to know the acute toxicity of these compounds. Based on this, peptides **BP100**, **BP76** and **BP15** were tested for oral acute toxicity in mice, with the median lethal dose (LD$_{50}$) and the lower limit lethal dose (LLD) being higher than 1000–2000 mg/Kg of body weight, which is considered a very low oral toxicity [66].

5. Control of Infections in Plant Material Caused by Phytopathogenic Bacteria in Controlled Environment Conditions

To assess the activity of peptides in vegetal material in controlled environment conditions, two strategies were performed. One strategy consists of using detached vegetal material, such as leaves, flowers or immature fruits and the other consists of using the whole plant under greenhouse conditions.

For example, while developing analogues of **BP100** that contained a D-amino acid in their sequence, a first experiment was performed with all the analogues using pear tree leaves for Pss, pepper leaves for Xav and pear immature fruits for Ea. This experiment allowed us to select the analogues **BP143** and **BP145**. Afterwards, these peptides were applied to whole plants using the same pathosystems, which demonstrated that **BP143** was as effective as streptomycin. Its activity was also greater than **BP100** in controlling Pss and Xav [40]. Aside from these analogues based on D-amino acid substitutions, the linear peptide **BP13** and the cyclic peptide **BPC200W** were also tested in pear trees, since they had antibacterial activity and were able to elicit the plant defense response. These peptides were applied using two different treatment schedules (two or three treatments before the inoculation of the pathogen); **BP100** and **BP143** were added to the experiment as reference peptides. The reference peptides only showed a significative reduction in the disease when three treatments were used, whereas **BP13** and **BP200W** were effective in all treatments [61].

Regarding the bifunctional peptide conjugates, an experiment was performed with **flg15-BP16** to assess their ability to control Ea infections on pear plants, when applied 48 h before the inoculation of the pathogen. In this experiment, the plants were also treated with the monomers of the conjugate, **flg15** and **BP16**, independently and combined for comparative purposes. The results showed that **flg15-BP16** and its monomers applied together (**BP16** and **flg15**) presented similar values to those of the treatment with kasugamycin [62]. In the case of other peptide conjugates, the effect of changing the elicitor peptide (**flg15** or **PIPI**), but not modifying the antimicrobial peptide (**BP475**), and the other way around by changing the antimicrobial peptide (**BP475** or **BP387**), but keeping the elicitor peptide (**flg15**), was studied for the control of infections caused by Xav in pepper plants. Peptide conjugate **flg15-BP475** showed an efficacy similar or higher than copper oxychloride (Figure 5) [63]. All these results are summarized in Table 2. The peptide conjugate **BP178** was tested in tomato plants to control Xav and Pst infections. **BP178** was applied using a spray and the results were compared to those obtained for **BP100**. When **BP178** was applied preventively 1 day before the inoculation of the pathogen, 70% efficacy in the control of the disease severity was observed. In addition, tomato plants with **BP178** showed significantly increased expression of about 100 genes of which 74.4% were related to plant defense and, particularly, to biotic stress responses using microarray analysis (Figure 3C) [10].

Table 2. Peptides assayed in planta for elicitation properties (overexpresed genes in tomato) and in detached plant material or plants for disease control in different pathosystems. Data extracted from several papers cited in the main text.

Peptide	Sequence [a]	Pathosystem	Gene Overexpression [b]
BP13	FKLFKKILKVL-NH$_2$	Ea/pear	3/11
BP16	KKLFKKILKKL-NH$_2$	Ea/pear	1/11
BP100	KKLFKKILKYL-NH$_2$	Xav/pepper, Pss/pear, Ea/pear	2/11
BP143	KKLfKKILKYL-NH$_2$	Xav/pepper, Pss/pear, Ea/pear	2/11
BP378	Ac-KKLFKKILKYK(COC$_5$H$_{11}$)-NH$_2$	-	3/11
BPC200W	c(LLLLKWKKLQ)	Ea/pear	4/11
BP178	KKLFKKILKYL-AGPA-GIGKFLHSAK-KDEL-OH	Pst/Xav/tomato	8/11
flg15-BP16	RINSAKDDAAGLQIA-KKLFKKILKKL-NH$_2$	Ea/pear	6/11
flg15-BP475	Ac-RINSAKDDAAGLQIA-KKLfKKILKK(COC$_3$H$_7$)L-NH$_2$	Xav/tomato	7/11

[a] Lower case letters correspond to D-amino acids. COC$_3$H$_7$, butanoyl. COC$_5$H$_{11}$, hexanoyl. [b] Number of overexpresed genes compared to the number of genes studied.

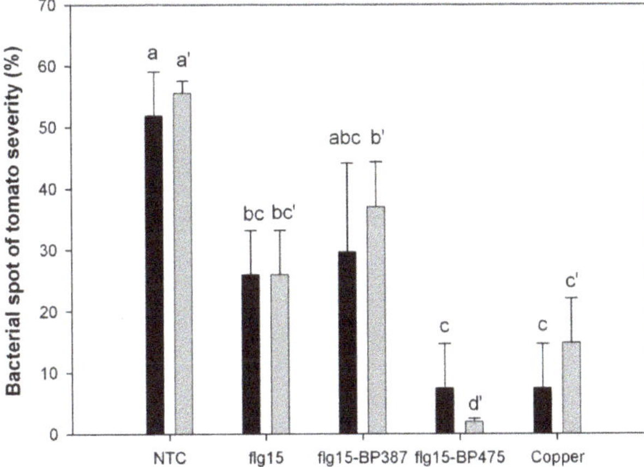

Figure 5. Effect of the application of **flg15** and of the peptide conjugates **flg15-BP387** and **flg15-BP475** on the disease severity of bacterial leaf spot blight caused by *X. axonopodis* pv. vesicatoria in tomato plants. Black and grey are the two independent experiments. The confidence intervals for the means are indicated on top of the bars. Different letters show significant differences between the treatments according to Duncan's test ($p < 0.05$, ANOVA, LSD) for the two experiments performed (with and without apostrophe). Data from Caravaca-Fuentes et al. [62]. Picture shows the disease symptoms in tomato leaves (red arrows).

6. Concluding Remarks

In summary, significant research has been carried out regarding the development of peptides as agents to control plant pathogenic bacteria of economic importance. Linear and cyclic peptides have been designed and synthesized, which have provided leads that have been further optimized by incorporating several structural modifications. This approach led to the identification of sequences with high antibacterial activity, low hemolysis and high stability to protease degradation. In addition, our research has not only focused on peptides that kill bacteria, but also on sequences with the ability to elicit plant defense responses. The achievement of these purposes has required the improvement of methods and technologies for the in vitro and/or in planta screening of a large number of compounds for their antibacterial, hemolytic and phytotoxic activities, as well as for their defense elicitor activity in plants.

The research along these lines has resulted in sequences with distinct biological activities, including antibacterial peptides, plant defense elicitor peptides and sequences with both activities. Thus, we have developed a versatile platform that can be useful at finding, with a certain guarantee of success, peptides that are able to control plant diseases caused by other bacteria. Moreover, the best peptides constitute good candidates with suitable properties to be further developed as effective phytosanitary products for their use in agriculture. Therefore, these peptides represent an alternative strategy that should be taken into account in the management of plant diseases in fields. Multidisciplinary work that involves different research areas, such as microbiology, organic chemistry and combinatorial chemistry, and phytopathology, is crucial in achieving our goal.

7. Ongoing Research

Nowadays, our research on the use of peptides as biopesticides is focused on the control of the xylem-restricted quarantine phytopathogen *Xylella fastidiosa* (Xf), which is the cause of several quarantine diseases found in the EU, such as the olive quick decline (OQDS), and on the control of the floem-restricted *Candidatus* Liberibacter asiaticus (CLas), causing citrus *huanglongbing* (HLB), previously called citrus greening disease, which is one of the most destructive diseases of citrus worldwide. The assessment of the antimicrobial activity of peptides against Xf and CLas is challenging because the former is a fastidious phytopathogen microorganism with a slow growth rate and special nutritional requirements, and the latter is considered a non-culturable microorganism. In the case of Xf, the previously mentioned conventional methods to assess the activity of peptides, such as the in vitro contact test, can be used but their antimicrobial activity can be overestimated [67]. Therefore, methodologies based on molecular methods had to be developed. These methods allow for the evaluation of bacterial viability, without taking into account its culturability [67,68]. Regarding CLas, researchers are using *Liberibacter crescens*, which is the closest cultured relative of this important uncultured crop pathogen.

The formation of biofilm in the xylem vessels of the host is one of the main pathogenicity mechanisms of *X. fastidiosa*; thus, we adapted a methodology using crystal violet [69] to identify peptides with antibiofilm activity against these bacteria [70]. We also focused our attention on the design of peptides that targeted the pathogenicity functions of *X. fastidiosa*. Furthermore, we are currently studying the peptides that are able to elicit defense responses in almond plants, because it is one of the most important crops in our research.

In addition, we have performed *in planta* experiments in *Prunus dulcis* Avijor inoculated with *X. fastidiosa* and treated with the bifunctional peptide **BP178** using endotherapy. The peptide was applied using a preventive and a curative strategy, in addition to a combination of both. The results showed that the combined strategy (preventive and curative) is able to reduce the population levels of *X. fastidiosa* and the disease severity [71].

Author Contributions: Writing—review and editing: E.B., M.P., L.F., L.M., A.B. and E.M. All authors have read and agreed to the published version of the manuscript.

Funding: This research has been funded in the last five years by different grants from Spain, including those by the Ministerio de Ciencia, Innovación y Universidades RTI2018-099410-B-C21; RTI2018-099410-B-C22; and ERTA INIA 2017-00004-C06-03 and from the European Union H2020 Program XF-Actors cod. 727987.

Conflicts of Interest: The authors declare no conflict of interest.

References

1. Oerke, E.C. Crop losses to pests. *J. Agric. Sci.* **2006**, *144*, 31–43. [CrossRef]
2. Mansfield, J.; Genin, S.; Magori, S.; Citovsky, V.; Sriariyanum, M.; Ronald, P.; Dow, M.; Verdier, V.; Beer, S.V.; Machado, M.A.; et al. Top 10 plant pathogenic bacteria in molecular plant pathology. *Mol. Plant Pathol.* **2012**, *13*, 614–629. [CrossRef] [PubMed]
3. López, M.M.; Jesús, M.; Montesinos, E.; Palacio-Bielsa, A. *Enfermedades de Plantas Causadas por Bacterias*; Sociedad Española de Fitopatología (SEF) y Bubok Publishing S.L.: Madrid, Spain, 2013. [CrossRef]
4. La Torre, A.; Jovino, V.; Caradonia, F. Copper in plant protection: Current situation and prospects. *Phytopathol. Mediterr.* **2018**, *57*, 201–236.
5. Taylor, P.; Reeder, R. Antibiotic use on crops in low and middle-income countries based on recommendations made by agricultural advisors. *CABI Agric. Biosci.* **2020**, *1*, 1. [CrossRef]
6. Manyi-Loh, C.; Mamphweli, S.; Meyer, E.; Okoh, A. Antibiotic Use in Agriculture and Its Consequential Resistance in Environmental Sources: Potential Public Health Implications. *Molecules* **2018**, *23*, 795. [CrossRef]
7. European Food Safety Authority; Parnell, S.; Camilleri, M.; Diakaki, M.; Schrader Gand Vos, S. Pest survey card on huanglongbing and its vectors. *EFSA Supporting Publ.* **2019**, *16*, 1574E. [CrossRef]
8. Montesinos, E. Antimicrobial peptides and plant disease control. *FEMS Microbial. Lett.* **2007**, *270*, 1–11. [CrossRef]
9. Huan, Y.; Kong, Q.; Mou, H.; Yi, H. Antimicrobial Peptides: Classification, Design, Application and Research Progress in Multiple Fields. *Front. Microbiol.* **2020**, *11*, 582779. [CrossRef] [PubMed]
10. Montesinos, L.; Gascón, B.; Ruz, L.; Badosa, E.; Planas, M.; Feliu, L.; Montesinos, E. A Bifunctional Synthetic Peptide With Antimicrobial and Plant Elicitation Properties That Protect Tomato Plants From Bacterial and Fungal Infections. *Appl. Environ. Microbiol.* **2020**, *12*, 756357. [CrossRef]
11. Benfield, A.H.; Henriques, S.T. Mode-of-Action of Antimicrobial Peptides: Membrane Disruption vs. Intracellular Mechanisms. *Front. Med. Technol.* **2020**, *2*, 25–28. [CrossRef]
12. Schäfer, A.B.; Wenzel, M. A How-To Guide for Mode of Action Analysis of Antimicrobial Peptides. *Front. Cell. Infect. Microbiol.* **2020**, *10*, 540898. [CrossRef] [PubMed]
13. Czékus, Z.; Kukri, A.; Hamow, K.Á.; Szalai, G.; Tari, I.; Ördög, A.; Poór, P. Activation of local and systemic defence responses by flg22 is dependent on daytime and ethylene in intact tomato plants. *Int. J. Mol. Sci.* **2021**, *22*, 8354. [CrossRef] [PubMed]

14. Ruiz, C.; Nadal, A.; Montesinos, E.; Pla, M. Novel Rosaceae plant elicitor peptides as sustainable tools to control *Xanthomonas arboricola* pv. pruni in *Prunus* spp. *Mol. Plant Pathol.* **2018**, *19*, 418–431, Erratum in *Mol. Plant Pathol.* **2021**, *22*, 896. [CrossRef] [PubMed]

15. Foix, L.; Nadal, A.; Zagorščak, M.; Ramšak, Ž.; Esteve-Codina, A.; Gruden, K.; Pla, M. *Prunus persica* plant endogenous peptides PpPep1 and PpPep2 cause PTI-like transcriptome reprogramming in peach and enhance resistance to *Xanthomonas arboricola* pv. pruni. *BMC Genom.* **2021**, *22*, 360. [CrossRef]

16. Santos-Silva, C.A.D.; Zupin, L.; Oliveira-Lima, M.; Vilela, L.M.B.; Bezerra-Neto, J.P.; Ferreira-Neto, J.R.; Ferreira, J.D.C.; Oliveira-Silva, R.L.D.; Pires, C.D.J.; Aburjaile, F.F.; et al. Plant Antimicrobial Peptides: State of the Art, In Silico Prediction and Perspectives in the Omics Era. *Bioinform. Biol. Insights* **2020**, *14*, 1177932220952739. [CrossRef]

17. Iqbal, A.; Khan, R.S.; Shehryar, K.; Imran, A.; Ali, F.; Attia, S.; Shah, S.; Mii, M. Antimicrobial Peptides as Effective Tools for Enhanced Disease Resistance in Plants. *Plant Cell. Tissue Organ Cult.* **2019**, *139*, 1–15. [CrossRef]

18. Phazang, P.; Negi, N.P.; Raina, M.; Kumar, D. Plant Antimicrobial Peptides: Next-Generation Bioactive Molecules for Plant Protection. In *Phyto-Microbiome in Stress Regulation*; Springer: Singapore, 2020; pp. 281–293. [CrossRef]

19. Li, J.; Hu, S.; Jian, W.; Xie, C.; Yang, X. Plant Antimicrobial Peptides: Structures, Functions, and Applications. *Bot. Stud.* **2021**, *62*, 5. [CrossRef]

20. Wu, Q.; Patočka, J.; Kuča, K. Insect Antimicrobial Peptides, a Mini Review. *Toxins* **2018**, *10*, 461. [CrossRef]

21. Józefiak, A.; Engberg, R.M. Insect Proteins as a Potential Source of Antimicrobial Peptides in Livestock Production. A Review. *J. Anim. Feed Sci.* **2017**, *26*, 87–99. [CrossRef]

22. Lee, H.; Lee, D.G. Mode of Action of Antimicrobial Peptides Identified from Insects. *J. Life Sci.* **2015**, *25*, 715–723. [CrossRef]

23. Kumar, R.; Ali, S.A.; Singh, S.K.; Bhushan, V.; Mathur, M.; Jamwal, S.; Mohanty, A.K.; Kaushik, J.K.; Kumar, S. Antimicrobial Peptides in Farm Animals: An Updated Review on Its Diversity, Function, Modes of Action and Therapeutic Prospects. *Vet. Sci.* **2020**, *7*, 206. [CrossRef] [PubMed]

24. Lee, D.W.; Kim, B.S. Antimicrobial Cyclic Peptides for Plant Disease Control. *Plant Pathol. J.* **2015**, *31*, 1–11. [CrossRef] [PubMed]

25. Kundu, R. Cationic Amphiphilic Peptides: Synthetic Antimicrobial Agents Inspired by Nature. *ChemMedChem* **2020**, *15*, 1887–1896. [CrossRef] [PubMed]

26. Kishi, R.N.I.; Stach-Machado, D.; de Lacorte Singulani, J.; Santos, C.T.D.; Fusco-Almeida, A.M.; Cilli, E.M.; Freitas-Astúa, J.; Picchi, S.C.; Machado, M.A. Evaluation of Cytotoxicity Features of Antimicrobial Peptides with Potential to Control Bacterial Diseases of Citrus. *PLoS ONE* **2018**, *13*, e0203451. [CrossRef]

27. Kumar, P.; Kizhakkedathu, J.N.; Straus, S.K. Antimicrobial Peptides: Diversity, Mechanism of Action and Strategies to Improve the Activity and Biocompatibility in Vivo. *Biomolecules* **2018**, *8*, 4. [CrossRef]

28. Manniello, M.D.; Moretta, A.; Salvia, R.; Scieuzo, C.; Lucchetti, D.; Vogel, H.; Sgambato, A.; Falabella, P. Insect Antimicrobial Peptides: Potential Weapons to Counteract the Antibiotic Resistance. *Cell. Mol. Life Sci.* **2021**, *78*, 4259–4282. [CrossRef]

29. Zhang, C.; Yang, M. Antimicrobial Peptides: From Design to Clinical Application. *Antibiotics* **2022**, *11*, 349. [CrossRef]

30. EFSA. Update of the *Xylella* spp. host plant database—Systematic literature search up to 31 December 2020. *EFSA J.* **2021**, *19*, 6674.

31. Baldi, P.; La Porta, N. *Xylella fastidiosa*: Host range and advance in molecular identification techniques. *Front. Plant Sci.* **2017**, *8*, 944. [CrossRef]

32. Gottwald, T.R.; Bassanezi, R.B. Citrus Huanglongbing: The Pathogen and Its Impact. *Plant Health Prog.* **2007**, *8*, 31. [CrossRef]

33. Blaustein, R.A.; Lorca, G.L.; Teplitski, M. Challenges for Managing Candidatus *Liberibacter* spp. (Huanglongbing Disease Pathogen): Current Control Measures and Future Directions. *Phytopathology* **2019**, *108*, 424–435. [CrossRef] [PubMed]

34. Cavallarin, L.; Andreu, D.; San Segundo, B. Cecropin A-derived peptides are potent inhibitors of fungal plant pathogens. *Mol. Plant Microbe Interact* **1998**, *11*, 218–227. [PubMed]

35. Ali, G.S.; Reddy, A.S.N. Inhibition of fungal and bacterial plant pathogens by synthetic peptides: In vitro growth inhibition, interaction between peptides, and inhibition of disease progression. *Mol. Plant Microbe Interact* **2000**, *13*, 847–859. [CrossRef] [PubMed]

36. Ferre, R.; Badosa, E.; Feliu, L.; Planas, M.; Montesino, E.; Bardají, E. Inhibition of plant-pathogenic bacteria by short synthetic cecropin A-melittin hybrid peptides. *Appl. Environ. Microbiol.* **2006**, *72*, 3302–3308. [CrossRef] [PubMed]

37. Badosa, E.; Ferre, R.; Planas, M.; Feliu, L.; Besalú, E.; Cabrefiga, J.; Bardají, E.; Montesinos, E. A library of lineal undecapeptides with bactericidal activity against phytopathogenic bacteria. *Peptides* **2007**, *28*, 2276–2285. [CrossRef]

38. Montesinos, E.; Bardají, E. Synthetic antimicrobial peptides as agricultural pesticides for plant-disease control. *Chem. Biodivers.* **2008**, *5*, 1225–1237. [CrossRef]

39. Ng-Choi, I.; Soler, M.; Güell, I.; Badosa, E.; Cabrefiga, J.; Bardají, E.; Montesinos, E.; Planas, M.; Feliu, L. Antimicrobial peptides incorporating non-natural amino acids as agents for plant protection. *Protein Pept. Lett.* **2014**, *21*, 357–367. [CrossRef]

40. Güell, I.; Cabrefiga, J.; Badosa, E.; Ferre, R.; Talleda, M.; Bardají, E.; Planas, M.; Feliu, L.; Montesinos, E. Improvement of the efficacy of linear undecapeptides against plant-pathogenic bacteria by incorporation of D-amino acids. *Appl. Environ. Microbiol.* **2011**, *77*, 2667–2675. [CrossRef]

41. Güell, I.; Micaló, L.; Cano, L.; Badosa, E.; Ferre, R.; Montesinos, E.; Bardají, E.; Feliu, L.; Planas, M. Peptidotriazoles with antimicrobial activity against bacterial and fungal plant pathogens. *Peptides* **2012**, *33*, 9–17. [CrossRef]

42. Ng-Choi, I.; Soler, M.; Cerezo, V.; Badosa, E.; Montesinos, E.; Planas, M.; Feliu, L. Solid-phase synthesis of 5-arylhistidine-containing peptides with antimicrobial activity through a microwave-assisted Suzuki-Miyaura cross-coupling. *Eur. J. Org. Chem.* **2012**, *2012*, 4321–4332. [CrossRef]

43. Oliveras, À.; Baró, A.; Montesinos, L.; Badosa, E.; Montesinos, E.; Feliu, L.; Planas, M. Antimicrobial activity of linear lipopeptides derived from BP100 towards plant pathogens. *PLoS ONE.* **2018**, *13*, e0201571. [CrossRef] [PubMed]

44. Oliveras, À.; Moll, L.; Riesco-Llach, G.; Tolosa-Canudas, A.; Gil-Caballero, S.; Badosa, E.; Bonaterra, A.; Montesinos, E.; Planas, M.; Feliu, L. D-amino acid-containing lipopeptides derived from the lead peptide bp100 with activity against plant pathogens. *Inter. J. Mol. Sci.* **2021**, *22*, 6631. [CrossRef] [PubMed]

45. Zamora-Carreras, H.; Strandberg, E.; Muhlhauser, P.; Burch, J.; Wadhwani, P.; Jimenez, M.A.; Bruix, M.; Ulrich, A.S. Alanine scan and 2H NMR analysis of the membrane-active peptide BP100 point to a distinct carpet mechanism of action. *Biochim. Biophys. Acta* **2016**, *1858*, 1328–1338. [CrossRef] [PubMed]

46. Badosa, E.; Moiset, M.; Montesinos, L.; Talleda, M.; Bardají, E.; Feliu, L.; Planas, M.; Montesinos, E. Derivatives of the antimicrobial peptide BP100 for expression in plant systems. *PLoS ONE* **2013**, *8*, e85515. [CrossRef]

47. Montesinos, L.; Bundó, M.; Badosa, E.; San Segundo, B.; Coca, M.; Montesinos, E. Production of BP178, a derivative of the synthetic antibacterial peptide BP100, in the rice seed endosperm. *BMC Plant Biol.* **2017**, *17*, 63. [CrossRef]

48. Monroc, S.; Badosa, E.; Feliu, L.; Planas, M.; Montesinos, E.; Bardají, E. De novo designed cyclic cationic peptides as inhibitors of plant pathogenic bacteria. *Peptides* **2006**, *27*, 2567–2574. [CrossRef]

49. Monroc, S.; Badosa, E.; Besalú, E.; Planas, M.; Bardají, E.; Montesinos, E.; Feliu, L. Improvement of cyclic decapeptides against plant pathogenic bacteria using a combinatorial chemistry approach. *Peptides* **2006**, *27*, 2575–2584. [CrossRef]

50. Camo, C.; Torné, M.; Besalú, E.; Rosés, C.; Cirac, A.D.; Moiset, G.; Badosa, E.; Bardaji, E.; Montesinos, E.; Planas, M.; et al. Tryptophan-containing cyclic decapeptides with activity against plant pathogenic bacteria. *Molecules* **2017**, *22*, 1817. [CrossRef]

51. Cirac, A.D.; Torné, M.; Badosa, E.; Montesinos, E.; Salvador, P.; Feliu, L.; Planas, M. Rational design of cyclic antimicrobial peptides based on BPC194 and BPC198. *Molecules* **2017**, *22*, 1054. [CrossRef]

52. Güell, I.; Vilà, S.; Micaló, L.; Badosa, E.; Montesinos, E.; Planas, M.; Feliu, L. Synthesis of cyclic peptidotriazoles with activity against phytopathogenic bacteria. *Eur. J. Org. Chem.* **2013**, *22*, 4933–4943. [CrossRef]

53. Güell, I.; Vilà, S.; Badosa, E.; Montesinos, E.; Feliu, L.; Planas, M. Design, synthesis, and biological evaluation of cyclic peptidotriazoles derived from BPC194 as novel agents for plant protection. *Pept. Sci.* **2017**, *108*, e23012. [CrossRef] [PubMed]

54. Vilà, S.; Badosa, E.; Montesinos, E.; Feliu, L.; Planas, M. A convenient solid-phase strategy for the synthesis of antimicrobial cyclic lipopeptides. *Org. Biomol. Chem.* **2013**, *11*, 3365–3374. [CrossRef] [PubMed]

55. Vilà, S.; Badosa, E.; Montesinos, E.; Planas, M.; Feliu, L. Synthetic Cyclolipopeptides Selective against Microbial, Plant and Animal Cell Targets by Incorporation of D-Amino Acids or Histidine. *PLoS ONE* **2016**, *11*, e0151639. [CrossRef] [PubMed]

56. Malik, N.A.A.; Kumar, I.S.; Nadarajah, K. Elicitor and receptor molecules: Orchestrators of plant defense and immunity. *Int. J. Mol. Sci.* **2020**, *21*, 963. [CrossRef]

57. Brunner, Â.; Rosahl, S.; Lee, J.; Rudd, J.J.; Geiler, C.; Scheel, D.; Nu, T. Pep-13, a plant defense-inducing pathogen associated pattern from *Phytophthora* transglutaminases. *EMBO J.* **2002**, *21*, 6681–6688. [CrossRef]

58. Miyashita, M.; Oda, M.; Ono, Y.; Komoda, E.; Miyagawa, H. Discovery of a Small Peptide from Combinatorial Libraries That Can Activate the Plant Immune System by a Jasmonic Acid Signaling Pathway. *Chem. Biol. Chem.* **2011**, *8502*, 1323–1329. [CrossRef]

59. Pieterse, C.M.J.; Van der Does, D.; Zamioudis, C.; Leon-Reyes, A.; Van Wees, S.C.M. Hormonal modulation of plant immunity. *Ann. Rev. Cell Dev. Biol.* **2012**, *28*, 489–521. [CrossRef]

60. Hajam, I.; Dar, P.; Shahnawaz, I.; Jaume, J.C.; Hwa Lee, J. Bacterial flagellin—A potent immunomodulatory agent. *Exp. Mol. Med.* **2017**, *49*, e373. [CrossRef]

61. Badosa, E.; Montesinos, L.; Camó, C.; Ruz, L.; Cabrefiga, J.; Francés, J.; Gascón, B.; Planas, M.; Feliu, L.; Montesinos, E. Control of fire blight infections with synthetic peptides that elicit plant defense responses. *J. Plant Pathol.* **2017**, *99*, 65–73. [CrossRef]

62. Caravaca-Fuentes, P.; Camó, C.; Oliveras, À.; Baró, A.; Francés, J.; Badosa, E.; Planas, M.; Feliu, L.; Montesinos, E.; Bonaterra, A. A bifunctional peptide conjugate that controls infections of erwinia amylovora in pear plants. *Molecules* **2021**, *26*, 3426. [CrossRef]

63. Oliveras, À.; Camó, C.; Caravaca-fuentes, P.; Moll, L.; Riesco-llach, G.; Badosa, E.; Bonaterra, A.; Montesinos, E.; Feliu, L. Peptide conjugates derived from flg15, Pep13 and PIP1 that are active against plant pathogenic bacteria and trigger plant defense responses. *Appl. Environ. Microbiol.* **2022**, *88*, e00574-22. [CrossRef] [PubMed]

64. Livak, K.; Schmittgen, T. Analysis of relative gene expression data using real-time quantitative PCR and the $2^{-\Delta\Delta CT}$ method. *Methods* **2001**, *25*, 402–408. [CrossRef] [PubMed]

65. Camó, C.; Bonaterra, A.; Badosa, E.; Baró, A.; Montesinos, L.; Montesinos, E.; Planas, M.; Feliu, L. Antimicrobial peptide KSL-W and analogues: Promising agents to control plant diseases. *Peptides* **2019**, *112*, 85–95. [CrossRef] [PubMed]

66. Montesinos, E.; Badosa, E.; Cabrefiga, J.; Planas, M.; Feliu, L.; Bardají, E. Antimicrobial Peptides for Plant Disease Control. From Discovery to Application. In *Small Wonders: Peptides for Disease Control*; ACS Symposium Series; American Chemical Society: Washington, DC, USA, 2012; pp. 235–261. [CrossRef]

67. Baró, A.; Mora, I.; Montesinos, L.; Montesinos, E. Differential susceptibility of *Xylella fastidiosa* strains to synthetic bactericidal peptides. *Phytopathology* **2020**, *110*, 1018–1026. [CrossRef] [PubMed]

68. Baró, A.; Badosa, E.; Montesinos, L.; Feliu, L.; Planas, M.; Montesinos, E.; Bonaterra, A. Screening and identification of BP100 peptide conjugates active against *Xylella fastidiosa* using a viability-qPCR method. *BMC Microbiol.* **2020**, *20*, 229. [CrossRef] [PubMed]
69. Zaini, P.A.; Fuente, L.D.L.; Hoch, H.C.; Burr, T.J. Grapevine xylem sap enhances biofilm development by *Xylella fastidiosa*. *FEMS Microb. Lett.* **2009**, *295*, 129–134. [CrossRef] [PubMed]
70. Moll, L.; Badosa, E.; Planas, M.; Feliu, L.; Montesinos, E.; Bonaterra, A. Antimicrobial Peptides With Antibiofilm Activity Against *Xylella fastidiosa*. *Front. Microbiol.* **2021**, *12*, 3243. [CrossRef]
71. Moll, L.; Baró, A.; Montesinos, L.; Badosa, E.; Bonaterra, A.; Montesinos, E. Induction of defense responses and protection of almond plants against *Xylella fastidiosa* by endotherapy with a bifunctional peptide. *Phytopathology* **2022**, *112*, 1907–1916. [CrossRef]

Review

Bacteria as Biological Control Agents of Plant Diseases

Anna Bonaterra *, Esther Badosa, Núria Daranas, Jesús Francés, Gemma Roselló and Emilio Montesinos

Laboratory of Plant Pathology, Institute of Food and Agricultural Technology-CIDSAV-Xarta,
University of Girona, Campus Montilivi, 17071 Girona, Spain
* Correspondence: anna.bonaterra@udg.edu

Abstract: Biological control is an effective and sustainable alternative or complement to conventional pesticides for fungal and bacterial plant disease management. Some of the most intensively studied biological control agents are bacteria that can use multiple mechanisms implicated in the limitation of plant disease development, and several bacterial-based products have been already registered and marketed as biopesticides. However, efforts are still required to increase the commercially available microbial biopesticides. The inconsistency in the performance of bacterial biocontrol agents in the biological control has limited their extensive use in commercial agriculture. Pathosystem factors and environmental conditions have been shown to be key factors involved in the final levels of disease control achieved by bacteria. Several biotic and abiotic factors can influence the performance of the biocontrol agents, affecting their mechanisms of action or the multitrophic interaction between the plant, the pathogen, and the bacteria. This review shows some relevant examples of known bacterial biocontrol agents, with especial emphasis on research carried out by Spanish groups. In addition, the importance of the screening process and of the key steps in the development of bacterial biocontrol agents is highlighted. Besides, some improvement approaches and future trends are considered.

Keywords: bacterial biological control agents; bacterial and fungal plant diseases; screening; improvement

Citation: Bonaterra, A.; Badosa, E.; Daranas, N.; Francés, J.; Roselló, G.; Montesinos, E. Bacteria as Biological Control Agents of Plant Diseases. *Microorganisms* **2022**, *10*, 1759. https://doi.org/10.3390/microorganisms10091759

Academic Editor: Ana Palacio-Bielsa

Received: 30 July 2022
Accepted: 24 August 2022
Published: 31 August 2022

Publisher's Note: MDPI stays neutral with regard to jurisdictional claims in published maps and institutional affiliations.

1. Introduction

Plant pathogens constitute a great threat to agricultural and forestry production since they cause diseases with important economic and environmental impact [1,2]. Currently, their effect has worsened due to globalization of markets and global climate change that facilitate the appearance of emerging diseases and their rapid spread [3]. New trends in crop protection have been oriented toward a reduction of reliance on conventional pesticides together with the compulsory implementation of integrated pest management (IPM) principles program addressed in the regulations of different countries [4,5]. Consequently, the interest in effective and sustainable alternative strategies to conventional pesticides has increased. Biological control is regarded as a promising alternative and a wide array of microbial biocontrol agents (BCA) have been developed in the past decades for the management of fungal and bacterial diseases. Some of the most intensively studied are bacteria belonging of the genus *Pseudomonas* spp., *Bacillus* spp., and *Streptomyces* spp., that have been already registered as commercial products and marketed. Nowadays, in EU there are 13 bacterial-based biocontrol agents (BCA) registered as biopesticides for the control of bacterial and fungal diseases (*Bacillus amyloliquefaciens* strains: QST 713, AH2, MBI 600, FZB24 and IT 45, *Bacillus amyloliquefaciens* subsp. *plantarum* strain D747, *Bacillus firmus* I-1582, *Bacillus pumilus* strain QST 2808, *Bacillus subtilis* strain IAB/BS03, *Pseudomonas* sp. strain DSMZ 13134, *Pseudomonas chlororaphis* strain MA 342, *Streptomyces* K61 and *Streptomyces lydicus* strain WYEC 108) (https://food.ec.europa.eu/plants/pesticides/eu-pesticides-database_en, accessed on 1 June 2022). However, efforts are still required to increase the commercially available microbial biopesticides for plant disease management [6].

The efficacy of a bacterial biocontrol agent against plant diseases depends on the microbial agent (mechanism of action, conditioning, dose, methods of application), plant pathogens targets (sensitivity), host (cultivar type, physical properties), and environmental conditions (biotic and abiotic factors, chemical residues, nutrient availability, temperature, moisture) [7]. Numerous interactions may affect the efficacy of biocontrol such as the variability from plant to plant, orchard, and year, and often lack of efficacy and inconsistent field performance have been reported. Therefore, it is necessary to know the efficacy and consistency of biological control in comparison to standard chemical fungicide and bactericide treatments under sufficiently wide production conditions in orchards representing different environments and agricultural practices [8,9].

Bacterial biocontrol agents use a great variety of mechanisms to protect plants from pathogen infections. They may use one or a combination of mechanisms to prevent or reduce plant disease, interacting directly or indirectly with the pathogen [10,11] (Figure 1). BCA can interact directly with the pathogen through the secretion of antimicrobial compounds, interfering with the pathogen virulence and competing for nutrients and space. Many BCA synthesize and release metabolites such as lipopeptides, bacteriocins, antibiotics, biosurfactants, cell-wall degrading enzymes or microbial volatile compounds which have antimicrobial activity by reducing growth or metabolic activity of pathogens. BCA may also interfere with the quorum sensing (QS) system of the pathogens, enzymatically degrading or inhibiting the synthesis of signal molecules used to initiate infections. For instance, producing QS inhibitors such as lactonases, pectinases, and chitinases that degrade QS signal molecules impairing pathogen infection and reducing the symptoms of plant diseases [12]. Moreover, BCA can diminish pathogen infection pressure through competitive exclusion over pathogens by reducing their growth without killing them. Highly competitive bacterial BCA may colonize and survive in the infection site and have a more efficient nutrient uptake system than the pathogen, such as low-molecular-weight siderophores with affinity for ferric iron. Besides direct interactions, BCA can protect plants indirectly, by triggering the defense response or promoting plant growth [10,11,13]. They may enhance host defense mechanism eliciting systemic resistance. This results in an accumulation of structural barriers and triggers many biochemical and molecular defense responses in the host, conferring a protective system against a wide range of pathogens. Moreover, BCA can promote plant growth by enhancing mineral and water absorption or producing plant growth stimulating compounds, such as hormones, and thereby improving plant health and fitness. In many cases, various mechanisms are involved in the complex interactions between plants, BCA, and pathogens. Therefore, identifying the mechanisms responsible for biocontrol is a great challenge. Understanding the mode of action responsible for the protective effect of a BCA will facilitate the optimization of biocontrol and allow the establishment of optimal conditions for the interaction between the BCA, the pathogen, and the host, and the design of appropriate formulations and methods of application to enhance plant health and sustainable agriculture.

This review shows some relevant examples of known bacterial BCA, and presents their main modes of action, including details concerning the mechanisms and molecules involved in the biocontrol activity with especial emphasis on research carried out by Spanish groups. In addition, the importance of the isolation, screening process, characterization of the key steps in the development of BCA is highlighted. Moreover, some improvement approaches and future trends are considered.

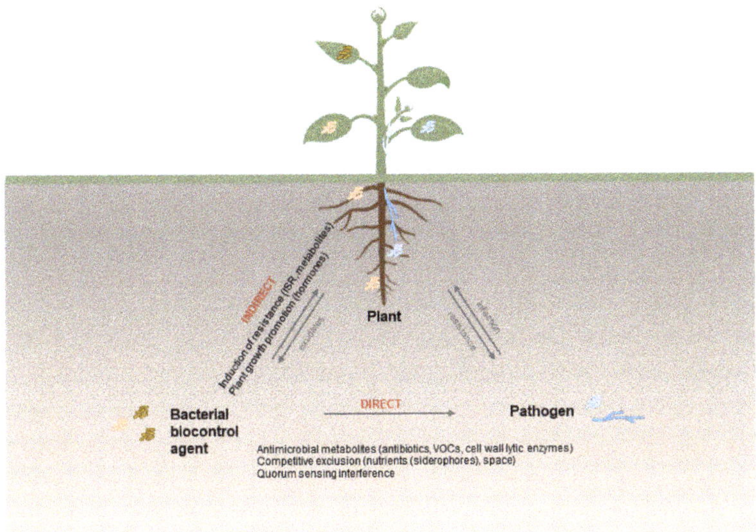

Figure 1. Overview of the direct and indirect mechanisms of biocontrol involving interaction between bacterial biocontrol agent, pathogen, and plant (created with BioRender.com).

2. Bacteria as Biological Control Agents of Plant Diseases

A wide variety of bacterial genera, including Agrobacterium, Alcaligenes, Arthrobacter, Bacillus, Enterobacter, Erwinia, Pseudomonas, Rhizobium, Serratia, Stenotrophomonas, Streptomyces, and Xanthomonas have been described to have plant disease protection activity against fungal and bacterial pathogens. These bacteria can use multiple mechanisms implicated in the limitation of plant pathogens development. These mechanisms of action include colonization of infection sites and competitive exclusion of the pathogen, antagonistic activity based on the secretion of highly active antimicrobials such as antibiotics or cell wall lytic enzymes and induction of plant resistance [7,14,15].

Several bacterial BCA of bacterial and fungal pathogens have been developed in research carried out within the framework of Spanish groups and some examples are highlighted (Table 1).

Table 1. Selected bacterial biocontrol agents [1] of plant diseases.

Microorganism and Strain	Target Pathogen or Disease [2]	In Vivo/In Planta Trials	Disease Reduction (%)/Application Dose/(CFU mL^{-1})	Mechanism Involved/Trait [3]	Reference
B. amyloliquefaciens PPCB004	Ac, B, Cg, Fa, Lt, Pc, Pp	orange fruits	20–70/10^8	Ab-fengycin, iturin A, surfactin	[16]
B. amyloliquefaciens CPA-8	Bc, Mf, Ml	cherry fruits	24–62/10^7	Ab-fengycin-like, VOCs	[17]
Bacillus subtilis UMAF6614 and UMAF6639	Pf	detached melon leaves	67–74/10^8	Ab-bacillomycin, fengycin, iturin A	[18]

Table 1. *Cont.*

Microorganism and Strain	Target Pathogen or Disease [2]	In Vivo/In Planta Trials	Disease Reduction (%)/Application Dose/(CFU mL^{-1})	Mechanism Involved/Trait [3]	Reference
Bacillus velezensis A17	Ea, Ps, Xa	-	-	Ab-bacillomycin, fengycin, iturin, surfactin,	[19,20]
Lactobacillus plantarum TC92, PM411	Ea, Psk, Xf	pear, kiwi, and strawberry plants	45–75/10^8	CE	[21,22]
Leuconostoc mesenteroides CM160	BFV	-	-	Ab-mesentericin	[23]
Pantoea agglomerans EPS125	PF	apricot, peach, and nectarine fruits	49–61/10^7	CE	[24,25]
P. agglomerans CPA-2	PF	pear fruits	50–95/10^7	CE	[26]
Pseudomonas chlororaphis PCL1606	Rn	avocado plants	40/10^9	Ab-2-hexyl, 5-propyl resorcinol	[27]
Pseudomonas fluorescens MVW1-2, MVP 1-4	Fop, Gt	-	-	Ab-phloroglucinol (DAPG)	[28]
P. fluorescens EPS62e	Ea	detached flowers, and pear plants	31–98/10^8	CE, NC	[29]
P. fluorescens EPS817, EPS894	Pc	strawberry plants	76–80/10^8	Ab-phenazines (PCA)	[30]
Pseudomonas simiae PICF7	Vd	olive plants	20–28/10^8	CE/IR-local and systemic defenses	[31,32]
Pseudomonas pseudoalcaligenes AVO110	Rn	-	-	CE	[33]
Streptomyces strains CBQ-EA-2, CBQ-B-8	Mp, Rs	bean plants	60–75/10^8	Extracellular enzyme activities	[34]
Streptomyces sp. VV/E1, VV/R1, VV/R4	GTD	grapevine plants	25–35/10^7	-	[35]
Weissella cibaria TM128	PBF	apple fruits	50/10^8	Ab-organic acids	[36]

[1] Only examples of studies performed by Spanish groups are selected. [2] BFV, bioprotection of fresh fruits and vegetables; GTD, grapevine trunk diseases; PBF, phytopathogenic bacteria and fungi; PF, postharvest fungi; Ac, *Alternaria citri*; B, *Botryosphaeria* sp.; Bc, *Botrytis cinerea*; Cg, *Colletotrichum gloesporioides*; Ea, *Erwinia amylovora*; Fa, *Fusicoccum aromaticum*; Fop, *Fusarium oxysporum* f. sp. pisi; Gt, *Gaeumannomyces tritici*; Lt, *Lasidiplodia theobromae*; Mp, *Macrophomina phaseolina*; Mf, *Monilia fructicola*; Ml, *Monilia laxa*; Pc, *Penicillium crustosum*; Pp, *Phomopsis perse*; Pc, *Phytophthora cactorum*; Pf, *Podosphaera fusca*; Ps, *Pseudomonas syringae*; Psk, *Pseudomonas syringae* pv kiwi; Rn, *Rosellinia necatrix*; Rs, *Rhizoctonia solani*; Vd, *Verticillium dahliae*; Xa, *Xanthomonas arboricola*; Xf, *Xanthomonas fragariae*. [3] Ab, antibiosis; CE, competitive exclusion; IR, induced resistance; NC, nutrient competition.

2.1. *Pseudomonas* spp.

Fluorescent pseudomonads are ubiquitously present in plant environments and possess several relevant traits for their effectiveness in the reduction of plant diseases. These traits include a high ecological fitness, a strong antagonistic activity toward various plant pathogens, and a potent ability to trigger an immune reaction in plant.

Many *Pseudomonas* spp. are efficient colonizers of the plant surface (rhizosphere and phyllosphere) and the endosphere. They can use many plant exudates as nutrients and have a high growth rate, which are prerequisites to efficiently compete with other microorganisms for space and nutrients in the plant environment [37–39]. For example, the activity of *P. fluorescens* EPS62e and *P. pseudoalcaligenes* AVO110 in the reduction of *Erwinia amylovora* or *Rosellinia necatrix* infections, respectively, is based on their strong fitness in colonizing plant tissues as they have higher growth potential and nutrient use efficiency than the target pathogens [29,33]. In addition, competition for limited nutrients has been described as an important mechanism of *Pseudomonas* spp., but it is only relevant when the concentration of a given limited nutrients is low, such as in the biological control of *Pythium ultimum* by *P. fluorescens* 54/96 [40] or in the case of siderophore-mediated competition for iron in the reduction of *Fusarium* wilt of carnation by *P. putida* WCS358 [41].

Another relevant trait of *Pseudomonas* spp. is that they are major producers of bioactive metabolites, such as antibiotics, cyclic peptides, or enzymes that play important ecological roles. Specifically, they produce different antimicrobial compounds such as phenazines, phloroglucinols, dialkylresorcinols, pyoluteorin, and pyrrolnitrin, whose involvement as a mechanism of action in biological control has been well documented [38,42]. Phenazines such as phenazine-1-carboxamide (PCN) or phenazine-1-carboxylate (PCA) are nitrogen-containing heterocyclic compounds with broad antifungal and antibacterial activities. These compounds are involved in the reduction of fungal pathogens infections of plants. For example, PCN produced by *P. chlororaphis* subsp. *aurantiaca* strain Pcho10 shows strong inhibitory activity against *Fusarium graminearum* [43] and PCA produced by *P. fluorescens* EPS894 inhibits *Phytophthora cactorum* in strawberry plants [30]. The phloroglucinols are phenolic broad-spectrum antibiotics produced by a wide variety of bacterial strains. Specifically, 2,4-diacetyl phloroglucinol (DAPG), produced by different strains of *Pseudomonas* spp., has a broad-spectrum action, and contributes to the biological control of plant disease, especially soil-borne plant diseases [28,44]. Dialkylresorcinols exhibit antifungal and antibacterial activities such as the compound 2-hexyl-5-propyl resorcinol produced by *P. chlororaphis* PCL 1606 is responsible for the biocontrol of *R. necatrix* [27]. Pyrrolnitrin have also been involved in the biocontrol of the *Fusarium* head blight by *P. chlororaphis* G05 [45]. Pyoluteorin, as well as the volatile compound hydrogen cyanide are other compounds produced by different strains of *Pseudomonas* spp. that have been involved in the biocontrol of some pathogens [46].

Moreover, pseudomonads produce cyclic lipopeptides (CLPs) that are amphiphilic molecules containing chains of 7–25 aminoacids of which several form a lactone ring coupled to a fatty acid tail. Many of the CLPs are biosurfactants, which can damage cell membranes, thereby causing leakage and cytolysis and are a common feature of both plant beneficial and pathogenic bacteria [46,47]. Interestingly, some of them such as orfamides synthesized by *P. protegens* have antimicrobial activity against a variety of organisms, including the pathogenic oomycetes *Pythium* and *Phytophthora*, and the fungus *Rhizoctonia* [48]. Other examples that show antifungal activity are the cyclic depsipeptide viscosinamide produced by *P. fluorescens* DR54 [49] or the peptide tensin produced by *P. fluorescens* 96.578 [50].

Pseudomonads can also produce lytic extracellular enzymes such as chitinases, β-1,3 glucanases, cellulases that have important roles in biocontrol activity by their degradative activities of cell wall compounds, such as chitin, glucan, and glucosidic bridges. For example, hydrolytic enzymes produced by *Pseudomonas* sp. have in vitro antifungal activity against *Pythium aphanidermatum* and *Rhizoctonia solani* and promote growth in chickpea [51].

Pseudomonas spp., can trigger defense responses of host plants through different pathways, conferring plants with resistance to multiple pathogens. In many cases they confer resistance to plant upon the activation of induced systemic resistance (ISR) that involves activation of immune response and priming state for a more efficient activation of defenses. For example, in *Vitis*, *P. fluorescens* PTA-CT2 induces ISR to *Plasmopara viticola* and *Botrytis cinerea* that depends on the activation of SA or JA and ABA defensive pathways [52]. In another case, the biocontrol endophytic bacterium *Pseudomonas simiae* PICF7 induces systemic defense responses in aerial tissues upon colonization of olive roots [31,32]. In addition, some compounds such as CLPs or phenazines have been reported to trigger defense responses in plants. For example, massetolide A of *P. fluorescens* enhanced resistance to infection by *Phytophthora infestans* in tomato plants [53] and phenazines from *Pseudomonas* sp. CMR12a induced systemic resistance on rice and bean [54].

2.2. Bacillus spp.

Bacillus species are among the most exploited beneficial bacteria as biopesticides. They are widely distributed in several habitats such as soil and plant surfaces, have broad physiological ability and capability to form endospores that confers resistance to adverse environmental conditions. They can develop antagonism against a wide range of bacterial and fungal plant pathogens. The most remarkable trait of *Bacillus* spp. is the ability to produce a wide variety of bioactive compounds valuable for agricultural applications, including metabolites with antimicrobial activity, surface-active, and implicated in the induction of plant defense responses [55,56].

Bacteriocins and bacteriocin-like substances are ribosomally synthesized peptides that act against target cells by interfering with the synthesis of the cell wall or by forming pores in the cell membrane. *Bacillus* spp. produce several bacteriocins with antimicrobial activity such as amylolysin, amylocyclicin, amysin, subtilin, subtilosin A, subtilosin B, thuricin [57]. Some of them have been involved in biocontrol of plant pathogens. For example, Bac-GM17 produced by *B. clausii* GM17 have activity against *Agrobacterium tumefaciens* [58] or thuricin Bn1 from *B. thuringiensis* subsp. *kurstaki* Bn1 against *Pseudomonas savastanoi* and *Pseudomonas syringae* [59].

Cyclic lipopeptides (CLPs) are non-ribosomally synthetized amphiphilic compounds, composed of a fatty acid tail linked to a short oligopeptide which form a macrocyclic ring structure that are widely spread in *Bacillus* spp. The most important CLPs produced by *Bacillus* are represented by iturins, fengicins, and surfactins. They interact with cell membrane of target pathogens forming pores and leading to an imbalance in transmembrane ion fluxes [60]. There are several examples of *Bacillus* spp. strains producing CLPs, that are responsible for the antifungal activity that protect plants from diseases. The fengycin, iturin A, and surfactin produced by *B. amyloliquefaciens* PPCB004 and bacillomycin, fengycin, and iturin A produced by *B. subtilis* UMAF6614 and UMAF6639 are key factors in the antagonism against fungal pathogens [16,18]. In addition, *Bacillus* strains producing CLPs have also antibacterial activity such as *B. amyloliquefaciens* A17 (currently *B. velezensis*) that produces bacillomycin, fengycin, iturin, and surfactin which act synergistically against several bacterial plant pathogens [19,20], or *B. amyloliquefaciens* KPS46 that produces surfactin, required to reduce infections by *Xanthomonas axonopodis* pv. glycines [61]. In many cases, lipopeptides and other peptides or volatile organic compounds (VOCs) act in a synergistic manner to improve their activity. For example, *B. amyloliquefaciens* CPA-8 produces fengycin and VOCs that are involved in the antifungal activity against *Monilinia* and *Botrytis* [17]. Besides their antimicrobial activity, some of these compounds act indirectly as elicitors of defense mechanism in the host plant or play an important role in favoring colonization [62].

Hydrolytic enzymes such as chitinases, chitosanases, glucanases, cellulases, lipases, and proteases, are also extensively produced by *Bacillus* spp. strains. These compounds efficiently hydrolyze the major components of the fungal and bacterial cell walls and have been involved in plant pathogen suppression. For example, a protease produced by *B. amyloliquefaciens* SP1 showed efficacy in biocontrol of *Fusarium oxysporum* [63] and the hydrolase activity (protease, chitinase, cellulase, glucanase) was identified as the key factor of *B. velezensis* in controlling pepper gray mold caused by *Botrytis cinerea* [64].

Various *Bacillus* spp. strains can elicit ISR in different plants and confer an enhanced defense mechanism against a range of pathogens. Several studies have shown that VOCs and CLPs, such as surfactin and fengycin, are involved in the immune response of plants elicitation [65,66]. For example, *B. amyloliquefaciens* FZB42 produced secondary metabolites (surfactin, fengycin, and bacillomycin D) that trigger plant defense gene expression and contribute to lettuce bottom rot reduction [67]. In another example, *Bacillus subtilis* OTPB1 increased the levels of growth hormones and defense-related enzymes in tomato, conferring protection against early and late blight [68].

2.3. Other Relevant Bacteria as BCA

There are other relevant species/strains which can be used to develop microbial biopesticides. Some are distributed among the Gram-negative bacteria of the families Rhizobiaceae, Enterobacteriaceae, and Xanthomonadaceae. Others can be found among Gram-positive bacteria such as Lactobacillaceae, Leuconostocaceae, and Streptomycetaceae [69]. Some examples, since they reduce plant pathogenic bacteria and fungi infections, include species of *Streptomyces* spp., *Pantoea* spp., and *Lactobacillus* spp.

Streptomyces spp. is one of the most studied genus of bacteria, since they produce bioactive compounds that inhibit plant pathogens in vitro and are effective in the controlling various bacterial and fungal plant diseases [70]. Examples of such metabolites include macrolide benzoquinones, aminoglycosides, polyenes, and nucleosides. *Streptomyces* strains are also known for their ability to produce extracellular enzymes active in fungal cell wall degradation. These hydrolases may be responsible for the mycoparasitic potential of some strains and the limitation of plant diseases, such as in the strains *Streptomyces* CBQ-EA-2 and CBQ-B-8 that have chitinolytic, cellulolytic, and proteolytic activity and reduced *Macrophomina phaseolina* and *Rhizoctonia solani* infections in *Phaseolus vulgaris* [34]. Other bioactive metabolites are produced, including VOCs, as signaling molecules to regulate plant growth and immunity in response to biotic and abiotic stresses. In addition, some strains can limit plant disease development through the induction of systemic resistance (ISR) in plants. ISR elicited by *Streptomyces* strains occurs via the activation of the jasmonic acid/ethylene and salicylic acid pathways. For example, *S. lydicus* M01 treatment reduced the reactive oxygen species (ROS) accumulation and increased the activities of antioxidases related with ROS scavenging, which indicated an enhanced resistance of cucumbers against *Alternaria alternata* foliar disease [71]. Predominantly, these bacteria are obtained from the soil, and from the endosphere and rhizosphere of plants. As an example, *Streptomyces* sp. endophytic strain VV/E1 and rhizosphere VV/R1 and VV/R4 strains exhibited antifungal activity and reduced nursery fungal graft infections on grapevine plants [35].

Many strains of *Pantoea* spp. have aptitudes as BCA because they are ubiquitous and produce antimicrobial compounds. Biopesticides based on *Pantoea* spp. are registered and commercially available in Canada, USA, and New Zealand. They have biocontrol activity through various mechanisms, including competitive colonization, production of antimicrobials, and/or induction of host systemic defense. Some strains of *Pantoea* species have been shown to target a wide spectrum of plant pathogens including bacteria, fungi, and oomycetes via secretion of antimicrobial compounds such as pantocins, herbicolins, microcins, and phenazines [72,73]. Other strains such as *P. agglomerans* EPS125 or strain CPA-2 require direct cell-to-cell interaction to combat postharvest fungal pathogens, without relying on the production of antibiotic substances or nutrient competition [24–26]. In another example, *Pantoea* species can also produce N-acyl-homoserine lactone (AHL), affecting quorum sensing in pathogens which, coupled with promoting environmental fitness in plants, may contribute to limit pathogen development [74].

Lactic acid bacteria (LAB) are good candidates as BCA because they include some strains categorized as Generally Regarded as Safe (GRAS) by the U.S. Food and Drug Administration (FDA) and as having Qualified Presumption of Safety (QPS) status by European Food Safety Authority (EFSA) and have been widely reported as biopreservatives of vegetables and fruits [23]. LAB show antimicrobial activity due to the production of one or more antimicrobial metabolites. These include organic acids, carbon dioxide, diacetyl, hydroxide peroxide and proteinaceous compounds such as bacteriocins and antifungal peptides. They may also exclude pathogens by pre-emptively colonizing plant tissues susceptible to infection, by competition for nutrients and space, or by inducing defense responses in plants. For example, *L. plantarum* PM411 and TC92 are effective in preventing bacterial plant diseases. Their broad spectrum of antagonism against plant pathogenic bacteria is based on antimicrobial metabolites, together with the reduction of infections by inhibition of pathogen population on plant surfaces [21,22,75]. Moreover, *Weissella cibaria* TM128 exhibited antimicrobial activity and prevented blue mold, mainly due to the production of organic acids and hydrogen peroxide [36].

3. Bacterial Biocontrol Agent's Development—Flowchart of Actions

The development of bacterial BCA requires several steps (Figure 2). It includes: (i) The isolation and selection of strains by means of screening methods able to analyze a high number of microorganisms; (ii) the characterization of the BCA, including the identification, the determination of phenotypic and genotypic traits, and the mechanisms of action, biocontrol efficacy in pilot tests and improvement; (iii) mass production and an appropriate formulation, which allow increasing biocontrol activity and ensuring its stability. Finally, the development of a monitoring system to detect and quantify the BCA in the environment and to make more extensive toxicology tests or environmental impact studies with the aim to register for use is required.

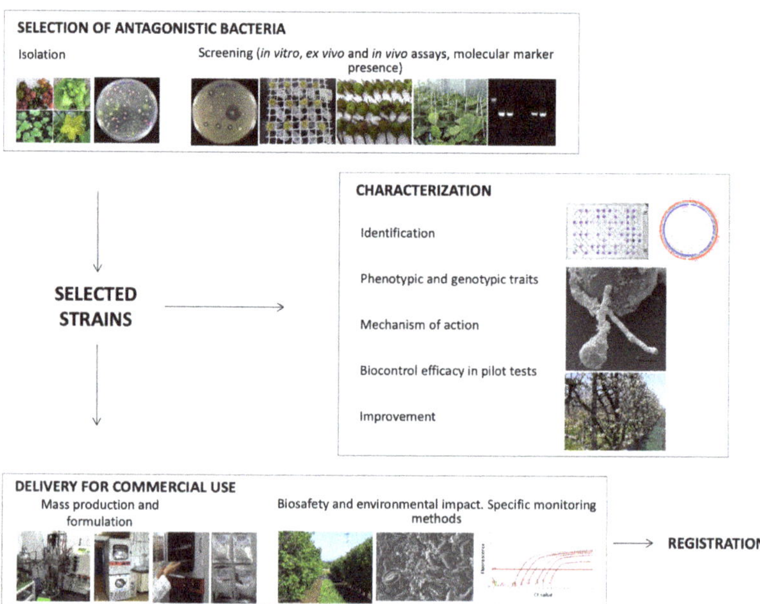

Figure 2. Flowchart of actions for bacterial biocontrol agents development.

3.1. Isolation and Screening for Strain Selection

The first stage of BCA development consists of the isolation and screening of isolates able of limiting the development of the targeted plant pathogen and reducing disease levels. Proper sampling at adequate niches can increase the probability of obtaining useful strains, therefore careful selection of the origin of samples, culture media composition, and enrichment-isolation techniques is very decisive [8]. Bacterial antagonists that prevent or limit disease development are naturally present in the plant environment (phyllosphere, rhizosphere, and endosphere) or in bare soil. Different habitats can be used as suitable sources to obtain candidates as BCA.

For example, samples may be taken from suppressive soils or healthy plants from epidemic areas, where there is evidence of presence of beneficial microorganisms, or near the pathogen infection site [8,9]. In addition, other habitats different from the plant environment can also allow to obtain beneficial bacteria. As the presence of microorganisms with suitable properties as BCA is relatively rare in a strain collection, the isolation of a high number of candidates is recommended. The choice of the isolation technique using selective and enrichment culture media allows for the successful isolation of microorganisms of interest. However, this approach restricts the type of microorganisms obtained and few bacteria genera have been systematically evaluated as BCA. Another approach deals with the use of molecular markers to prospect BCA candidates by means of the specific detection of genes involved in the biocontrol and can be used as a good strategy to increase the efficiency of screening procedures [7,49,50]. The advances in genome sequencing and annotation, and the understanding of the mechanisms of action of BCA have greatly increased the availability of marker genes as tools for the screening [76]. Moreover, considering that a wide array of bacteria from different taxonomic group that studies the structure and function of plant microbiome have been identified [77,78], in-depth study of genetic diversity of microbial communities associated with plants can allow finding new bacteria with relevant traits related to biocontrol which can extend the candidates for plant diseases management [79].

Once a collection of isolates has been made, the putative BCA will be selected based on their attributes. The screening for appropriate candidates is a critical step in the development of novel bacterial BCA and determines the type of microorganism selected [7,9,80]. Rapid-throughput in vitro assays are widely used. In these assays, the target pathogen and candidate biocontrol agents are grown together in solid or liquid media to test for direct reduction of pathogen growth. These assays are fast, reproducible, and reliable, and allow the analysis of many isolates. However, they only permit the selection of bacteria with antagonistic activity, and they may not identify microorganisms with other mechanisms of action such as competitive exclusion or induction of plant resistance [10,81]. Screening procedures such as small-scale whole-plant bioassays in which pathogen and antagonists interact with the host in controlled conditions allow the selection of microorganisms with other mechanisms of action and have a good correlation with biocontrol efficacy in the field. However, these assays are time-consuming and require significant number of resources. The development of ex vivo bioassays on seeds, detached leaves, flowers, and fruits reduces plant material size and permits faster, reliable, and efficient screening [82,83]. A multi-pathogen approach is recommended to select strains with a broad spectrum of activity [74,84].

3.2. Characterization of Selected Strains

The deep characterization of the selected strains is an important stage of BCA development since it provides relevant information about strains for their exploitation as biopesticides. The identification, and phenotypic and genotypic characterization of the strains reveals key attributes in their activity as biocontrol agents. Some of these traits include the synthesis of compounds related to the antimicrobial activity such as enzymes, antibiotics, bacteriocins, or toxins that have detrimental activity against other microorganisms, or to their ability to trigger an immune reaction in plant tissues. Moreover, other traits contribute to the ability of a bacterial strain to colonize plant environment such as the efficient use and uptake of nutrients from exudates (amino acids, organic acids, sugars), motility (flagella), fast growth rate, ability to synthesize amino acids and vitamins, and presence of different structures for adhesion to plant surfaces, such as pili, fimbriae, major outer membrane proteins, or the O-antigen chain of lipopolysaccharides [85,86]. Understanding the traits that are involved as the mechanism of action of a BCA may help finding optimum conditions for implementing biocontrol in each pathosystem. However, the assessment of the mechanisms is a complex and difficult task because of the need of prospective studies to reveal the implication of a given process (e.g., antibiosis, nutrient competition, host colonization, induction of plant defense) and because, in most cases, there are several mechanisms involved and the importance of each one depends on the particular biotic and abiotic conditions.

Nowadays, the genome sequencing of BCA and its comparison with related published genomes will provide a framework for further functional studies of their colonization of plant environment competence and biocontrol effectiveness [87]. Comparative genomics between bacterial strains of varying biocontrol activities allow the identification of new candidate genes putatively involved in the biocontrol. This analysis will unravel novel insights into the biocontrol mechanisms of bacterial BCA and provide new resources for disease control [88,89].

Before bacterial strain is seriously considered for a microbial biopesticide development, pilot trials (greenhouse and field bioassays) must be conducted in several pathosystems and under diverse environmental conditions to ensure a wide range of applicability, as well as consistency in efficacy under real conditions [8]. Considering that the relative dose of pathogen and BCA is an important factor determining the efficacy and consistency of biological control, it is necessary to optimize the dose and frequency of applications. Dose–response models have been developed to obtain quantitative parameters that describe the efficacy of the BCA [90]. These parameters may give information on the dose range of the BCA needed to provide reliable, economical biological control, and allowing for the comparison of different BCA and pathosystems [8,69,90]. The required dose of BCA may be dependent on the mechanism by which a biocontrol agent performs its action. For a strain which acts via antibiosis or competitive exclusion it may be assumed that proper colonization is needed to deliver antimicrobial compounds or to compete with the pathogen, whereas for a strain which acts through ISR a smaller number of bacteria during a restricted period may be sufficient to elicit a successful response in the host plant [17].

3.3. Formulation and Delivery for Commercial Use

The final stages of B-BCA development include industrial scale production, formulation, and preservation. Suitable and cost-effective mass production at the industrial scale system must be carefully developed to obtain the highest number of cells in the shortest period. Moreover, it must be guaranteed that the production method does not alter the characteristics of the strains responsible for biocontrol. Culturing conditions determine population densities at the time of harvest and influence the viability and fitness of the microbes during formulation, storage, and application. These are however specific for each microbial strain and need to be screened carefully for improving final performance of microorganisms in the field [91]. Subsequently, developing an appropriate formulation (dry or liquid) is fundamental to increasing shelf-life, improving delivery, enhancing persistence in the field, and maintaining the viability and biocontrol efficacy [92]. Thus, the use of protective additives and adjuvants compatible with the BCA is common and they can be incorporated at different points of the production-formulation process. Classical protective substances (sucrose, glycerol, Arabic gum) improve survival of the microorganisms and adjuvants (surfactants, emulsifiers, dispersants, coupling agents, stabilizing agents) facilitate mixing, handling, application, and effectiveness [91].

In addition, biosafety studies must be undertaken to guarantee the lack of adverse effects of the active ingredient and the formulated product in plants and non-target organisms, including humans. It is also required to perform risk assessment studies on traceability, residue analysis, and environmental impact [8]. Thus, the development of reliable monitoring methods that accurately identify the released microorganism at strain level and track its population dynamics over time is a registration requirement [93]. Examples of strain specific quantitative monitoring methods developed for BCAs are real-time PCR for *P. fluorescens* EPS62e [94–96] or viable qPCR for *L. plantarum* PM411 [97]. These methods are useful for monitoring the fate and behavior of a released strain in the environment and for the quality control during production and formulation of the microbial biopesticide.

For placing the microbial biopesticide on the EU market, the active substance (i.e., bacterial BCA strain) needs to be approved at EU level and the formulated product must be authorized at Member State level (Regulation (EC) No 2009/1107 and (EC) 2017/1432). The registration procedure generally requires detailed dossiers accounting for scientific data on microorganism identity, biological properties, efficacy, specific analytical methods, residues, traceability, and potential adverse effects on human health and non-target organisms [8,93]. Microorganisms categorized as safe are highly appreciated for the development of microbial biopesticides. For example, bacteria designated with the GRAS and QPS status by the FDA and the EFSA, respectively, have a history of safe use in agriculture and in food and feed crops and lack known toxic or allergenic properties. These microorganisms are considered non-pathogenic to humans, or non-deleterious to the environment according. Therefore, the fact of belonging to this group facilitates the registration process for marketing.

4. Improvement of Biocontrol and Future Trends

The inconsistency in the performance of BCA in the biological control of phytopathogenic fungi and bacteria has limited their extensive use in commercial agriculture. Pathosystem factors such as host genotype, intrinsic characteristics of the pathogen, pathogen inoculum density, and environmental conditions have been shown to be key factors involved in the final levels of disease control achieved by bacteria. Multitude of biotic and abiotic factors can negatively influence the performance of the BCA, affecting their mechanisms of action or the multitrophic interaction between the plant, the pathogen, and the bacteria. However, some strategies can be adopted to improve the performance of BCA consisting of nutritional enhancement, physiological adaptation of BCA to stress and improvement of formulation (Table 2), as well as genetic manipulation of microorganisms. In addition, another challenge is to develop specific delivery systems that favor the success of biocontrol programs. Delivery methods must be carefully selected based on the characteristics of a particular BCA against a specific pathogen. Bacteria can be applied directly to seeds by different methods such as biopriming, encapsulation, or fluid drilling, to soil by drenching, mixing, or microbigation, and on plant aerial parts by foliar spraying or directly into the vascular system by means of endotherapy [98].

Table 2. Some strategies for the physiological improvement of bacterial biocontrol agents.

Microorganism and Strain	Approach for the Improvement	Effect Observed on B-BCA	Reference
Lactobacillus plantarum PM411	Combined hyperosmotic and acid stress adaptation	Increased survival on plant surfaces and overexpression of stress-related genes.	[99]
L. plantarum TC92 and PM411	Mixed bacteria combined with lactic acid	Improvement of efficiency and reliability of biocontrol of fire blight.	[100]
Pantoea agglomerans EPS125	Combined saline osmotic stress and osmolyte amendment	Intracellular accumulation of trehalose and glycine betaine and higher tolerance to desiccation.	[101]
Pseudomonas fluorescens EPS62e	Combined saline osmotic stress and osmolyte amendment	Intracellular accumulation of trehalose, glucosyl-glycerol, and N-acetylglutaminylglutamine amide and improvement of cell survival on plant surfaces and after formulation.	[102,103]
P. fluorescens EPS62e	Nutritional enhancement combined with osmoadaptation	Improvement of fitness in plant surfaces and efficacy in biocontrol of fire blight.	[104]
P. fluorescens EPS817 and EPS894	Mixed bacteria producing different bioactive metabolites	Improvement of efficiency and reliability of biocontrol of *Phytophthora* root.	[30]

An improvement strategy of BCAs is based on nutritional enhancement, which consists of adding nutrients to the formulation that are more efficiently used by the biocontrol agent than by the pathogen. For example, the addition of glycine and Tween 80 to the formulation of *P. fluorescens* EPS62e improved its survival and adaptability in the plant environment [104] or the glucose analog, 2-deoxy-D-glucose enhanced biocontrol of blue mold on apples and pears [105]. Another effective approach to enhance the epiphytic establishment of BCA on plant surfaces is the physiological adaptation by osmoadaptation. This procedure based on the combination of saline osmotic stress and osmolyte amendment of the growth medium has been used to increase intracellular accumulation of osmolytes and drought stress tolerance. This strategy improved epiphytic survival and biocontrol efficacy of the apple blue mold biocontrol agent *P. agglomerans* EPS125 [101] and CPA-2 [106] and the fire blight biocontrol agents *P. fluorescens* EPS62e [102–104], *P. agglomerans* E325 [107] and *L. plantarum* PM411 [99].

The improvement of biocontrol can be achieved by application of mixtures of BCAs, the so-called consortia. This approach consists of designing mixtures of compatible strains that complement each other in terms of the mechanism of action and ecological attributes. This strategy may increase the efficacy and reliability of biocontrol in different environmental conditions, as well as provide a broader spectrum activity due to the synergistic effect of different mechanisms of action of the introduced biocontrol strains. Some examples are, dual mixtures of *P. fluorescens* and *Pantoea* sp. that enhanced the biocontrol of fire blight of pear [108], or mixtures of *P. fluorescens* producing different bioactive metabolites that improved the biocontrol of *P. cactorum* root rot in strawberry plants [30] and *P. infestans* in potato plants [109]. In some cases, the consortia include a high number of bacteria such in a consortium of seven different bacterial species used to protect maize against *Fusarium* [110] or a mixture of eight *Pseudomonas* strains that enhanced protection of tomato against bacterial wilt [111]. In addition, another possible strategy to improve the biocontrol efficacy is the amendment of BCAs with low toxic antimicrobial compounds. Several studies reported the combination with compounds such as bioregulators, organic acids, or essential oils. Improved biological control was reported by combining *L. plantarum* strains PM411 and TC92 with lactic acid [100], and *Bacillus amyloliquefaciens* or *L. plantarum* strains with essential oils [112,113]. Or in another approach, improved bioformulations containing living bacteria and concentrated culture supernatants with antimicrobial metabolites have also been reported [114]. Moreover, BCA performance can be improved by genetic alterations to enhance the efficacy of selected strains for biological control. This may be achieved by conventional approaches as well as through recombinant DNA techniques. However, regulation restrictions to apply and release genetically modified organisms (GMO) into the environment must be considered since genetic manipulation is an impediment for registration of a GM-biological control agent. Genetic engineered bacteria for development of improved bioformulations may offer a good opportunity for future. This approach may include engineered strains without foreign genes but containing useful mutations in genes affecting the biocontrol or strains containing genes from other bacteria. There are several examples of genetic improvement, such as the overproduction of the antimicrobial polyketides, pyoluteorin and 2,4-diacetylphloroglucinol, in *P. fluorescens* CHA0 [115] or the enhancement of mycosubtilin production in *B. subtilis* ATCC 6633 [116].

In conclusion, in recent years there have been important advances in the knowledge of BCA for the development of commercial products for bacterial and fungal disease management. However, large-scale implementation of biological control is hampered by the limitation of commercially available and efficient BCA. Future trends should include the identification of novel BCA and require rapid and robust screening methods suitable to evaluate high numbers of candidates. Moreover, a deep study of model BCA using comparative genome analysis, and genome, transcriptome and proteome analysis will provide a valuable framework allowing for a detailed analysis of the biological mechanisms of BCA and to design strategies enhancing its beneficial action. In addition, this multi-omics approach will allow to analyze the impact of field application of bacteria on the indigenous microbiome of plants. This study would allow analyzing the environmental impact of BCA, to ensure its biosafety, and understand how to modulate the microbiome to improve the efficacy of biocontrol.

Author Contributions: Writing—original draft preparation, A.B., E.B., N.D., G.R., J.F. and E.M. Writing—review and editing, A.B., E.B., N.D., G.R., J.F. and E.M. All authors have read and agreed to the published version of the manuscript.

Funding: This research was funded by different grants from Spain Ministerio de Ciencia, Innovación y Universidades AGL2015-69876-C2-1-R, RTI2018–099410-B-C21, and from the European Union FP7-KBBE.2013.1.2-04 613678 DROPSA.

Data Availability Statement: Not applicable.

Conflicts of Interest: The authors declare no conflict of interest.

References

1. FAO. *The Future of Food and Agriculture—Alternative Pathways to 2050*, Summary version; FAO: Rome, Italy, 2018.
2. FAO. *The Future of Food and Agriculture—Trends and Challenges*; FAO: Rome, Italy, 2017.
3. Chakraborty, S.; Newton, A.C. Climate change, plant diseases and food security: An overview. *Plant Pathol.* **2011**, *60*, 2–14. [CrossRef]
4. Directive of the European Parliament and of the Council of 21 October 2009 Establishing a Framework for Community Action to Achieve the Sustainable Use of Pesticides, 2009/128/EC, OJ L 309:71–86. 2009. Available online: https://eur-lex.europa.eu/LexUriServ/LexUriServ.do?uri=OJ:L:2009:309:0071:0086:en:PDF (accessed on 15 June 2022).

5. Lamichhane, J.R.; Dachbrodt-Saaydeh, S.; Kudsk, P.; Messéan, A. Toward a reduced reliance on conventional pesticides in European agriculture. *Plant Dis.* **2016**, *100*, 10–24. [CrossRef] [PubMed]

6. Robin, D.C.; Marchand, P.A. Evolution of the biocontrol active substances in the framework of the European Pesticide Regulation (EC) No. 1107/2009. *Pest. Manag. Sci.* **2019**, *75*, 950–958. [CrossRef] [PubMed]

7. Montesinos, E.; Bonaterra, A. Microbial Pesticides. In *Encyclopedia of Microbiology*, 3rd ed.; Schaechter, M., Ed.; Elsevier: Amsterdam, The Netherlands, 2009; pp. 110–120.

8. Montesinos, E. Development, registration and commercialization of microbial pesticides for plant protection. *Int. Microbiol.* **2003**, *6*, 245–252. [CrossRef]

9. Köhl, J.; Postma, J.; Nicot, P.; Ruocco, M.; Blum, B. Stepwise screening of microorganisms for commercial use in biological control of plant-pathogenic fungi and bacteria. *Biol. Control* **2011**, *57*, 1–12.

10. Köhl, J.; Kolnaar, R.; Ravensberg, W.J. Mode of action of microbial biological control agents against plant diseases: Relevance beyond efficacy. *Front. Plant Sci.* **2019**, *10*, 845. [CrossRef]

11. Legein, M.; Smets, W.; Vandenheuvel, D.; Eilers, T.; Muyshondt, B.; Prinsen, E.; Samson, R.; Lebeer, S. Modes of action of microbial biocontrol in the phyllosphere. *Front. Microbiol.* **2020**, *11*, 1619. [CrossRef]

12. Kalia, V.C.; Patel, S.K.S.; Kang, Y.C.; Lee, J.K. Quorum sensing inhibitors as antipathogens: Biotechnological applications. *Biotechnol. Adv.* **2019**, *37*, 68–90. [CrossRef]

13. Elnahal, A.S.M.; El-Saadony, M.T.; Saad, A.M.; Desoky, E.S.M.; El-Tahan, A.M.; Rady, M.M.; AbuQamar, S.F.; El-Tarabily, K.A. The use of microbial inoculants for biological control, plant growth promotion, and sustainable agriculture: A review. *Eur. J. Plant Pathol.* **2022**, *162*, 759–792. [CrossRef]

14. Lugtenberg, B.; Kamilova, F. Plant-growth-promoting rhizobacteria. *Annu. Rev. Microbiol.* **2009**, *63*, 541–556. [CrossRef]

15. Berendsen, R.L.; Pieterse, C.M.J.; Bakker, P.A.H.M. The rhizosphere microbiome and plant health. *Trends Plant Sci.* **2012**, *17*, 478–486. [CrossRef] [PubMed]

16. Arrebola, E.; Jacobs, R.; Korsten, L. Iturin A is the principal inhibitor in the biocontrol activity of *Bacillus amyloliquefaciens* PPCB004 against postharvest fungal pathogens. *J. Appl. Microbiol.* **2010**, *108*, 386–395. [CrossRef] [PubMed]

17. Gotor-Vila, A.; Teixidó, N.; Di Francesco, A.; Usall, J.; Ugolini, L.; Torres, R.; Mari, M. Antifungal effect of volatile organic compounds produced by *Bacillus amyloliquefaciens* CPA-8 against fruit pathogen decays of cherry. *Food Microbiol.* **2017**, *64*, 219–225. [CrossRef] [PubMed]

18. Zeriouh, H.; Romero, D.; Garcia-Gutierrez, L.; Cazorla, F.M.; de Vicente, A.; Perez-Garcia, A. The iturin-like lipopeptides are essential components in the biological control arsenal of *Bacillus subtilis* against bacterial diseases of cucurbits. *Mol. Plant Microbe Interact.* **2011**, *24*, 1540–1552. [CrossRef] [PubMed]

19. Mora, I.; Cabrefiga, J.; Montesinos, E. Antimicrobial peptide genes in *Bacillus* strains from plant environments. *Int. Microbiol.* **2011**, *14*, 213–223.

20. Mora, I.; Cabrefiga, J.; Montesinos, E. Cyclic lipopeptide biosynthetic genes and products, and inhibitory activity of plant-associated *Bacillus* against phytopathogenic bacteria. *PLoS ONE* **2015**, *10*, e0127738. [CrossRef]

21. Roselló, G.; Bonaterra, A.; Francés, J.; Montesinos, L.; Badosa, E.; Montesinos, E. Biological control of fire blight of apple and pear with antagonistic *Lactobacillus plantarum*. *Eur. J. Plant. Pathol.* **2013**, *137*, 621–633. [CrossRef]

22. Daranas, N.; Roselló, G.; Cabrefiga, J.; Donati, I.; Francés, J.; Badosa, E.; Spinelli, F.; Montesinos, E.; Bonaterra, A. Biological control of bacterial plant diseases with *Lactobacillus plantarum* strains selected for their broad-spectrum activity. *Ann. Appl. Biol.* **2019**, *174*, 92–105. [CrossRef]

23. Trias, R.; Badosa, E.; Montesinos, E.; Bañeras, L. Bioprotective *Leuconostoc* strains against *Listeria monocytogenes* in fresh fruits and vegetables. *Int. J. Food Microbiol.* **2008**, *127*, 91–98. [CrossRef]

24. Bonaterra, A.; Mari, M.; Casalini, L.; Montesinos, E. Biological control of *Monilinia laxa* and *Rhizopus stolonifer* in postharvest of stone fruit by *Pantoea agglomerans* EPS125 and putative mechanisms of antagonism. *Int. J. Food Microbiol.* **2003**, *84*, 93–104. [CrossRef]

25. Francés, J.; Bonaterra, A.; Moreno, M.C.; Cabrefiga, J.; Badosa, E.; Montesinos, E. Pathogen aggressiveness and postharvest biocontrol efficiency in *Pantoea agglomerans*. *Postharvest Biol. Technol.* **2006**, *39*, 299–307. [CrossRef]

26. Nunes, C.; Usall, J.; Teixidó, N.; Viñas, I. Biological control of postharvest pear diseases using a bacterium, *Pantoea agglomerans* CPA-2. *Int. J. Food. Microbiol.* **2001**, *70*, 53–61. [CrossRef]

27. Calderón, C.E.; Pérez-García, A.; de Vicente, A.; Cazorla, F.M. The dar genes of *Pseudomonas chlororaphis* PCL1606 are crucial for biocontrol activity via production of the antifungal compound 2-hexyl, 5-propyl resorcinol. *Mol. Plant Microbe Interact.* **2013**, *26*, 554–565. [CrossRef] [PubMed]

28. Landa, B.B.; Mavrodi, O.V.; Raaijmakers, J.M.; McSpadden-Gardener, B.B.; Thomashow, L.S.; Weller, D.M. Differential ability of genotypes of 2, 4-diacetylphloroglucinol-producing *Pseudomonas fluorescens* strains to colonize the roots of pea plants. *Appl. Environ. Microbiol.* **2002**, *68*, 3226–3237. [CrossRef] [PubMed]

29. Cabrefiga, J.; Bonaterra, A.; Montesinos, E. Mechanisms of antagonism of *Pseudomonas fluorescens* EPS62e against *Erwinia amylovora*, the causal agent of fire blight. *Int. Microbiol.* **2007**, *10*, 123–132.

30. Agustí, L.; Bonaterra, A.; Moragrega, C.; Camps, J.; Montesinos, E. Biocontrol of root rot of strawberry caused by *Phytophthora cactorum* with a combination of two *Pseudomonas fluorescens* strains. *J. Plant Pathol.* **2011**, *93*, 363–372.

31. Gómez-Lama Cabanás, C.; Schilirò, E.; Valverde-Corredor, A.; Mercado-Blanco, J. The biocontrol endophytic bacterium *Pseudomonas fluorescens* PICF7 induces systemic defense responses in aerial tissues upon colonization of olive roots. *Front Microbiol.* **2014**, *5*, 427. [CrossRef]

32. Montes-Osuna, N.; Gómez-Lama Cabanás, C.; Valverde-Corredor, A.; Berendsen, R.L.; Prieto, P.; Mercado-Blanco, J. Assessing the involvement of selected phenotypes of *Pseudomonas simiae* PICF7 in olive root colonization and biological control of *Verticillium dahliae*. *Plants* **2021**, *10*, 412. [CrossRef]

33. Pliego, C.; de Weert, S.; Lamers, G.; de Vicente, A.; Bloemberg, G.V.; Cazorla, F.M.; Ramos, C. Two similar enhanced root colonizing *Pseudomonas* strains differ largely in their colonization strategies of avocado roots and *Rosellinia necatrix* hyphae. *Environ. Microbiol.* **2008**, *10*, 3295–3304. [CrossRef]

34. Díaz-Díaz, M.; Bernal-Cabrera, A.; Trapero, A.; Medina-Marrero, R.; Sifontes-Rodríguez, S.; Cupull-Santana, R.D.; García-Bernal, M.; Agustí-Brisach, C. Characterization of actinobacterial strains as potential biocontrol agents against *Macrophomina phaseolina* and *Rhizoctonia solani*, the main soil-borne pathogens of *Phaseolus vulgaris* in Cuba. *Plants* **2022**, *11*, 645. [CrossRef]

35. Álvarez-Pérez, J.M.; González-García, S.; Cobos, R.; Olego, M.Á.; Ibañez, A.; Díez-Galán, A.; Garzón-Jimeno, E.; Coque, J.J.R. Use of endophytic and rhizosphere actinobacteria from grapevine plants to reduce nursery fungal graft infections that lead to young grapevine decline. *Appl. Environ. Microbiol.* **2017**, *83*, e01564-17. [CrossRef] [PubMed]

36. Trias, R.; Bañeras, L.; Montesinos, E.; Badosa, E. Lactic acid bacteria from fresh fruit and vegetables as biocontrol agents of phytopathogenic bacteria and fungi. *Int. Microbiol.* **2008**, *11*, 231–236. [PubMed]

37. Lugtenberg, B.J.J.; Dekkers, L.; Bloemberg, G.V. Molecular determinants of rhizosphere colonization by *Pseudomonas*. *Annu. Rev. Phytopathol.* **2001**, *39*, 461–490. [CrossRef] [PubMed]

38. Chin-A-Woeng, T.F.; Bloemberg, G.V.; Lugtenberg, B.J.J. Phenazines and their role in biocontrol by *Pseudomonas* bacteria. *New Phytol.* **2003**, *157*, 503–523. [CrossRef] [PubMed]

39. Oso, S.; Walters, M.; Schlechter, R.O.; Remus-Emsermann, M.N.P. Utilisation of hydrocarbons and production of surfactants by bacteria isolated from plant leaf surfaces. *FEMS Microbiol. Lett.* **2019**, *366*, fnz061. [CrossRef]

40. Ellis, R.J.; Timms-Wilson, T.M.; Beringer, J.E.; Rhodes, D.; Renwick, A.; Stevenson, L.; Bailey, M.J. Ecological basis for biocontrol of damping-off disease by *Pseudomonas fluorescens* 54/96. *J. Appl. Microbiol.* **1999**, *87*, 454–463. [CrossRef] [PubMed]

41. Duijff, B.J.; Bakker, P.A.H.M.; Schippers, B. Suppression of fusarium wilt of carnation by *Pseudomonas putida* WCS358 at different levels of disease incidence and iron availability. *Biocontrol. Sci. Technol.* **1994**, *4*, 279–288. [CrossRef]

42. Haas, D.; Keel, C. Regulation of antibiotic production in root-colonizing *Pseudomonas* spp. and relevance for biological control of plant disease. *Annu. Rev. Phytopathol.* **2003**, *41*, 117–153. [CrossRef]

43. Hu, W.; Gao, Q.; Hamada, M.S.; Dawood, D.H.; Zheng, J.; Chen, Y.; Ma, Z. Potential of *Pseudomonas chlororaphis* subsp. *aurantiaca* strain Pcho10 as a biocontrol agent against *Fusarium graminearum*. *Phytopathology* **2014**, *104*, 1289–1297. [CrossRef]

44. Rezzonico, F.; Zala, M.; Keel, C.; Duffy, B.; Moënne-Loccoz, Y.; Défago, G. Is the ability of biocontrol fluorescent pseudomonads to produce the antifungal metabolite 2,4-diacetylphloroglucinol really synonymous with higher plant protection? *New Phytol.* **2007**, *173*, 861–872. [CrossRef]

45. Huang, R.; Feng, Z.; Chi, X.; Sun, X.; Lu, Y.; Zhang, B.; Lu, R.; Luo, W.; Wang, Y.; Miao, J.; et al. Pyrrolnitrin is more essential than phenazines for *Pseudomonas chlororaphis* G05 in its suppression of *Fusarium graminearum*. *Microbiol. Res.* **2018**, *215*, 55–64. [CrossRef] [PubMed]

46. Flury, P.; Vesga, P.; Péchy-Tarr, M.; Aellen, N.; Dennert, F.; Hofer, N.; Kupferschmied, K.P.; Kupferschmied, P.; Metla, Z.; Ma, Z.; et al. Antimicrobial and insecticidal: Cyclic lipopeptides and hydrogen cyanide produced by plant-beneficial *Pseudomonas* Strains CHA0, CMR12a, and PCL1391 contribute to insect killing. *Front. Microbiol.* **2017**, *8*, 100. [CrossRef] [PubMed]

47. Raaijmakers, J.M.; De Bruijn, I.; de Kock, M.J. Cyclic lipopeptide production by plant-associated *Pseudomonas* spp.: Diversity, activity, biosynthesis, and regulation. *Mol. Plant Microbe Int.* **2006**, *19*, 699–710. [CrossRef]

48. Ma, Z.; Geudens, N.; Kieu, N.P.; Sinnaeve, D.; Ongena, M.; Martins, J.C.; Höfte, M. Biosynthesis, chemical structure, andstructure-activity relationship of orfamide lipopeptides produced by *Pseudomonas protegens* and related species. *Front. Microbiol.* **2016**, *7*, 382. [CrossRef] [PubMed]

49. Nielsen, T.H.; Christophersen, C.; Anthoni, U.; Sorensen, J. Viscosinamide, a new cyclic depsipeptide with surfactant and antifungal properties produced by *Pseudomonas fluorescens* DR54. *J. Appl. Microbiol.* **1999**, *87*, 80–90. [CrossRef] [PubMed]

50. Nielsen, T.H.; Thrane, C.; Christophersen, C.; Anthoni, U.; Sørensen, J. Structure, production characteristics and fungal antagonism of tensin—A new antifungal cyclic lipopeptide from *Pseudomonas fluorescens* strain 96.578. *J. Appl. Microbiol.* **2000**, *89*, 992–1001. [CrossRef] [PubMed]

51. Sindhu, S.S.; Dadarwal, K.R. Chitinolytic and cellulolytic *Pseudomonas* sp. antagonistic to fungal pathogens enhances nodulation by *Mesorhizobium* sp. *Cicer* in chickpea. *Microbiol. Res.* **2001**, *156*, 353–358. [CrossRef]

52. Lakkis, S.; Trotel-Aziz, P.; Rabenoelina, F.; Schwarzenberg, A.; Nguema-Ona, E.; Clément, C.; Aziz, A. Strengthening grapevine resistance by *Pseudomonas fluorescens* PTA-CT2 relies on distinct defense pathways in susceptible and partially resistant genotypes to downy mildew and gray mold diseases. *Front. Plant Sci.* **2019**, *10*, 1112. [CrossRef]

53. Tran, H.; Ficke, A.; Asiimwe, T.; Hofte, M.; Raaijmakers, J.M. Role of the cyclic lipopeptide massetolide A in biological control of *Phytophthora infestans* and in colonization of tomato plants by *Pseudomonas fluorescens*. *New Phytol.* **2007**, *175*, 731–742. [CrossRef]

54. Ma, Z.; Hua, G.K.; Ongena, M.; Hofte, M. Role of phenazines and cyclic lipopeptides produced by *Pseudomonas* sp. CMR12a in induced systemic resistance on rice and bean. *Environ. Microbiol.* **2016**, *8*, 896–904. [CrossRef]

55. McSpadden Gardener, B.B. Ecology of *Bacillus* and *Paenibacillus* spp. in agricultural systems. *Phytopathology* **2004**, *94*, 1252–1258. [CrossRef]

56. Ongena, M.; Jacques, P. *Bacillus* lipopeptides: Versatile weapons for plant disease biocontrol. *Trends Microbiol.* **2008**, *16*, 115–125. [CrossRef] [PubMed]

57. Abriouel, H.; Franz, C.M.A.P.; Omar, N.B.; Gálvez, A. Diversity and applications of *Bacillus* bacteriocins. *FEMS Microbiol. Rev.* **2011**, *35*, 201–232. [CrossRef] [PubMed]

58. Mouloud, G.; Daoud, H.; Bassem, J.; Laribi Atef, I.; Hani, B. New bacteriocin from *Bacillus clausii* strainGM17: Purification, characterization, and biological activity. *Appl. Biochem. Biotechnol.* **2013**, *171*, 2186–2200. [CrossRef] [PubMed]

59. Ugras, S.; Sezen, K.; Kati, H.; Demirbag, Z. Purification and characterization of the bacteriocin Thuricin Bn1 produced by *Bacillus thuringiensis* subsp. kurstaki Bn1 isolated from a hazelnut pest. *J. Microbiol. Biotechnol.* **2013**, *23*, 167–176. [CrossRef]

60. Raaijmakers, J.M.; De Bruijn, I.; Nybroe, O.; Ongena, M. Natural functions of lipopeptides from *Bacillus* and *Pseudomonas*: More than surfactants and antibiotics. *FEMS Microbiol. Rev.* **2010**, *34*, 1037–1062. [CrossRef]

61. Preecha, C.; Sadowsky, M.J.; Prathuangwong, S. Lipopeptide surfactin produced by *Bacillus amyloliquefaciens* KPS46 is required for biocontrol efficacy against *Xanthomonas axonopodis* pv. glycines. *Kasetsart J. Nat. Sci.* **2010**, *44*, 84–99.

62. Ongena, M.; Duby, F.; Jourdan, E.; Beaudry, T.; Jadin, V.; Dommes, J.; Thonart, P. *Bacillus subtilis* M4 decreases plant susceptibility towards fungal pathogens by increasing host resistance associated with differential gene expression. *Appl. Microbiol. Biot.* **2005**, *67*, 692–698. [CrossRef]

63. Guleria, S.; Walia, A.; Chauhan, A.; Shirkot, C.K. Molecular characterization of alkaline protease of *Bacillus amyloliquefaciens* SP1 involved in biocontrol of *Fusarium oxysporum*. *Int. J. Food Microbiol.* **2016**, *232*, 134–143. [CrossRef]

64. Jiang, C.-H.; Liao, M.-J.; Wang, H.-K.; Zheng, M.-Z.; Xu, J.-J.; Guo, J.-H. *Bacillus velezensis* a potential and efficient biocontrol agent in control of pepper gray mold caused by *Botrytis cinerea*. *Biol. Control* **2018**, *126*, 147–157. [CrossRef]

65. Ongena, M.; Jourdan, E.; Adam, A.; Paquot, M.; Brans, A.; Joris, B.; Arpigny, J.L.; Thonart, P. Surfactin and fengycin lipopeptides of *Bacillus subtilis* as elicitors of induced systemic resistance in plants. *Environ. Microbiol.* **2007**, *9*, 1084–1090. [CrossRef] [PubMed]

66. Cawoy, H.; Debois, D.; Franzil, L.; De Pauw, E.; Thonart, P.; Ongena, M. Lipopeptides as main ingredients for inhibition of fungal phytopathogens by *Bacillus subtilis/amyloliquefaciens*. *Microb. Biotechnol.* **2015**, *8*, 281–295. [CrossRef] [PubMed]

67. Chowdhury, S.P.; Uhl, J.; Grosch, R.; Alquéres, S.; Pittroff, S.; Dietel, K.; Schmitt-Kopplin, P.; Borriss, R.; Hartmann, A. Cyclic Lipopeptides of *Bacillus amyloliquefaciens* subsp. *Plantarum* colonizing the lettuce rhizosphere enhance plant defense responses toward the bottom rot pathogen *Rhizoctonia solani*. *Mol. Plant Microbe Interact.* **2015**, *28*, 984–995. [CrossRef] [PubMed]

68. Chowdappa, P.; Kumar, S.P.M.; Lakshmi, M.J.K.; Upreti, K. Growth stimulation and induction of systemic resistance in tomato against early and late blight by *Bacillus subtilis* OTPB1 or *Trichoderma harzianum* OTPB3. *Biol. Control.* **2013**, *65*, 109–117. [CrossRef]

69. Montesinos, E.; Bonaterra, A. Are there bacterial bioprotectants besides *Bacillus* and *Pseudomonas* species? In *Microbial Bioprotectants for Plant Disease Management*; Köhl, J., Ravensberg, W.J., Eds.; Burleigh Dodds Series Publishing: London, UK, 2022; p. 734.

70. Viaene, T.; Langendries, S.; Beirinckx, S.; Maes, M.; Goormachtig, S. *Streptomyces* as a plant's best friend? *FEMS Microbiol. Ecol.* **2016**, *92*, fiw119. [CrossRef]

71. Wang, M.; Xue, J.; Ma, J.; Feng, X.; Ying, H.; Xu, H. *Streptomyces lydicus* M01 regulates soil microbial community and alleviates foliar disease caused by *Alternaria alternata* on cucumbers. *Front. Microbiol.* **2020**, *11*, 942. [CrossRef]

72. Pusey, P.L.; Stockwell, V.O.; Reardon, C.L.; Smits, T.H.M.; Duffy, B. Antibiosis activity of *Pantoea agglomerans* biocontrol strain E325 against *Erwinia amylovora* on apple flower stigmas. *Phytopathology* **2011**, *101*, 1234–1241. [CrossRef]

73. Walterson, A.M.; Stavrinides, J. *Pantoea*: Insights into a highly versatile and diverse genus within the Enterobacteriaceae. *FEMS Microbiol. Rev.* **2015**, *39*, 968–984. [CrossRef]

74. Smits, T.H.M.; Rezzonico, F.; Pelludat, C.; Goesmann, A.; Frey, J.E.; Duffy, B. Genomic and phenotypic characterization of a non-pigmented variant of *Pantoea vagans* biocontrol strain C9-1 lacking the 530 kb megaplasmid pPag3. *FEMS Microbiol. Lett.* **2010**, *308*, 48–54. [CrossRef]

75. Trias, R.; Bañeras, L.; Badosa, E.; Montesinos, E. Bioprotection of Golden Delicious apples and Iceberg lettuce against foodborne bacterial pathogens by lactic acid bacteria. *Int. J. Food Microbiol.* **2008**, *123*, 50–60. [CrossRef]

76. Joshi, R.; McSpadden Gardener, B.B. Identification and characterization of novel genetic markers associated with biological control activities in *Bacillus subtilis*. *Phytopathology* **2006**, *96*, 145–154. [CrossRef] [PubMed]

77. Mendes, R.; Kruijt, M.; de Bruijn, I.; Dekkers, E.; van der Voort, M.; Schneider, J.H.M.; Piceno, Y.M.; DeSantis, T.Z.; Andersen, G.L.; Bakker, P.A.; et al. Deciphering the rhizosphere microbiome for disease-suppressive bacteria. *Science* **2011**, *332*, 1097–1100. [CrossRef] [PubMed]

78. Berg, G.; Grube, M.; Schloter, M.; Smalla, K. Unraveling the plant microbiome: Looking back and future perspectives. *Front Microbiol.* **2014**, *5*, 148. [CrossRef] [PubMed]

79. Aranda, S.; Montes-Borrego, M.; Jiménez-Díaz, R.M.; Landa, B.B. Microbial communities associated with the root system of wild olives (*Olea europaea* L. subsp. *europaea* var. *sylvestris*) are good reservoirs of bacteria with antagonistic potential against *Verticillium dahliae*. *Plant Soil.* **2011**, *343*, 329–345. [CrossRef]

80. Raymaekers, K.; Ponet, L.; Holtappels, D.; Berckmans, B.; Cammue, B.P.A. Screening for novel biocontrol agents applicable in plant disease management—A review. *Biol. Control* **2020**, *144*, 104240. [CrossRef]

81. Pliego, C.; Ramos, C.; de Vicente, A.; Cazorla, F.M. Screening for candidate bacterial biocontrol agents against soilborne fungal plant pathogens. *Plant Soil* **2011**, *340*, 505–520. [CrossRef]

82. Montesinos, E.; Bonaterra, A.; Ophir, Y.; Beer, S.V. Antagonism of selected bacterial strains to *Stemphylium vesicarium* and biological control of brown spot of pear under controlled environment conditions. *Phytopathology* **1996**, *86*, 856–863. [CrossRef]

83. Pusey, P.L. Crab apple blossoms as a model for research on biological control of fire blight. *Phytopathology* **1997**, *87*, 1096–1102. [CrossRef]

84. Wang, L.Y.; Xie, Y.S.; Cui, Y.Y.; Xu, J.; He, W.; Chen, H.G.; Guo, J.H. Conjunctively screening of biocontrol agents (BCAs) against fusarium root rot and fusarium head blight caused by *Fusarium graminearum*. *Microbiol. Res.* **2015**, *177*, 34–42. [CrossRef]

85. Lugtenberg, B.J.J.; Dekkers, L.C. What makes *Pseudomonas* bacteria rhizosphere competent? *Environ. Microbiol.* **1999**, *1*, 9–13. [CrossRef]

86. Compant, S.; Duffy, B.; Nowak, J.; Clément, C.; Barka, E.A. Use of plant growth-promoting bacteria for biocontrol of plant diseases: Principles, mechanisms of action, and future prospects. *Appl. Environ. Microbiol.* **2005**, *71*, 4951–4959. [CrossRef] [PubMed]

87. Martínez-García, P.M.; Ruano-Rosa, D.; Schilirò, E.; Prieto, P.; Ramos, C.; Rodríguez-Palenzuela, P.; Mercado-Blanco, J. Complete genome sequence of *Pseudomonas fluorescens* strain PICF7, an indigenous root endophyte from olive (*Olea europaea* L.) and effective biocontrol agent against *Verticillium dahliae*. *Stand. Genomic Sci.* **2015**, *10*, 10. [CrossRef] [PubMed]

88. De Vrieze, M.; Varadarajan, A.R.; Schneeberger, K.; Bailly, A.; Rohr, R.P.; Ahrens, C.H.; Weisskopf, L. Linking comparative genomics of nine potato-associated *Pseudomonas* isolates with their differing biocontrol potential against late blight. *Front. Microbiol.* **2020**, *11*, 857. [CrossRef] [PubMed]

89. Xu, S.; Zhang, Z.; Xie, X.; Shi, Y.; Chai, A.; Fan, T.; Li, B.; Li, L. Comparative genomics provides insights into the potential biocontrol mechanism of two *Lysobacter enzymogenes* strains with distinct antagonistic activities. *Front. Microbiol.* **2022**, *13*, 966986. [CrossRef]

90. Montesinos, E.; Bonaterra, A. Dose-response models in biological control of plant pathogens: An empirical verification. *Phytopathology* **1996**, *86*, 464–472. [CrossRef]

91. Bejarano, A.; Puopolo, G. Bioformulation of microbial biocontrol agents for a sustainable agriculture. In *How Research Can Stimulate the Development of Commercial Biological Control against Plant Diseases*; De Cal, A., Melgarejo, P., Magan, N., Eds.; Progress in Biological Control; Springer: Cham, Switzerland, 2020; Volume 21, pp. 275–293.

92. Segarra, G.; Puopolo, G.; Giovannini, O.; Pertot, I. Stepwise flow diagram for the development of formulations of non spore-forming bacteria against foliar pathogens: The case of *Lysobacter capsici* AZ78. *J. Biotechnol.* **2015**, *216*, 56–64. [CrossRef]

93. Bonaterra, A.; Badosa, E.; Cabrefiga, J.; Francés, J.; Montesinos, E. Prospects and limitations of microbial pesticides for control of bacterial and fungal pomefruit tree diseases. *Trees-Struct. Funct.* **2012**, *26*, 215–226. [CrossRef]

94. Pujol, M.; Badosa, E.; Cabrefiga, J.; Montesinos, E. Development of a strain-specific quantitative method for monitoring *Pseudomonas fluorescens* EPS62e, a novel biocontrol agent of fire blight. *FEMS Microbiol. Lett.* **2005**, *249*, 343–352. [CrossRef]

95. Pujol, M.; Badosa, E.; Manceau, C.; Montesinos, E. Assessment of the environmental fate of the biological control agent of fire blight, *Pseudomonas fluorescens* EPS62e, on apple by culture and real-time PCR methods. *Appl. Environ. Microbiol.* **2006**, *72*, 2421–2427. [CrossRef]

96. Pujol, M.; Badosa, E.; Montesinos, E. Epiphytic fitness of a biological control agent of fire blight in apple and pear orchards under Mediterranean weather conditions. *FEMS Microbiol. Ecol.* **2007**, *59*, 186–193. [CrossRef]

97. Daranas, N.; Bonaterra, A.; Francés, J.; Montesinos, E.; Badosa, E. Monitoring viable cells of the biological control agent *Lactobacillus plantarum* PM411 in aerial plant surfaces by means of a strain-specific. *Appl. Environ. Microbiol.* **2018**, *84*, e00107–e00118. [CrossRef] [PubMed]

98. Jambhulkar, P.P.; Sharma, P.; Yadav, R. Delivery systems for introduction of microbial inoculants in the field. In *Microbial Inoculants in Sustainable Agricultural Productivity*; Singh, D., Singh, H., Prabha, R., Eds.; Springer: New Delhi, India, 2016.

99. Daranas, N.; Badosa, E.; Frances, J.; Montesinos, E.; Bonaterra, A. Enhancing water stress tolerance improves fitness in biological control strains of Lactobacillus plantarum in plant environments. *PLoS ONE* **2018**, *13*, e0190931. [CrossRef]

100. Roselló, G.; Francés, J.; Daranas, N.; Montesinos, E.; Bonaterra, A. Control of fire blight of pear trees with mixed inocula of two *Lactobacillus plantarum* strains and lactic acid. *J. Plant Pathol.* **2017**, *99*, 111–120.

101. Bonaterra, A.; Camps, J.; Montesinos, E. Osmotically induced trehalose and glycine betaine accumulation improves tolerance to desiccation, survival and efficacy of the postharvest biocontrol agent *Pantoea agglomerans* EPS125. *FEMS Microbiol. Lett.* **2005**, *250*, 1–8. [CrossRef] [PubMed]

102. Bonaterra, A.; Cabrefiga, J.; Camps, J.; Montesinos, E. Increasing survival and efficacy of a bacterial biocontrol agent of fire blight of rosaceous plants by means of osmoadaptation. *FEMS Microbiol. Ecol.* **2007**, *61*, 185–195. [CrossRef] [PubMed]

103. Cabrefiga, J.; Francés, J.; Montesinos, E.; Bonaterra, A. Improvement of a dry formulation of *Pseudomonas fluorescens* EPS62e for fire blight disease biocontrol by combination of culture osmoadaptation with a freeze-drying lyoprotectant. *J. Appl. Microbiol.* **2014**, *117*, 1122–1131. [CrossRef] [PubMed]

104. Cabrefiga, J.; Francés, J.; Montesinos, E.; Bonaterra, A. Improvement of fitness and efficacy of a fire blight biocontrol agent via nutritional enhancement combined with osmoadaptation. *Appl. Environ. Microbiol.* **2011**, *77*, 3174–3181. [CrossRef] [PubMed]

105. Janisiewicz, W.J. Enhancement of biocontrol of blue mold with nutrient analog 2-deoxy-D-glucose on apples and pears. *Appl. Environ. Microbiol.* **1994**, *60*, 2671–2676. [CrossRef]

106. Teixidó, N.; Cañamás, T.P.; Abadias, M.; Usall, J.; Solsona, C.; Casals, C.; Viñas, I. Improving low water activity and desiccation tolerance of the biocontrol agent *Pantoea agglomerans* CPA-2 by osmotic treatments. *J. Appl. Microbiol.* **2006**, *101*, 927–937. [CrossRef]

107. Pusey, P.L.; Wend, C. Potential of osmoadaptation for improving *Pantoea agglomerans* E325 as biocontrol agent for fire blight of apple and pear. *Biol. Control* **2012**, *62*, 29–37. [CrossRef]
108. Stockwell, V.O.; Johnson, K.B.; Sugar, D.; Loper, J.E. Control of fire blight by *Pseudomonas fluorescens* A506 and *Pantoea vagans* C9-1 applied as single strains and mixed inocula. *Phytopathology* **2010**, *100*, 1330–1339. [CrossRef] [PubMed]
109. De Vrieze, M.; Germanier, F.; Vuille, N.; Weisskopf, L. Combining different potato-associated *Pseudomonas* strains for improved biocontrol of *Phytophthora infestans*. *Front. Microbiol.* **2018**, *9*, 2573. [CrossRef] [PubMed]
110. Niu, B.; Paulson, J.N.; Zheng, X.; Kolter, R.; Lindow, S.E. Simplified and representative bacterial community of maize roots. *Proc. Natl. Acad. Sci. USA* **2017**, *14*, E2450–E2459. [CrossRef]
111. Hu, J.; Wei, Z.; Friman, V.P.; Gu, S.H.; Wang, X.F.; Eisenhauer, N.; Jousset, A. Probiotic diversity enhances rhizosphere microbiome function and plant disease suppression. *MBio* **2016**, *7*, e01790-16. [CrossRef] [PubMed]
112. Arrebola, E.; Sivakumar, D.; Bacigalupo, R.; Korsten, L. Combined application of antagonist *Bacillus amyloliquefaciens* and essential oils for the control of peach postharvest diseases. *Crop Protect.* **2010**, *29*, 369–377. [CrossRef]
113. Zamani-Zadeh, M.; Soleimanian-Zad, S.; Sheikh-Zeinoddin, M.; Amir, S.; Goli, H. Integration of *Lactobacillus plantarum* A7 with thyme and cumin essential oils as a potential biocontrol tool for gray mold rot on strawberry fruit. *Postharvest. Biol. Technol.* **2014**, *92*, 149–156. [CrossRef]
114. Wu, L.; Wu, H.-J.; Qiao, J.; Gao, X.; Borriss, R. Novel routes for improving biocontrol activity of *Bacillus* based bioinoculants. *Front. Microbiol.* **2015**, *6*, 1395. [CrossRef]
115. Girlanda, M.; Perotto, S.; Moenne-Loccoz, Y.; Bergero, R.; Lazzari, A.; Defago, G.; Bonfante, P.; Luppi, A.M. Impact of biocontrol *Pseudomonas fluorescens* CHA0 and a genetically modified derivative on the diversity of culturable fungi in the cucumber rhizosphere. *Appl. Environ. Microbiol.* **2001**, *67*, 1851–1864. [CrossRef]
116. Fickers, P.; Guez, J.E.; Damblon, C.; Leclère, V.; Béchet, M.; Jacques, P.; Joris, B. High-level biosynthesis of the anteiso-C_{17} isoform of the antibiotic mycosubtilin in *Bacillus subtilis* and characterization of its candidacidal activity. *Appl. Environ. Microbiol.* **2009**, *75*, 4636–4640. [CrossRef]

MDPI
St. Alban-Anlage 66
4052 Basel
Switzerland
Tel. +41 61 683 77 34
Fax +41 61 302 89 18
www.mdpi.com

Microorganisms Editorial Office
E-mail: microorganisms@mdpi.com
www.mdpi.com/journal/microorganisms